ESCAPE FROM EARTH

ESCAPE
FROM
EARTH

A Secret History of the Space Rocket

Fraser MacDonald

PUBLICAFFAIRS

New York

PublicAffairs
Hachette Book Group
1290 Avenue of the Americas, New York, NY 10104
www.publicaffairsbooks.com
@Public_Affairs

Printed in the United States of America

First published in Great Britain in 2019 by Profile Books Ltd.

First US Edition: June 2019

Published by PublicAffairs, an imprint of Perseus Books, LLC, a subsidiary of Hachette Book Group, Inc. The PublicAffairs name and logo is a trademark of the Hachette Book Group.

The Hachette Speakers Bureau provides a wide range of authors for speaking events. To find out more, go to www.hachettespeakersbureau.com or call (866) 376-6591.

The publisher is not responsible for websites (or their content) that are not owned by the publisher.

Typeset in Garamond by MacGuru Ltd.

Library of Congress Control Number: 2019936734

ISBNs: 978-1-61039-871-8 (hardcover), 978-1-61039-869-5 (ebook)

LSC-C

10 9 8 7 6 5 4 3 2 1

For Roger Bacon

'There is something addictive about secrets.'

J. Edgar Hoover

CONTENTS

PREFACE

SPACE FLIGHT MAY SEEM like a transcendent theme – the stuff of soaring visions and azure skies – but its history is grounded in the dirt. This book is the unearthing.

It is a story that I've reconstructed from archives buried in obscure places. Perhaps that's why the research has so often felt like an exhumation. It's not just that the principal characters in this book are dead, which they are, but that their reputations have followed them down to the grave. This is about people who have, for the most part, been forgotten, even though their lives are central to the achievements of the twentieth century. I only found out about them through an accident of geography.

In the closing years of the twentieth century, I was conducting some doctoral fieldwork on the island of North Uist, in Scotland's Outer Hebrides. I didn't go there to study rockets but my interest in the cultural landscape made me curious about the one place on the island to which I was denied access: a hilltop called Cleatrabhal, 'hill of the ridge' in Old Norse. Its militarised summit gives a commanding view over the irregular carpet of moor and loch; there are even traces of Neolithic and Iron Age communities. But it's the Cold War infrastructure that still dominates Cleatrabhal, and it was there that I first started to dig into the story of the Space Age.

I learned that this site had been part of a rocket testing range built on the neighbouring island of South Uist in the late 1950s. The next time I was down in London, I dredged the National Archives to find declassified military files about the planning of the range. I discovered that it had been built to test a type of American rocket. And not just any rocket: the Corporal was the first guided missile authorised to carry a nuclear warhead. To my mild shame, I had never heard of it.

I wrote a few dry academic papers about missile testing and Cold War geopolitics, but the origins of this technology remained a bit hazy. I knew that the Corporal had been designed at the Jet Propulsion Laboratory and I knew, too, that JPL was at the forefront of space exploration today. But I thought it a bit odd that the key engineer behind both the rocket and the laboratory should be such a distant figure. His name was Frank J. Malina.

In 2006, I noticed a new Wikipedia article about Malina – a one-sentence entry that described him as an 'aeronautical engineer and painter'. Ploughing through a few books and oral histories turned up more information. I read that when he was little more than a graduate student, with the help of friends whose credentials were even less impressive than his own, he developed the first US rocket to reach an extreme altitude. It's called the WAC Corporal, the precursor to the Corporal. These days 'rocket science' is a cliché for complexity, a shorthand for engineering brilliance. In the 1930s, however, the opposite was the case: rocketry was so discredited that it didn't belong anywhere near the word 'science'. Yet it was Frank Malina, arguably more than anyone else in the United States, who made it respectable. Why then was his name absent from histories of space flight? There were rumours about his politics, and even more outlandish stories about his colleagues.

Years passed. I was invited to give a paper at the International Astronautical Congress, where by chance I ran into the astronomer Roger Malina, Frank's son. I had no particular plans to write about his father but I was intrigued by why he wasn't better known. Why did he walk away from practical rocketry? Why did he leave the United States? 'You should come to our home in Paris,' Roger suggested. 'We have a family archive there.'

It took me a few more years – life happens; I wasn't in a hurry – but eventually I made it to the Malina house. Roger opened the gate and welcomed me through the courtyard into the home where he grew up. Tucked away off a quiet back street in Boulogne-Billancourt, the house exudes a kind of homely modernism: simple concrete lines, a quirky spiral staircase, high ceilings and low furniture. In the study was a panorama of books, photographs and paintings, preserved in a state of lifelike disorder.

In the adjacent office I scanned the shelves. Each was laden with box files of letters, drawings, photos, sketches, more letters, magazines, exhibition catalogues, receipts, so many letters. There were documents of every conceivable kind – many of them intimate rather than institutional. Love letters. Letters to his mother. There were more formal papers too: a thick correspondence with lawyers, an archive box on which was written 'Box V: Witchhunt file'. Frank's life felt so close at hand, it was as if he had just stepped out to the patisserie. On his bedside table I spotted his wristwatch, a tiny calendar clipped to the strap: November 1981.

I had only been in Paris for a few hours when I realised that what had been an idle curiosity on Cleatrabhal, then an academic interest in London's National Archives, was now something urgent and personal. Here was an extraordinary life. I didn't know the full story then, not even half of it, but I felt certain that there *was* a story.

With Roger's permission, I photographed everything I could find, page by page, and read the material back in Scotland. I filled notebooks with details of Malina's friends and colleagues. I pieced together his relationships from the letters, working out who he trusted and who he didn't. I started to find gaps: letters missing; things that didn't add up. I found Frank Malina's FBI file and blinked at some of the allegations it contained. I submitted my own Freedom of Information requests to declassify the FBI files on Malina's friends. There were thousands of pages to examine in this house, but it was only a starting point; the search took me spiralling outwards, into other circuits of association.

The momentum I built up in Frank's archive was dragged by the search for FBI files. First you have to prove that the subject is dead and provide enough information (social security number, dates of birth, death, marriage) to identify the relevant files. If any are found you then join the declassification queue; that can take five years. Released files have many of the names redacted – blacked out – so that although you have some idea of what has happened, it's difficult to know *who* it happened to. It requires endless comparisons with other files and other archives. Much of this is repetitive and boring. Now and again, I'd find a little nugget. In the spring of 2016, a new file arrived. And with a single name on a single page, mistakenly left unredacted, I found the motherlode.

The trouble with FBI files as sources is that they're only as reliable as the agents and their informants. They can be useful, but they aren't the Truth. On reading them I still needed wider evidence – letters, diaries, oral histories – to give a more nuanced picture. Foremost here were the papers and journals kept by Frank's first wife, Liljan. Even with all this material, getting the measure of this story depended on getting to know its characters; that in turn meant getting to know their children, even grandchildren. These conversations weren't always relaxed. I was asking about past membership in the US Communist Party, not the kind of talk that puts anyone at ease. But in time, unseen and often unknown sources began to emerge, sometimes dramatically changing the story: the people who remembered FBI agents sitting in cars at the bottom of their driveway; or those who recalled the suited men watching as they bought ice cream as children; the family that came home one day to find a nail driven into a door frame, preventing its tight closure. Some of these recollections cut deep.

Half a century after the moon landings, we have inherited a particular image of America's Space Age pioneers: the steely-eyed missile man facing the great unknown. In the mass of papers and testimonies piling up in my own study, I saw something else: something repressed and unspeakable, something hidden and shameful. Something secret.

This is the story of the birth of the space rocket. In the pages that follow, you'll learn how humankind first reached beyond the atmosphere of Earth to worlds beyond. But it is more than that. It's about what we will allow ourselves to know about the darker legacies of the twentieth century, and the dangerous ideas that won't stay buried. I didn't expect this story to turn out as it did. Then again, I didn't think I'd be the one to uncover it.

PROLOGUE

THE PHONE RANG. The caller didn't identify herself; she just began speaking. At first Liljan Malina was too busy trying to absorb the information to recognise the voice. Then it clicked: this was Katja Liepmann. She could tell from the German accent. Not a close friend – their husbands had worked together at Caltech – but she knew Katja understood political danger. Katja had twice fled from the Nazis.[1] And, yes, it was definitely Katja who was saying that the FBI were on their way to the Malinas' house, that they'd be there within the hour.

Liljan snapped into alertness. It was a scenario she and her husband Frank had feared; they'd half planned for it, though neither of them quite expected it would happen like this. Katja, in a matter-of-fact tone, said that other homes were being raided even as they spoke, although she gave no clue as to how she knew that the Malinas were next. She calmly instructed Liljan to grab and destroy anything that might incriminate them.

After Liljan hung up she tried the office number at the Jet Propulsion Laboratory. No answer. She started to panic. She ran through the house from room to room scanning for books, pamphlets, fliers, magazines. There didn't seem to be much lying around, but then they'd both become a bit more careful recently. Earlier that year, in

April 1945, Frank had told her about a long train journey during which he found himself chatting to another passenger who, it turned out, worked for the FBI. 'His ideas were most discouraging,' Frank wrote; he 'said he would like Southern California if there weren't so many radical inhabitants'.[2] The whole episode had seemed quite funny, inconsequential.

Liljan calmed herself down and remembered the spring clean that Frank had undertaken months before. He had burned a lot of material. What was left he had placed in two cartons, which were put out of the way in the attic. Sure enough, behind some old furniture, Liljan found the cartons. It took two trips to heave them to the car, racing up and down steps two at a time, after which she carefully locked the front door of their Pasadena house.

Liljan drove towards Los Angeles – in the 1940s it was still quite separate from Pasadena – and at a quiet stretch she pulled off the road to call a friend, Saki Dikran, from a payphone. Saki's mother, Helen Blair, was part of the scene, and so was Helen's partner, Jack Frankel, an attorney to a number of beleaguered leftists. He would soon be handling Liljan and Frank's divorce. Liljan skirted LA and drove west, winding her way up to Saki's Spanish-style villa in Laurel Canyon, high in the Hollywood Hills.[3] It was a steep, narrow road with houses dug right into the hillside, all steps and walls and terraces. The place was tucked back from the road and nestled in greenery. On the horizon, Italian cypress trees pointed skywards. When the car pulled in to the house, Saki was standing in the driveway. Liljan motioned to the garage. Saki opened it without a word. Once inside, Liljan spilled out everything that had happened. Saki poured her a drink and in her usual quiet way said, 'Well … we'll simply burn the whole damned mess.' Together they collected a few leaves and twigs from the back yard and started a fire that would consume, page by page, the contents of the two cartons, until nothing but white ash remained.

Early that evening, Liljan returned to Pasadena feeling as though she had disposed of a body. It was dark by the time she got back, and she was concerned to the see the house bright against the evening sky, a light on in every room. She found Frank in his study. The posture of his slight frame was tense, his face pale and angry.

They had interrogated him in his office. Books, paper and files were strewn about the floor. It was the same in Liljan's studio, and in their bedroom. Frank's only crumb of comfort was that it was Liljan, not the FBI, who had removed the two cartons from the attic.[4]

YOU MIGHT NOT HAVE RECOGNISED his name, but Frank J. Malina is among the most important figures in twentieth-century science. His life is central to the bigger story of how humans first reached beyond the boundary of Earth, yet it has been obscured by the politics of the mid-twentieth century. In part because of a decades-long campaign of surveillance and harassment, both Malina and the pioneering rocket he created have vanished from the pages of history.

Malina was the first US rocketeer to achieve the main purpose of rocketry – high-altitude flight. Of course, building such a complex system is never the work of just one individual, but Malina was the architect of America's first successful liquid-propellant rocket, the WAC Corporal. The reclusive Clark University physicist Robert Goddard is known as an early pioneer with liquid fuel, but he never came close to the 'extreme altitudes' that were his own stated aim.[5] Malina's WAC Corporal rocket owed no design debt to Goddard's, and it soared twenty-seven times higher. In collaboration with the self-taught chemist and occultist Jack Parsons and aerodynamicist Theodore von Kármán, Malina also helped develop solid-fuel rocketry, another mainstay of contemporary space flight. If we put aside the Nazi engineering team that built Hitler's V-2 (many of whom became born-again Americans), it was Malina's initiative and leadership that transformed early rocketry in the United States from fantasy to science.

Malina occasionally turns up as a bit-part player in other people's histories – not least those of the charismatic Parsons.[6] But he is weirdly absent from so many accounts of the Space Age. Take, for instance, Walter A. McDougall's Pulitzer Prize-winning *The Heavens and the Earth: A Political History of the Space Age*. It is one of the most authoritative histories of space exploration and yet, across five hundred pages, it has more references to Bob Dylan than to Frank Malina.[7] This is very strange. The institutions Malina founded have been crucial to human achievements in space for over seventy years. His Jet Propulsion Laboratory (JPL), for instance, is now a NASA research centre with 6,000 employees that daily pushes back the frontiers of autonomous space exploration. Those amazing hi-res pictures of Martian terrain? Many are from JPL's *Curiosity* rover; by 2020 a new JPL mission will assess the habitability of the red planet. It's precisely because these institutional legacies are so enduring that *our* curiosity might usefully explore the secret struggles that were part of their genesis.

One of Frank Malina's most important contributions was a paper he published in the *Journal of Aeronautical Sciences* with his colleague Martin Summerfield. 'The Problem of Escape from the Earth by Rocket' detailed for the first time the mathematical criteria for 'multi-staged rocketry', a strategy by which a rocket achieves high altitude by dropping pieces once it has finished with them, so that it gets lighter as it ascends. Though the efficiency of rockets has since improved immeasurably, staged rocketry has remained the standard means of reaching orbit for over half a century.

But the problem of escape from Earth turned out to be as much political as mathematical. For Malina and Summerfield, it was the problem of how to make the rocket the bearer of hope rather than fear, and how their bid for the stars could be part of a progressive politics here on Earth. Closer to home, it was the problem of how to maintain a personal and professional life in the face of suspicion that reached the highest levels of America's security apparatus.

Malina didn't especially like to put a name on his politics. He thought of his views as being common sense, even 'scientific'. During the early years of his work at Caltech, he campaigned against racial segregation and raised money for the republicans in the Spanish Civil War. Some called him a radical, a socialist and a communist. These labels aren't wrong. Nevertheless, after years examining his private archive, it's clear to me that the core of his politics was anti-fascist. That was enough to put him in a difficult position.

As his rocket work became more successful in the interregnum between World War and Cold War, a refined version of his WAC Corporal was anointed as the Corporal missile, the first rocket authorised to carry a nuclear warhead. 'Four battalions of Corporal missiles alone are equivalent in fire power to all the artillery used in World War II,' boasted President Eisenhower.[8] The Corporal offered what US Secretary of Defence Charles Erwin Wilson memorably called 'a bigger bang for the buck'.[9] In other words, the rocket that Malina developed as a vessel for scientific exploration became the progenitor of contemporary weapons of mass destruction. He found himself making instruments of terror that were intended to destroy the very political movement he believed in.

AROUND THE TIME that Malina's WAC Corporal was soaring into the upper atmosphere, another group of rocketeers were settling into a new life in America. Wernher von Braun, Arthur Rudolph and Walter Dornberger had been the driving forces behind the V-2 that terrorised London and Antwerp in the final year of World War II. Their rocket wasn't much of a killing machine: the 3,000 or so British victims can be contrasted with the 43,000 that died in the German Blitz four years earlier. The real horrors of the V-2 were located at the production sites rather than at the targets. An assembly line, latterly housed in deep tunnels of the Mittelwerk factory on the outskirts of Nordhausen, used prisoners from the Mittelbau-Dora concentration camp, drawn from among Jews, Roma and Sinti, as well as French resistance fighters. Even the more conservative estimate puts the number of deaths at Mittelbau-Dora at around 20,000, of which about half can be attributed to V-2 production.[10] Words cannot describe the suffering.[11] Yet this was the reality of fascism. It's what Malina and his friends were fighting while von Braun was trying not to notice the executions of enslaved prisoners who were deemed a threat to output or quality.[12]

Wernher von Braun, whose smiling face would personify the Space Age to millions of Disney viewers, was not what you'd call a career Nazi. He was happy to work with whoever could propel his engineering ambitions, including the SS. Eventually, as the Third Reich disintegrated, he and his colleagues would seek a new sponsor. There was only one state that really fitted the bill. As another German engineer put it, 'we despise the French; we are mortally afraid of the

Soviets; we do not believe that the British can afford us, so that leaves the Americans'.[13]

Under the auspices of 'Operation Paperclip', 1,600 German engineers, including von Braun, Dornberger and Rudolph, transferred to the United States, many with laundered war records. Wernher von Braun wasn't the worst; he was not among the architects of Nazi death. But while the extent of his culpability for war crimes was never tested in court, survivor accounts indicate he wasn't above dishing out a bit of his own personal brutality.[14] When he got to America, he would still use anti-Semitic slurs for his military minder, Arno J. Mayer, later a Princeton historian.[15] Even with all this political baggage, von Braun had nothing like the difficulty in the US that Frank Malina encountered. Then again, Hitler's Boy Wonder had the advantage that he was a fervent anti-communist.[16]

Perhaps this comparison of Malina and von Braun is a little too neat. It will become clear that this book isn't about caricaturing heroes and villains of the Space Age. Still, we lose a great deal if don't attend to the Cold War political circumstances that have allowed von Braun's story to prosper while the legacies of Frank Malina have been obscured.

Here's one example. In 1958, when the United States launched its first satellite, a beaming Wernher von Braun was pictured at JPL holding aloft the successful Explorer 1. We might have expected Malina to share the limelight, not only as JPL's founder, but as the proponent of an earlier satellite that the US chose not to fund. Then again, he received no plaudits when, back in February 1949, his WAC Corporal reached the record altitude of 244 miles as the second stage on top of a captured V-2 – the so-called BUMPER WAC Corporal. Though it lasted just 390 seconds, that flight completed a voyage as remarkable in its own way as those of Columbus, Magellan and Cook. It was the first human object to reach into extraterrestrial space as it was then understood, and the first vehicle to achieve hypersonic flight (Mach 5).[17] 'If those who publicise such matters had not been asleep in our country,' Malina later complained, then 'a reasonable claim could have been made ... that the Bumper WAC project opened up the Space Age well before the *Sputnik*.'[18] But when Explorer 1 orbited the Earth, the US had rescinded Malina's passport and FBI agents sat in a car outside his house. By the time their interest in him finally waned, humans would be walking on the moon.

THIS IS THE STORY of how Malina and a close circle of friends pursued two strains of twentieth-century optimism: space flight and socialism. These were connected in remarkable ways, not least in that the purposes of both movements became corrupted, and many of their advocates were persecuted and forgotten. If this has been a largely hidden story, the responsibility for its concealment lies as much with the radical rocketeers as with the US space and security establishment. It has taken time for the truth to emerge from the fabric of finely woven secrets, accusations, denials and evasions.

At the dark heart of the Red Scare lay a straightforward question – 'the question that we have for so long been worried about', as Malina once put it.[19] Malina avoided it where he could, but he could not escape it forever. In 1958, he faced it directly on Form DSP-11, which he was completing in the hope that the US might eventually return his passport. It was the question that made the full rights of citizenship dependent on one's political thought being acceptable to the government – a question that negated any kind of achievement, no matter how far out of this world.

'Have you ever been a member of the Communist Party?'

Lest this question seem open, Form DSP-11 followed it up with a binary instruction: '(WRITE YES OR NO)'.

Malina filled the box with his clear capitals: 'NOT TO MY KNOWLEDGE'.[20]

The evasion in these four words is at the core of Malina's adult life, of how his life has been elided from the history of the twentieth century, and of what, finally, we can know about our escape from Earth.

CALIFORNIA

IT TOOK TWO DAYS and two trains to get to Los Angeles, travers-
ing the southern United States from east to west. The rail cars were
busy. Frank slept for a bit and chatted to a few other passengers. His
folksy politeness was part of the baggage he carried with him from
Brenham, Texas, even as he pushed against the gravity of its conventions, its religion, its small-town politics. Most of the journey was
spent staring out of the window, gazing at the subtle gradations of
scenery as the railroad bisected the Panhandle, crossed through the
sands and scrub of New Mexico and Arizona, where the scale of the
desert was hard to take in. There was just so much space.[1]

At nearly 22 Frank Joseph Malina was slim, with close-set brown
eyes, angled brows and swept-back hair that was just beginning to
recede. He looked every inch an American youth, yet he was very
proud of his Czech identity. Born in Brenham in 1912, Frank's family
had moved back to Moravia, Czechoslovakia, when he was seven,
only to return to Brenham when Frank was 12. It was at school in
Moravia that he had first read Jules Verne's *From the Earth to the
Moon* – in Czech, of course. The range and seriousness of Frank's
present interests, from science and engineering to art and philosophy, reflected this eclectic gymnázium background. It felt to him as
if 'Czecho' was as close as California. Even now, looking out of the

window, he observed that the mountains of Arizona were slightly bigger than those of Rožnov.[2]

It was dark when he finally arrived at the station in Los Angeles. As he collected his luggage and stepped off the train, Frank found his path down the platform blocked by a lieutenant from the Los Angeles Police Department. That was unexpected. Frank's frame stiffened until he recognised the cop as an older cousin. He simply wanted to welcome Frank and make sure that he settled into the big city.[3]

There was quite a clan of Malinas in the US and a few in LA, but mostly they had stayed in or around Brenham, where Frank's father, Frank Malina Sr, was the Leader of Band Instruments at the local high school. Music tuition wasn't well paid and the family had little money, but the house was a lively place for Frank and his younger sister Carolyn to grow up – a centre of Czech conversation, singing, four-handed piano-playing, philosophising and politicking, while their mother Caroline, and grandmother, Baba, would prepare elaborate lunches.[4]

Frank loved his family, but he was different from them. He was a confirmed atheist, which troubled his mother, who played the piano in the Lutheran church.[5] Moving to Caltech, however, was the bigger rebellion: his father had wanted him to follow a military career, as much out of a commitment to music as to warfare. The brass instruments they all played had a martial provenance, and Frank had even paid his way through Texas A&M by playing trumpet in the college cadet band. But the unquestioning obedience to authority unsettled him. Frank's insistence on civilian life became a raw family argument. 'I follow the dictates of my own conscience,' declared Frank in his high-school yearbook. His father was at times an adversary: 'very impatient if I could not carry out his desires'. He 'made cynical remarks that I bore very painfully … he was never really brutal about it explicitly, but I knew he carried many disappointments about his boy as he dreamt he should be'.[6]

His cousin drove him around the city, up to Hollywood, Beverly Hills and, of course, to Pasadena for his first look at Caltech. 'It is a fine school,' he reassured his parents in his regular letter home, 'it is not as large as A&M but you can see that it has quality'.[7] Even on this first visit to campus, before Frank had formally enrolled in his Master's in Mechanical Engineering, his interest was piqued by the sights of the Guggenheim Aeronautical Laboratory (GALCIT), which housed the first wind tunnel in the United States.[8] It looked like a rectilinear doughnut, consisting of a series of ten-foot diameter tubes that could give a controlled 'return' airflow, isolated from the fluctuations of the external atmosphere. The capacity and accuracy of this vast tool was unprecedented in the United States. It transformed airfoil design, and anyone who knew about aerodynamics regarded it as a kind of marvel. But as the months went by and Frank progressed through his degree he came to learn that GALCIT's outstanding reputation was founded less on the wind tunnel than on its eccentric Hungarian designer, Professor Theodore von Kármán.

BY EARLY 1935, Frank had secured a job as a technical assistant on the wind tunnel. At just 25 cents an hour, it wasn't well paid, but it was interesting to test new propeller designs, like that of Howard Hughes' H-I racer.[9] Frank worked nearly every waking hour, yet somehow still convinced himself that he was lazy.[10] 'You probably gather I have done little but flit around,' he wrote to his parents, before adding that 'I worked a total of 90 or more hours during the holidays in the wind tunnel'.[11] The fact that the theoretical work was so difficult was part of what made it interesting. Frank started to think that he could use a wind tunnel assistantship to fund another master's degree, this time in aeronautical engineering, to work with the great Theodore von Kármán.

Though small in stature, the professor possessed a kind of patri-archal magnetism, a supreme confidence in his own ability that was inspiring, even contagious. Von Kármán ranked himself third in the pantheon of scientific greats, behind Isaac Newton and von Kármán's friend Albert Einstein.[12] Yet to his students, this sky-high self-esteem was almost justified by the relentless rigour of his mathematical approach to aerodynamic problems. Caltech President Robert A. Millikan, himself a Nobel laureate in physics, was certainly among those who believed in the genius of Theodore von Kármán.[13] Millikan had sought him out as part of a wider recruitment of outstanding

European scientists that already included Niels Bohr, Paul Dirac, Erwin Schrödinger and Werner Heisenberg.[14] Thanks to Millikan, even Einstein could be found bicycling around Caltech in the early 1930s. The appointment of von Kármán, however, was about more than prestige. Millikan knew that the United States was behind in airplane design. He figured that von Kármán, then at Aachen University in Germany, could help America catch up.[15]

Von Kármán fitted this role, and was perfectly comfortable as an applied aeronautical engineer. But he also felt that he was cut from a different cloth to his colleagues at GALCIT. Despite his unwavering commitment to mathematical theory, he had an appetite for 'unconventional ideas'.[16] Frank loved this combination of boldness and orthodoxy, and felt a sort of kinship with von Kármán's broad interests across the arts. Not that it was easy to keep up with von Kármán in a lecture. The professor spoke in heavily accented English, such that students mystified by his emphasis on the physics of 'cows' were relieved to discover that he meant 'chaos'.[17] Those who could make out his emphatic diction still struggled to follow the train of equations that he'd chalk up on the board. Even Frank, who was plainly a gifted student, had to work hard. But the more he understood, the more it seemed that this kind of analysis could prise open the world.

It all started at an ordinary seminar in March 1935. As it happened, another of von Kármán's graduate assistants, William Bollay, was presenting. Everyone knew that von Kármán, who was slightly deaf, would switch off his hearing aid if the speaker didn't hold his attention.[18] Bollay, however, had the advantage of speaking about rocket-powered aircraft, the sort of topic that was guaranteed to raise eyebrows. GALCIT might have been one of the world's leading centres on the problems of high-speed flight, but most GALCIT faculty thought rockets were for cranks and fantasists. Frank listened intently as Bollay discussed the work of Eugen Sänger, an Austrian engineer who had witnessed Jewish colleagues like von Kármán flee Germany for the safety of the US, and then joined the Nazi Party anyway.

Bollay's seminar caught the attention of the *Los Angeles Times*, which covered it under the headline: 'Rocket Plane Visualised Flying

1,200 Miles an Hour'.[19] People loved this stuff. Aviation and speed had become part of the metabolism of the city – why not rockets next? That was the view of two young amateur rocketeers from Pasadena who were so inspired by this news report that they turned up on the jasmine-scented campus, looking to tap some of Caltech's famed expertise. Bill Bollay was either too busy to deal with these visitors or he didn't take them seriously. 'They were sent to me,' recalled Frank Malina, 'and then this story began.'[20]

JACK PARSONS STOOD TALL, rakish, with curly dark hair and, as von Kármán would later put it, 'penetrating black eyes which appealed to the ladies'.[21] It's a description that would have pleased its subject. He sported the sort of thin moustache that might belong to a silver-screen cad; he invariably wore a full suit, tie and waistcoat, and his presence was announced by an eau de cologne that failed to mask his body odour.[22] Born John Whiteside Parsons in Los Angeles on 2 October 1914 – he and Malina would discover they shared a birthday – Jack was the son of an unhappy union between Ruth Whiteside and Marvel Parsons. After his parents divorced in an acrimonious dispute involving Marvel's relationships with prostitutes, Jack was primarily raised by his mother and maternal grandparents, Walter and Carrie Whiteside.[23] In 1916, the Whitesides moved west from Massachusetts to be closer to Ruth, settling in Pasadena, California, a growing city of panoramic views and enviable climate.[24] Young Jack's grandfather, automobile executive Walter Whiteside, was a man who believed wealth was there to be spent: he bought a rambling Italianate mansion on Orange Grove Avenue (Pasadena's most affluent address), surrounded by one and a half acres of ornamental gardens. They all moved in together.

This Millionaires' Row was home to some of the household names of American consumer culture. The chewing-gum industrialist William J. Wrigley had his mansion nearby; so did Adolphus Busch, the Budweiser baron. Keeping up with this company was an expensive affair. Young Jack was chauffeured to Washington Junior High School in a limousine, but this lordly treatment stopped at the gate. Inside, the environment was unforgiving and often violent, an experience that Parsons made part of his own personal mythology. In later life he would write a memoir in the second person: 'your father separated from your mother in order that you might grow up with a hatred of authority and a spirit of revolution … your isolation as a child developed the necessary background of literature and scholarship.'[25]

When other people entered his life, Parsons sometimes gave the impression that they existed, much like his servants at home, in order to further his personal vocation. Foremost in the supporting cast from these early years was his friend Edward S. Forman, whose first acquaintance came when Parsons was being badly beaten by another student. The powerfully built Forman dragged off the aggressor, and deftly broke his nose, before turning to help the bloodied victim.[26] By any measure this sort of introduction breeds loyalty, but Parsons and Forman also bonded over a love of science fiction and a shared curiosity about rockets. Parsons had money and a poetic inclination. Forman was good with his hands and respectful of his friend's passions. If he had passions of his own, they tended to be eclipsed. He didn't seem to mind. Together they became a couple of 'powder monkeys', initially messing around with fireworks bought with the $20 note that Walter Whiteside dished out daily to his only grandson.[27] With this kind of support the 'two fellows' soon graduated to crude home-made rockets, which in 1928 pitted and scarred the lawns of the Whiteside mansion.[28]

Damage to the gardens was cosmetic, but the crises that engulfed the household at the end of the Roaring Twenties were far more existential. Walter Whiteside did not fare well in the Wall Street Crash. His prosperity and health ebbed in tandem, and his death in 1931 not only forced the sale of the house but also meant that its residents needed to find work: Ruth as a shop assistant, and Jack at the Hercules Powder Company, a job which he combined with trying to finish high school. It was a formative time: as Parsons would put it, 'the loss of family fortune developed your sense of self-reliance at a critical period'.[29] Forman had dropped out of school altogether, doing casual jobs until he, too, found work at the same Hercules Powder Company, an explosives manufacturer supplying the mining and construction industries.

In addition to a regular income, Hercules gave Parsons an education in practical chemistry – an introduction to the world of ammonium nitrate, nitroglycerin, ammonia dynamite and trinitro-toluene (TNT). Here was a whole new language for exploring order and its catastrophic breakdown. He soaked up the differences between high explosives and low explosives, between detonation and deflagration, between those reactions that occured faster than sound and those, like rocket propellants, that could be artfully slowed.

Forman, meanwhile, had become a machinist's apprentice at Hercules, where he learned the craft of metal-working on a lathe, creating shells and repairing guns.[30] When Parsons came up with a potential new rocket design, Forman could often manufacture it at work. Where before they had worked with cardboard and wood, now they graduated to metal. Parsons, too, was open to experimenting with company resources, improvising with different types of solid fuel. And yet despite their developing knowledge and ready access to materials, their ambitions were hampered by their inability to understand or measure the power output of their creations. It wasn't just that they lacked the instruments to measure thrust. They also lacked the mathematics necessary to conceive of such an object of measurement in the first place. To their credit, they had learned that a viable rocket could not be a firework; they appreciated that it would have to be a complex system that exceeded the capacities of any individual. In the absence of textbooks, they did what they could in the circumstances, and tried to find someone with relevant expertise.

THEY FIRST MET MALINA in February 1935. All three of them could see that there were benefits to combining their expertise: Malina had the mathematical knowledge while Parsons had an uncommon feeling for explosives. Forman had practical skills as a machinist, but he bristled at the way conversation between Malina and Parsons so readily turned to topics outside his ken. There was something about Malina that amplified the differences between the two old high-school friends, feeding Parsons' belief in himself as a man 'of literature and scholarship'. Forman's isolation was perhaps compounded by Parsons' marriage just a few weeks later to Helen Northrup, a secretary four years his senior (Parsons' engagement presents to Helen were a three-carat diamond ring and a .25 calibre pistol).[31]

Nothing between the aspirant rocketeers happened very quickly. By November 1935, having started his new aeronautics degree, Malina was considering whether he might pursue doctoral research at Caltech. But he also had a backup plan: 'if I see that things are forcing me to do otherwise,' he wrote to his parents, 'I shall try and get an exchange scholarship to Russia. I feel fairly certain that I am not brilliant in research, but I do believe I have determination to succeed in that work.'[32] It was nearly a year after the Bollay seminar when Malina admitted that he 'wouldn't be surprised if I got my fingers mixed into rocket propulsion'.[33] But even in this letter there is a sense that he was keeping his options open, in part by learning Russian: 'a few more months of study and I should be able to use the language for my purposes'.[34]

Part of the delay in getting a rocket research project off the ground was a difference in approach between himself and Parsons and Forman. It became apparent, for example, that Malina had no desire to start firing rockets just for the thrill of the launch. Under the influence of von Kármán's theoretical aerodynamics course, Frank cautiously drew up a programme of work to design a high-altitude 'sounding' rocket, propelled by either liquid or solid fuel.[35] A sounding rocket was a vehicle for scientific research, a bid to explore the outer boundary of Earth just as, on the same axis, nineteenth-century sounding voyages had ascertained the depths of its oceans. With existing balloon technology unable to exceed 100,000 feet, little was then known about conditions in the upper atmosphere. 'If we could develop a rocket to go up and come down again safely,' reasoned Malina, 'we would be able to get much data useful to the weather men and also for cosmic ray study.'[36]

That, however, was a distant goal. In the meantime Malina proposed a purely theoretical enquiry into the 'thermodynamic problems of the reaction principle and of flight performance requirements of a sounding rocket'.[37] The reaction principle is enshrined in Newton's third law of motion: for every action there is an equal but opposite reaction. At GALCIT this was well understood in relation to propeller dynamics but not in relation to rocketry. Only after this theoretical study was completed, insisted Malina, could experimental work begin; even then it would be static tests on a stationary rocket motor that would be facing the 'wrong' way. Even by Malina's own admission this was 'an austere program', and Parsons and Forman were not happy. What was the point of research if it didn't end up in a launch?[38]

'They could not resist the temptation,' Frank later complained, 'of firing some models with black powder motors during the next three years. Their attitude is symptomatic of the anxiety of pioneers … in order to obtain support for their dreams, they are under pressure to demonstrate them before they can be technically accomplished.'[39]

There was something else that made Frank uneasy. Jack's thrill-seeking curiosity – his desire to 'shoot' – was symptomatic of a wider lack of restraint. It didn't help that working in an explosives factory gave Jack ready access to a veritable candy store of combustibles. Though Frank had genuine admiration for Jack's 'free-wheeling' mind, he was deeply unnerved by the cavalier attitude to hazardous chemicals.[40] It became part of Parsons' schtick that, in an era when smoking was as ordinary as breathing, he liked to keep barrels of gunpowder, open, on his porch.[41] And it wasn't as if he observed any prohibition about taking explosives inside. One kitchen cupboard held a bottle of tetranitromethane, a highly sensitive oxidant that reacts violently with just about anything. Parsons thought it might have potential as a rocket propellant.[42]

Frank liked Jack and Helen: they were 'a pair of good intelligent friends'. Jack 'reads a lot,' Frank told his parents, 'and has similar viewpoints on social problems as I have'.[43] There was some truth to this, at least in these early days. But Frank did not enjoy the uncertain chemical hazards that awaited visitors to Jack's home. There was always a sense of background threat.

THE SUICIDE SQUAD

BETWEEN WORKING ON the wind tunnel, finishing his master's in aeronautical engineering and trying to get the rocket project off the ground, Malina didn't have much spare time.[1] But he still paid close attention to what was happening in the world. From what he could tell – from reading the leftist *New Republic* magazine, and from speaking to other colleagues at Caltech – 1936 was taking a frightening turn.

'Events in Europe are certainly leading to another war,' he warned his parents. 'There seems to be only one hope; overthrowing of the capitalist system in all countries and an economic union of all nations.'[2] This assessment held a particular importance for his Czech family. 'The European countries seem destined to bear the brunt as battlefields,' he wrote; 'I hope this is wrong, all wrong.'[3] The future held a curious mixture of dread and anticipation: 'after the war, there will probably be an exchange of power from the capitalists to a minority supported by the masses. An application of the scientific method in economic matters will certainly be tried.'[4]

The scientific method: for Frank Malina, it was the stuff of life. Science was modern and it was optimistic. It could renew agriculture and transport. It could transform industrial and domestic labour. And while a rocket might not seem like the most urgent proletarian necessity, it was nonetheless part of an avowedly civilian endeavour to explore the upper atmosphere. Caltech meteorologists such as Irving Krick had high hopes that Malina's project might have everyday applications in weather forecasting.

The essential precursor to any experimental enquiry was a review of the relevant literature, so Malina pursued an exhaustive study of other rocket researchers, including Hermann Oberth, Konstantin Tsiolkovosky and Robert Esnault-Pelterie, as well as Eugen Sänger. Closer to home, there was also America's own Robert Goddard, the well-known physics professor at Clark University. But surprisingly little could be gleaned from all this reading. Sänger's method wasn't clear. And ever since Goddard's famous 1920 paper 'A Method of Reaching Extreme Altitudes' had been unfairly ridiculed by the *New York Times*, he had proved secretive, showing little interest in publication.[5] But from looking at data published by the American Rocket Society, Malina deduced that it wouldn't be easy to design a workable rocket that could go higher than balloons, then the reigning technology for exploring the skies. The primary problem lay in building an engine with sufficient power to carry its own weight plus that of the shell, propellants and guidance equipment. There was little point fussing over any of these other details until an examination of the fundamental physical principles could confirm that an engine could, in theory, produce the necessary thrust.

In March 1936 Malina proposed that, working informally alongside Parsons and Forman, he would commence work for a PhD at Caltech to specifically study rocketry. This would allow all three of them access to the conceptual and mathematical supervision they might need, as well as some laboratory space and equipment. Malina's first port of call was Clark B. Millikan, an associate professor of aeronautics who commanded much respect and no little fear among the student body. He was a good teacher, albeit notorious for setting exams in which 95 per cent of the class failed to finish the paper.[6] It was as if he had something to prove, which, being the eldest son of Caltech President Robert A. Millikan (the most famous scientist in America, bar Albert Einstein), he sort of did.[7] Clark Millikan was von Kármán's safe pair of hands to supervise work on the wind tunnel, training the personnel and testing the models of the booming Los Angeles aviation industry.[8]

In the beginning, Malina and Millikan got on well. Frank described how at 'a fine buffet dinner [at the Millikans] ... everyone drank all the "Martinis" they cared for. I drank a number and found that they bothered me not at all. As long as I don't drink beer, I am safe.'[9] He remained sufficiently alert to observe that 'Millikan married rich and has a beautiful home ... all of Pasadena can be seen'. A few months later Malina was asked back to a dinner to mark Clark's 33rd birthday.[10] He then asked Millikan to supervise his rocketry PhD. 'I don't know if he was skeptical of the whole affair,' recalled Frank later, 'but ... he told me not to do it. He suggested that this was a good time to get my master's and go into the aircraft industry. Well, that was a rather bleak day.'[11]

Malina wasn't deterred. 'So I went to von Kármán and he overruled Clark, and said "Okay, you can stay and do it".'[12] That, at least, is how Malina told the story in public. Privately he brooded over the fact that Millikan talked about his PhD as 'Kármán's folly'.[13]

Von Kármán shared a sprawling Spanish-style house on South Marrengo Avenue with his mother and sister, Pipö.[14] Since his father's death during World War I, von Kármán had become the centre of this domestic universe. The famously absent-minded professor need do nothing for himself. Matters like what he should eat or wear were attended to by his mother and sister; the same went for diary scheduling or arranging social events. It was an infantilising arrangement – liberating in some ways, suffocating in others – but von Kármán benefited from the rich social circle of European émigrés maintained by Pipö.

Regular visitors to the household constituted an A-list of science, Hollywood and the arts: physicists such as Enrico Fermi and Albert Einstein, actors such as Paul Lukas and Bela Lugosi, musicians such as Hungarian pianist Ervin Nyireghházi and Russian opera star Feodor Chaliapin, and writers such as the American novelist Theodore Dreiser.[15] Each visitor would be welcomed personally by Kármán's mother from her wheelchair (he 'attracts the most peculiar types of people,' she'd say, even as her son emerged from his study wearing a kimono).[16] The von Kármáns were generous and hospitable; their house was often a place for parties and impromptu scientific meetings. It was into this rich social and intellectual atmosphere that the three rocketeers were now being introduced. Von Kármán not only agreed to supervise Malina but also allowed Parsons and Forman to use the GALCIT laboratory after hours. 'I was immediately captivated,' recalled the professor, 'by the earnestness and enthusiasm of these young men.'

MALINA, PARSONS AND FORMAN were by no means the only researchers interested in rockets. Ahead of the game in the United States was the reclusive physicist Robert Goddard, whose work was funded by the Daniel and Florence Guggenheim Foundation. As Robert Millikan sat on the board that oversaw this work, he took the opportunity to introduce Goddard to Frank Malina, who was brimful of the group's hopes and research plans. This short conversation in August 1936 must have dismayed the older rocketeer, who had not forgotten the vigour with which he first embarked on his own rocketry career at the end of the nineteenth century; he had since looked on the technology as his own private reserve. But the ridicule that followed the publication of his Smithsonian paper of 1920 had haunted Goddard, making him both secretive and possessive. Unwilling to give anything away, he mostly worked outside the scientific peer-review process, choosing instead the more defensive strategy of filing patents – often of uncertain and untested efficacy. The last thing Goddard needed was a Caltech team intimating a shared interest in liquid propulsion, far less that one of his own funders should encourage cooperation between the two projects. On the other hand, an absolute refusal to engage might not impress the foundation; caught in this bind, it was hard for Goddard to refuse an invitation to Malina to visit his ranch at Roswell, New Mexico the next time that the student was at home in Brenham, Texas.

One month later Robert and Esther Goddard were picking Malina up from the railroad station at Roswell. They took him to dinner at the Hotel Nickson and spent a cordial evening chatting rockets on the back porch of their Mescalero ranch.[17] Yet the hosts remained cagey. The next day, there was a quick tour of Goddard's workshop and static test stand, but any actual components were kept strictly under wraps. Then it was straight back to the train station.[18] By way of a conciliatory gesture, as Malina departed, Goddard floated the possibility that the student might return to work at Roswell after the completion of his PhD. But by the end of the 1930s that would not be necessary.[19]

From the outset, Malina took a different approach. As he later told Goddard's biographer, Milton Lehman, 'a successful high altitude rocket simply could not be built in a reasonable length of time by one scientist and a few helpers'.[20] Malina considered teamwork and openness to be a more promising model than Goddard's secrecy. In any case, a viable rocket was a bigger project than one mind could tackle. There were so many different elements, but at this stage they were just setting out to explore the design of a viable motor. The motor is the heart of the vehicle, where fuel and oxidant react in the combustion chamber to produce pressurised exhaust gases that, shaped by a nozzle, can vent at high velocity. It's a bit like letting go of an untied balloon. According to Newton's third law of motion, the velocity of the exhaust gases is an action that produces a corresponding reaction: the thrust that accelerates the rocket. But what size should the combustion chamber be? And what combination of propellants would give the greatest efficiency? Should they pursue solid-state propellants like the black powder of fireworks? (Simpler, but less controllable once ignited.) Or, like Goddard, try liquid propellants such as methyl alcohol with gaseous oxygen? (Greater thrust, but harder to handle and with the additional weight that comes with pressurised tanks, plumbing, pumps and so on.) What might be an optimal pressure inside the chamber? What materials might withstand the phenomenal heat generated? These questions scarcely scratched the surface of rocket motor design, and that's long before the contemplation of guidance, stabilisation and launching method – not to mention payload.

Goddard had tried to do everything himself, and it hadn't worked out well. He spent years, for instance, on an elaborate gyroscopic stabiliser before he had a viable motor that could even bear the weight of such a mechanism. The reason that 'rocket science' has become a byword for complexity is that the technical challenges are each individually so complex and delicate – from engine design to electronics and guidance – but they still have to work together as a system. And even then, it's a system that has to withstand the unprecedented mechanical and thermodynamic forces of supersonic motion. The calculation involved was vast. So when two other Caltech students told Frank they were interested in the rocket project, their contribution was warmly received. After all, Parsons and Forman could never provide mathematical assistance.

Apollo Milton Olin Smith, a Harvard graduate, was a master's student in aeronautical engineering, and could help with the theoretical analysis of flight performance. If his name was not enough to make his mark on campus (everyone called him by his initials: 'Amo'), Smith stood out from the crowd by wearing a modified pith helmet of his own design, with revolving ventilator and weather vane. His office mate cut an altogether more serious figure. Hsue-Shen Tsien (Qian Xuesen) was a Chinese PhD student who had come to study under the great von Kármán.[21] Tsien was quiet, refined, mathematically brilliant and, by all accounts, arrogant. If he taught you and you couldn't keep up, that was your problem, not his. But beneath his intimidating exterior was a luminous intellectual curiosity. After overhearing Malina and Smith working on a theoretical problem, Tsien also got sucked into the rocket group.[22]

Even with this additional help, the task of understanding propellant combustion had scarcely started. It would be even harder to bend that knowledge into a precise, workable design. That task is not called 'rocket science'. It is properly called engineering, and it works best as a team effort. Goddard, the lone wolf, has been hailed as America's pioneering rocketeer, but it is hard to make the case that he is, in any meaningful sense, a founding figure. Von Kármán's assessment was brutal but accurate. 'There is no direct line from Goddard to present-day rocketry,' shrugged the professor. 'He is on a branch that died.'[23]

THE FIRST HURDLE in the pursuit of space flight was the question of how to pay for it. The institutional support of von Kármán and

GALCIT was essential, but it did not cover the cost of materials. Malina was already working forty-eight hours a week on the wind tunnel; that income had to cover all his living expenses and his Caltech fees.[24] He always budgeted carefully and had managed earlier in the year to save enough to buy a second-hand car, a vehicle he christened Means of Transport – 'MOT' for short.[25] It allowed him to get out of Pasadena, a sleepy town with little going on, but it also helped the rocketeers find some of the materials they needed but struggled to afford. 'Parsons and I drove all over LA looking for high-pressure tanks and meters. Didn't have any luck. Two instruments we need cost $60 a piece and we are trying to find them second hand. I am convinced it is a hopeless task.'[26]

With experimental costs coming out of their own pockets, the rocketeers had to search for more creative ways of making money. Malina made a little on the side by providing the technical illustrations for an engineering textbook.[27] Parsons was at Hercules Powder Company but not earning much. Being bailed out by his wife, Helen, was not an ideal solution in the long term, nor, for that matter, was pawning her diamond engagement ring – though he was perfectly willing to do so when the occasion demanded.[28] The funding problem seemed particularly intractable because they already needed all their time for either paid work or rocket work. Even in these difficult circumstances, the solution they settled on looked desperate: they decided to write a movie.

The typescript doesn't have a name; it's not clear if it ever did. The words 'MGM Project' are scribbled in the top-right corner of what appears to be the synopsis of a screenplay written by Parsons and Malina.[29] Parsons was going to pitch it as a movie proposal that would then yield the riches with which they could build the rocket. Its forty-two pages started with this inauspicious opener:

> This story is to be built on the present stage of rocketry as the foundation, with a superstructure of the dynamic social problems now existing. The story in the latter contents is not to be propaganda, but the various characters are to express the true and sincere belief of the various classes of people in a capitalistic country.

Foundation and superstructure? A perceptive film director might recognise this concept from Karl Marx's *A Contribution to the Critique of Political Economy*. But for the MGM moguls who had recently snapped up the rights to *The Wizard of Oz* and *Gone with the Wind*, rocketry and Marxism weren't the most likely ingredients for box office success. These forbidding themes were given a baffling, sinuous plot in which a group of clever young men try to build a sounding rocket while taking on the dark forces of 1930s America: corrupt corporations, strike-breaking cops, compliant university authorities, sinister security services and a hysterical press. Behind all these elements are Nazis determined to get their hands on the rocketeers' engineering secrets. And of course the heroes struggle with a lack of funding: 'various schemes are thought out to little advantage'. Indeed. The circularity is dizzying, but precisely because the protagonists are so thinly veiled, the script also gives some insight into how Malina and Parsons saw themselves.

Though the key characters do not have identical analogues among the Caltech rocket group – some exhibit hybrid traits drawn from their wider circle – the similarities are more than passing. Franklin Hamilton, a hybrid of Parsons and von Kármán, is 'a typical brilliant scientist', though 'not the type usually thought to be scholastic', who was 'highly sexed', as 'is common among great men'. He had 'black, closely cut, slightly curly hair which tops a good looking, sharply chiselled face'; 'in any mixed crowd, he draws attention, especially of women'; his 'colleagues smilingly refer to him … as a "lady's man"'. Then there's Malina's alter ego: Jan Kavan (Malina had a relative with exactly this name), a trained mechanical engineer of Czech descent, 23 years old, who has also become interested in rocket research; his character was to 'express the ideas … of Socialism and Communism'. One wonders what Forman might have made of his portrait as George Pratt, who 'did not distinguish himself in school' but 'was always inventing something' and 'thought reading was a silly business … he could never understand Kavan's desire to read and was sceptical of all solutions of problems concerning social science, economics and so forth'. Tsien tactfully asked the authors to leave out the character of Lin Lao, 'a Chinese of medium height, in neat occidental clothes, who is lean in build ... with a light complexion and with eyes that are not very Chinese'. The authors, however, did not oblige ('I think we can make good use of him,' said Frank).[30]

The synopsis seems so outlandish that it's hard to imagine Malina and Parsons ever took it seriously. And yet they wrote it together every Monday night, mostly at Parsons' house.[31] One inspiration was probably a consulting job that Frank undertook for MGM studios, where he gave technical advice on an 'aerial adventure' film called *Shadow of the Wing* starring Clark Gable. (The role was 'to keep the writers from saying impossible things about airplanes', specifically those of the Royal Air Force; in the end the British Air Ministry refused to cooperate, leaving the film unfinished and preventing the writers from saying anything at all about airplanes.[32]) This brush with Hollywood likely helped Frank realise that an implausible plot was not in itself an obstacle to success.

Even so, this pitch was never going to fly. Perhaps the writing was driven as much by their own need to make sense of all that was going on. 'My brain is going round and round,' confessed Malina, 'aeroplanes, socialism, capitalism, logic, Napoleon, opera, thermodynamics – our slide rules are being worn out'.[33] In this light, the script may have been less about raising money than about anticipating the arc of their own story. What if their sounding rocket actually worked? What might that mean for themselves and for the world? As it turned out, their absurd fictional plot was uncannily prescient.

SHORTLY AFTER HIS RETURN from Roswell, Malina felt that the team had sufficiently grasped the theoretical issues to at least try a static test of a rocket motor. 'Go ahead and shoot,' von Kármán told them – even if it was a little late for Parsons' liking. On Friday 30 October 1936, after six months of preparation, Malina, Parsons, Forman and Smith drove around collecting tanks, fittings and instruments, which, to minimise damage or disturbance, they assembled at an out-of-the-way site on the dry bed of the Arroyo Seco canyon, just behind the Devil's Gate Dam. They laboured through the night lugging all the gear from the Caltech truck, digging trenches, filling sandbags and setting up the equipment. It was exhausting. Someone took a photograph of them all lying on the ground, each one reclining like Venus in a Renaissance painting.

It wasn't until 1 p.m. on the Saturday, Halloween, after a few hours of snatched sleep, that the test setup was ready: a water-cooled Duralumin motor that would run with methyl alcohol and gaseous oxygen. Four hoses ran to the motor: one for fuel; one for the oxidant; one for water to cool the motor jacket; and the fourth to monitor the pressure inside the chamber.[34] A simple fuse would ignite the motor. Thrust would be measured as the skyward-facing motor pressed down on a diamond-tipped arm, etching a trace onto a clock-driven glass drum. And in case the combustion pressure got too much, Malina and Forman were ready to shut off the supply of propellants. It was a well-conceived experiment – though it didn't quite work.

Each time Parsons lit the fuse and then dived behind the sandbags, the rush of gas into the motor just blew it out before combustion could take place, and fuel then flooded the motor. Each time they had to cautiously empty it and start again. A few other students had come to witness the spectacle. Bill Bollay and his wife were there, along with another student, Rudolf Schott; two others, Carlos Wood and William Rockefeller, brought a camera but, finding little worthy of their film, had decided to pack up and head home.[35] For the fourth and final attempt, Parsons decided to tie the fuse to the chamber so that it couldn't be blown out. He lit the fuse and retired as before.

Ignition!

For three full seconds the motor spat out a foot-long flame. Then: chaos. A line whipped round, hissing. It took a moment to realise what was happening, that an oxygen hose had broken free and was swinging out of control, igniting earlier fuel spills. The spectators were scarcely forty feet away. In the absence of a formal safety protocol, they did what they could: they ran like the Santa Ana wind through the dry canyon. 'We all tore out across the country wondering if our check valves would work,' Malina told his parents, but 'as a whole the test was successful'.[36] It was not the only time that success arrived in the guise of an instructive failure. And it's fitting that what is now regarded as the nativity scene of American rocketry should have taken place in such unprepossessing circumstances, amid dirt, sweat and confusion. Certainly the young men lying in the sand were far from despondent. 'Very many things happened that will teach us what to do next time,' was Malina's conclusion. The life of rocket-building would thereafter become a matter of planning for 'next time', a steady shuttling between analysis and experiment, where improvised modifications would recast their theoretical understanding. Even accidents would have their epistemic value.

Malina was exhausted by the efforts of the weekend. On Monday, he developed a cold and committed himself solely to routine work and to getting to bed by 11 p.m. every night.[37] On Wednesday he felt restored by an unexpected tonic: the landslide re-election of President Franklin D. Roosevelt. It was a strong endorsement of the New Deal with its emphasis on public investment (the Works Progress Administration gave jobs to millions of unemployed), pro-union policies and support for tenant farmers and migrant labour. 'For us that believe the capitalistic method is hopeless,' wrote Malina, 'a new hope is given.'[38] It was a sign that his political faith was shared by others. 'In the near future,' he predicted, 'we will probably start studying the Russian method to see if their good points can be adapted here.'[39]

Two weeks later, the three-second flame had extended to five seconds, albeit with another small fire as the oyxgen line ignited. At the end of the month a different oxidant, nitrogen dioxide (NO_2), enjoyed modest success. By January 1937, a test at the Arroyo with a copper motor ran for forty-four seconds uninterrupted – so long, in fact, that the entire apparatus glowed red.[40] Now that the group was generating data, von Kármán was properly attentive. He asked Malina to present a seminar on the work so far; this in turn brought out another interested party, Weld Arnold, a graduate student in meteorology, who was so confident about the potential of rocketry in his own field that he promised the group $1,000 if he could act as official photographer. They gratefully accepted without enquiring too hard as to where Arnold, a paid assistant in the astrophysics lab, could possibly lay hands on this kind of money. (It's since become a firm part of JPL's myth of origin, but nothing about this story makes sense. What kind of graduate student gives his peers $1,000 cash – more than six months' average salary – no questions asked?

It's doubly odd given that there's no record of Arnold taking photographs.) For Malina, the gift meant that less depended on the fate of the MGM script, which in turn meant less time amid the hazards of Parsons' kitchen. Arnold arrived by bicycle with the first $500 instalment in small denomination notes wrapped in newspaper.[41] Malina proudly parked these on the desk of a 'flabbergasted' Clark Millikan, asking, 'How do we open a fund at Caltech?'[42] The pleasure with which Malina would relate this encounter is one measure of their strained relationship.

According to Malina's letter home, there were five members of what was now being called the GALCIT Rocket Research Group: Parsons, Smith, Tsien, Arnold and himself (Forman's presence is less clear – Malina sometimes counts him as an active member and sometimes not).[43] The advent of new experimental data meant that Theodore von Kármán rewarded the rocketeers with permission to conduct small-scale tests on the GALCIT campus itself. This saved the laborious effort of setting up in the Arroyo, but the professor soon came to regret the decision.

For one experiment, the group mounted a small motor on a bob suspended fifty feet from the ceiling of the GALCIT laboratory. It was a neat idea: different combinations of liquid propellants could be tested and the deflection of the pendulum would provide a measure of thrust. The only problem was that, as they found in the Arroyo, combustion was often uncertain. This mattered less in the wilds of the chaparral. But handling the chemicals on a crowded campus wasn't easy. On one occasion, a valve on an NO_2 cylinder jammed open, causing a fountain of the toxic oxidant to spray over a Caltech lawn. The gardener was far from happy. And when the motor on the pendulum misfired, nor was the janitor. An acrid cloud of ill-combusted nitrogen dioxide and alcohol filled the building, leaving a thin layer of rust on much of the lab's permanent equipment.[44]

This little mishap brought losses and gains. A significant loss was the lab space, from which they were now evicted, though not before they were handed some oily rags and told to make good the damage. A dubious gain was a new campus notoriety that came with their unpredictable bangs and roars, flashes and fumes. They were dubbed

33

'the suicide squad' – not entirely without warrant. A subsequent static test with gaseous oxygen and ethylene caused an explosion that blasted a pressure gauge into some wood where Malina had been sitting moments before.[45] But the triumph of the brown-mist affair that enraged the janitor and tarnished the patina of GALCIT was that it led to tangible breakthroughs in developing storable liquid oxidants. Nitrogen dioxide (NO_2) anticipated experiments with the even less stable nitrogen tetroxide (N_2O_4) from which Parsons finally arrived at Red Fuming Nitric Acid (RFNA – nitric acid with 13 per cent nitrogen tetroxide), a storable oxidant that, though highly toxic, would in one form or another propel generations of rockets.

'With Malina's first explosion,' wrote one historian, 'GALCIT had lost her virginity'.[46] It's an odd feature of working on rockets, and also of writing about them, that sex – OK, male sex – so often seems to be the first analogy within reach. It isn't just the eyebrow-raising talk of 'thrust-to-weight ratios'; or even Defence Secretary Wilson's 'bigger bang for the buck'. Engineers would later talk about 'missile erection' and 'warhead mating'; the 'orifice position' of propellant supply; stealth 'penetration aids'; and about the problem of combustion instability widely known as 'throbbing' (one solution to throbbing was pioneered by the chemical engineer Ray C. Stiff).[47] In other words, some intimate aspects of masculinity were built into the idea and into the idiom of rocketry from the very outset.[48]

Reflecting on Parsons' inspired chemistry, Malina suggested that 'sex played a role, like it does in so many of these kind of activities'.[49] In this way, scientific knowledge becomes a kind of carnal knowledge – an identification that somehow slips from machines to bodies – and, with this knowledge, comes an attendant loss of innocence. From here, it's obviously a short patriarchal step to nativity, to birth and to the rocket that Frank Malina would sometimes call his 'baby'.[50] Little wonder then that one of the most potent forces to enter Malina's life, and to his monolithically masculine culture of engineering, should be a woman.

FACTS AND FANCIES

THEY MET IN THE LONG shadow of the revolution – Bastille Day, 14 July 1938 – which was the day Liljan Darcourt went to the ball. She wasn't a debutante in the strict sense (the party at L'Alliance Française of Los Angeles was not that kind of event), but she still was conscious of attending for the first time as an adult rather than as a child. René and Georgette Darcourt were French immigrants who thought it was about time that their daughter was properly introduced to the city's French society. 'Maman' bought some apricot organza to make a long dress for the occasion. Liljan loved the whole ensemble, from her new shoes to the satin ribbon that held back sun-streaked strands of hair.[1] At 17 she was superficially confident, 'neither surprised nor shy at the stares and glances' of 'the many young men who wanted a dance'.

At L'Alliance Française she was first introduced to Theodore and Pipö von Kármán, both noted Francophiles, and then to their guest, a young man who for all his evident lack of French seemed perfectly at ease in this crowd. The chatter about Frank Malina was that he was very clever – 'absolutely brilliant', 'von Kármán's protégé' – but Liljan also noticed his 'meltingly warm dark brown eyes, a beautiful mouth, a completely captivating smile'.[2] That Frank should be a guest of von Kármán was early evidence that the professor had become more than just a doctoral supervisor. Pipö and Theodore would sometimes claim to be Malina's aunt and uncle; 'they are both very kind to me and I only hope I do not disappoint their faith in me,' Frank told his parents.[3] Frank would fondly refer to von Kármán as 'the Boss' (sometimes 'Doctor'), but to see this as a mentor–protégé relationship, as many clearly did, doesn't quite convey the lasting depth of their interdependence. If von Kármán claimed him as a nephew, Frank saw him as a 'second father'. Without children of his own, the Boss would often be in the background for the big moments in Frank's life. Meeting Liljan Darcourt turned out to be one of them.

'It was just a very lovely evening, very simple,' she remembered, but it wasn't without its complications.[4] There was an asymmetry: 'he was smitten and I was flattered'.[5] And there was the fact that she was already in the midst of a 'passionate though proper' love affair with 'a handsome rambunctious Irish boy' called Carl.[6] Liljan and Carl had met at a church social under the watchful gaze of Father McCoy, who, she surmised, 'had plans of marrying us off as soon as possible'.[7] For six months, the pair had been inseparable: they'd go to beach parties, dance to the big band sounds of Benny Goodman or Artie Shaw, and 'melt in each other's arms smothered by unfulfilled passions'.[8] On meeting Frank, however, Liljan 'immediately sensed another level of intelligence from my tall, handsome Carl'. Frank's 'wild dreams of space rocketry' fascinated her; he was 'attentive, friendly, funny and also very serious'.[9] In the weeks and months after the ball, he was persistent.

Liljan didn't stop seeing Carl, but she found it hard to decline Frank's many invitations. He took her to concerts and to the opera; they played tennis and tried some archery.[10] He introduced her to his close friends, Jack Parsons and Tsien. Naturally, Frank drew Liljan into the social circle of the von Kármáns, a different milieu from any she had previously known, and for someone who had just graduated from Hollywood High School, a stimulating change. She noticed how enthusiastic Frank was about her desire to go to art school, even if he sometimes saw her painting as an object for his own mastery ('between the two of us we will manage to get your talent developed').[11] He would always arrive with a new book on philosophy, art or politics. Carl, on the other hand, seemed unmoved by Liljan's interests and ambitions. 'I began to understand that I had something of a problem,' Liljan later wrote: 'wildly physically attracted to Carl while equally being drawn to Frank's mind and work'. The 'situation could have continued just so, indefinitely, and I would have been very happy'.[12] That, however, wasn't part of Frank's plan.

It's not clear what, if anything, he knew about Carl, but after meeting Liljan Frank didn't hold back. He wanted to see her all the time. He even slackened his relentless pace of work, an interlude made possible by the unexpected dispersal of the Rocket Research Project. In June 1938, Amo Smith had left Caltech to work for the Douglas Aircraft Company; Tsien was primarily focused on his doctoral work with von Kármán; Weld Arnold, who had donated $1,000 to the rocket cause, had taken a job in New York; and Jack Parsons and Ed Forman were away for weeks at a time with a new employer, the Halifax Powder Company, making explosives out in the Mojave Desert.[13] Malina often felt he was left 'trying to keep this thing alive', but with progress slowing he was glad to find a different aspect to life.[14]

When Malina took the train south to Texas to see his parents at the end of August, his two days aboard a railway car were entirely occupied with thoughts of Liljan. On arriving at Brenham he immediately wrote to tell her so. Frank was unaccustomed to these intensities of feeling. It was as if they triggered some kind of confusion. On the one hand there was love and the delirium of their half-restrained erotic life ('every time I enjoy myself I think of you,' he wrote to her).[15] On the other hand there was work, and his quest for soaring success. Their years together would be beset by this abiding tension between Frank's libidinally charged professional life – he'd eventually call his rocket 'the little lady' – and a love affair with Liljan that he often conceived as a kind of collaboration.[16] Even at this early stage in their dating, he expected her to be sweetheart, secretary, research assistant, translator and artist – yet also subject to the occasional admonishment. 'Don't take work too seriously,' he told her, 'take it in your stride, and gracefully, as I know you can.'[17] He praised her to his parents for being 'pretty, serious and a hard worker'.[18]

Frank being in Texas for about a month meant that their relationship was, characteristically for him, mostly worked out on paper. In one sense, their letters were not untypical for young people in love. Frank sometimes concluded with a doodle or sketch of some kind, a thinly veiled sexual scene or motif: an arrow (rocket?) from Brenham heading for Hollywood, or a phallic locomotive streaking towards a cleft in a decorously arranged landscape. Sometimes he dispensed with the veil altogether: 'start preparing to receive a love that is being added to each day' – a daunting sign-off for even the most ardent addressee.[19] The expectations of Liljan's administrative assistance were no less onerous. By September he asked her to translate a paper – 'Characteristics of the rocket motor unit based on a theory of perfect gases' – which he submitted for the French REP-Hirsch medal in astronautics.[20] That worked so well that he jokingly enquired if she would like to help him finish another paper: 'are you a good computer – as good as you are at translating French?'[21] But he also wanted to work *with* Liljan, to have their own project, a desire that took the form of yet another literary collaboration, even before his MGM script with Parsons was complete. It would be a children's story about ants, 'different types of ants living in their anthill could represent the nations of the world'. 'The main thing,' said Malina, 'is to have a man of action.'[22]

Much later Liljan would understand 'that being "forceful" and "decisive" were qualities that Frank much believed in and admired'.[23] So too was a capacity for thorough planning. 'As yet I have not asked if you love me nor to marry me,' Frank wrote to her from Texas, before launching into detailed speculation about their potential domestic finances: 'I need a minimum of $50 per month to live on … two cannot live cheaper than one but two should be able to live on twice as much as one. At the end of June I should be able to meet this requirement.'[24] On the strength of these calculations, Frank returned from Texas with a proposal that they get married in June 1939. Liljan hesitated. He was eight years older; she felt too young and unready for marriage, and preferred things to remain as they were.[25] But her concerns were met with techniques of blunt persuasion: absolute confidence on his part ('I passionately believe I can make you happy') coupled with an ultimatum.[26] Frank's 'way of being forceful,' recalled Liljan, 'was to give me a week to make up my mind, and if I answered "no" I would never see him again.'[27] He had 'immediate answers for

Keep well and start preparing to receive a love that is being added to each day. I hope you get this letter in time to answer this week.

Love
Frank

HOLLYWOOD

BRENHAM

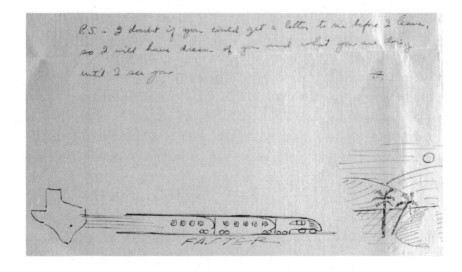

P.S. — I doubt if you could get a letter to me before I leave, so I will have dreams of you and what you are doing until I see you.

FASTER

everything and it was cut and dried like week-old toast'; he 'knew exactly how much we could spend for a house rental … how much for food, for pocket money etc. All of which I had nothing to do with. No surprises, no excitements.'[28]

This is how it seemed to her in retrospect, that there wasn't much room to negotiate. It didn't help that René and Georgette seemed much more excited about the prospective union than their daughter. Liljan felt pressured by a consensus between her parents and her boy-friend. And then there was Carl, who talked wistfully about quitting his place at Loyola College and eloping with her. But he was also drinking quite a lot and Liljan was wary about marrying a committed Catholic. 'There would be no birth control,' she supposed, 'and what of art school?'[29] Frank cared about her art, brought her books and promised security. She liked him. Didn't she? Heartfelt declarations of love did not happen overnight, but she was persuaded.

A FEW DAYS AFTER their engagement on 24 September, Hitler annexed Sudetenland. He did so with the acquiescence of Prime Minister Neville Chamberlain, an appeasement that in Britain was called the 'Munich Agreement', but for Czechs and Slovaks it was the 'Munich Betrayal' or 'Munich Diktat'. The arrival of fascism at home and the prospect of another war was a grim outlook for Frank's extended family. He was particularly scathing about Chamberlain, observing to his parents that 'the capitalists of England, fearing the success of Russia, nicely sold out as they always will do'.[30] Hitler may have been the obvious enemy, but Malina also felt that 'the British capitalists are the troublemakers together with the fascists', warning too that 'big business in Czecho will sell out the people just as big business sells out in every capitalist country'.[31] It's not clear that his parents entirely agreed, but in any case discussing events in Europe was safer territory for his letters home than his impending marriage. 'Your letter did not reflect very great happiness over the news of our engagement,' he noted, a little curtly. 'The words reminded me of a mother bear cautioning a bear cub.'[32]

The likelihood of war in Europe was unsettling enough, but Malina could also discern the impact it might have on his work at

Caltech. More precisely, he could see the impact that his rocketry work might have on warfare. 'A bigshot from the Army ordnance division was here today,' Malina wrote to his parents in October 1938. 'He ... thought there was little possibility of using them for military purposes. I silently rejoiced, however, Parsons who is about broke is not so happy about it as he hoped to get some funds for research from the army.'[33] But the rocketeers felt that this initial reticence would not last long amid talk of the next war. 'I have no doubt that war will break out sooner or later,' wrote Malina in November, but '[I] can't get very enthusiastic over making rockets for murdering purposes.'[34]

The very next week von Kármán asked Frank to present his work at a Sigma Xi luncheon at Caltech's Athenaeum faculty club.[35] He called the presentation 'Facts and Fancies of Rockets', a title that succinctly distinguished the secure ground of scientific knowledge from other excitable visions. 'Some say it was good' was all he told his parents about the lecture, but a clearly impressed von Kármán and Robert Millikan afterwards asked Malina if he would fly to Washington DC to present a report to the National Academy of Sciences Committee on Army Air Corps Research.[36] This was a committee established by Major General H. H. ('Hap') Arnold, the newly appointed Air Corps Chief, whose passion for harnessing the latest scientific developments had seen him drop in on the rocket work at Caltech back in May. Rocketry was still seen as an eccentric technology, but Malina's work was beginning to be noticed. His theoretical paper with Amo Smith, 'Flight Analysis of the Sounding Rocket', became the first publication on rocketry in the *Journal of Aeronautical Sciences*. Its concluding sentence was both sensational and unambiguous: 'if a rocket motor of high efficiency can be constructed, far greater altitudes can be reached than is possible by any other known means.'[37]

Malina was also in touch with the Consolidated Aircraft Company about using rocket units to give heavy aircraft a boost with take-off, an idea that may have prompted Arnold to make 'assisted take-off', along with the de-icing of windows, one of the priorities for his new committee.[38] Arnold was curious as to whether rocket units on heavy aircraft might advance US air supremacy over the Pacific by enabling heavy bombers to take off from short island runways. Rival engineering teams were scathing. 'I don't understand how a serious scientist or engineer can play around with rockets,' said the distinguished engineer Vannevar Bush to Robert Millikan, while Bush's colleague at MIT, Jerome Hunsaker, told Arnold and the committee: 'we'll take the problem of visibility. Kármán can have the Buck Rogers job.'[39]

The Buck Rogers job came with money: only $1,000 initially, but it represented the first investment in rocket flight by the US government, less than thirty years before the moon landings (it probably didn't hurt that both von Kármán and Robert Millikan sat on the assessing committee).[40] By July 1939, that had been extended under a new National Academy of Sciences Air Corps contract to $10,000 to investigate using rockets for the 'super-performance' of aircraft.[41] This support allowed Malina to work part-time while Parsons and Forman, delighted with military support, could work full-time. They 'can earn enough money to afford smoking ready rolled cigarettes,' Malina told his parents.[42] Malina himself had mixed feelings, as did Liljan. 'I know you're not very enthusiastic about it,' she consoled him, 'but it is still Science whether it be for the people or the army men. However it does seem that if the army gets a stronghold on it, it will be used against the people. In the meantime it will mean a lot to Jack and Ed.'[43]

Almost overnight the Air Corps involvement transformed a quirky student project into something much more professional. The GALCIT Rocket Research Project became the Air Corps Jet Propulsion Research Project, Malina and von Kármán dropping the R-word in favour of 'jet propulsion' as a sop to the cynics. A bigger adjustment, however, was the blanket of military secrecy that now enveloped their work. When Malina first appeared before the NAS committee he told his parents 'it is for the Army Air Corps so don't tell people why I am going to Washington. This secret and confidention [sic] business is very childish, but there is no point in bucking it at present.'[44] Once the money came through he again cautioned his parents, lest the good folk of Brenham should learn too much.[45] Extending this now classified information to his parents was itself a breach, but Malina was still working out the limits to what could and couldn't be shared. The problem with secrecy was that it came with an implicit threat – exposure – that in this case was less about leaking the object of secrecy (jet-assisted take-off) than about being revealed as someone who couldn't be trusted. The rocketeers were in the midst of learning, each of them in different ways, that secret knowledge could bring people together and also drive them apart.

ON THE EVENING BEFORE the wedding, Liljan had been over at a friend's house when Carl showed up. They had continued to see each other in the run-up to the wedding, 'only infrequently' according to Liljan. Sitting in Carl's car outside, he once again tried to persuade Liljan to abandon the wedding and leave with him. She didn't want to; they parted tearfully, Liljan telling him to leave her alone while he took off somewhere so he could drink. For all Frank's persuasiveness Liljan was still conflicted, less about a future with Carl – there wasn't one – than about marriage more generally. She was just 18. 'Frank, with all his intelligence and planning, was getting the wife he desired without realizing what a child she was,' she later wrote.[46]

The wedding reception was planned for 150 guests to be held on the rolling lawns of the von Kármán house at 1501 South Marengo Avenue (it sat atop the line of the Raymond Fault).[47] Notwithstanding their own quiet reservations, Frank's parents and sister had made the trip up from Texas. There had been a few tense exchanges in the months after the engagement. 'Have gotten some laughs from the pointed remarks that appear now and then in your letters,' wrote Frank to his mother. 'I suppose the psychologists are right when they say that women distrust women and that mothers hate to see their sons marry.'[48]

On the morning of the wedding, 24 June 1939, Liljan was getting ready. She had a new skirt suit, white hat and matching peep-toe heels. The phone rang – a concerned friend of Carl's asking if she knew where he was. Liljan had no idea, other than that last time she saw him he was 'already well sauced', and that they had parted in acrimony. This did not exactly allay Liljan's pre-wedding jitters. Here was 'Carl, the one I still half-loved … off some place drinking himself into a coma'. The official wedding photograph shows Liljan looking sullen and anxious. Frank manages a weak smile. It really hadn't been an easy morning. As it turned out, Carl was in rude health. He was seeing another woman with whom he had six kids ('when they weren't making children they were fighting,' was Liljan's caustic recollection).[49]

Liljan's feelings about Frank were complicated. 'Something about him disturbed and angered me.'[50] One source of conflict was the balance between work and intimacy, matters which needn't necessarily have been counterposed but in practice tended to be. For her part, Liljan couldn't have been prepared for the extent of Frank's underlying commitment to science, nor for the multiple ways in which it intruded on their domestic life. His devotion greatly exceeded the ordinary demands of undertaking a PhD or the basic need for paid employment as a research assistant. In quite a profound sense, though one different from Jack Parsons, Frank took science home.

One of his paid jobs at this time was to assist von Kármán in the preparation of technical diagrams for a textbook, *Mathematical Methods in Engineering*, which the Boss was writing with another Caltech engineer, Tony Biot.[51] In the run-up to the wedding and as the book was being finalised, von Kármán teased that he'd send Malina the proofs to check on his honeymoon as – nudge, nudge – he'd have nothing else to do.[52] It was a gentle, ribald joshing, a way for male scientists to acknowledge sex without having to talk about it. In anticipation of their life together Frank and Liljan *did* have to talk about sex (and about contraception), but this wasn't easy, not for Frank anyway. Liljan was less inhibited. In fact, she really didn't want to wait for their physical relationship but Frank was insistent that until the wedding they would do no more than 'fondle' ('his ability at self-discipline went a little too far').[53] She liked to tease him, like at Christmas when she joked about going to see Santa: 'I'm afraid he's going to say I've been a bad girl.'[54] Her future husband didn't especially enjoy these jokes. A few weeks before the wedding, he directed Liljan to a gynaecologist to be fitted for a diaphragm and gave her a marriage manual – with diagrams. There was no doubt that she was marrying a mechanical engineer.

In her journal Liljan recounted the engagement, wedding and honeymoon in unsparing detail:

> Meanwhile, the tiny cottage on Cordova St, Pasadena was all ready for us to start housekeeping, but not before our honeymoon at Lake Tahoe.
>
> I knew Frank was a virgin, but I assumed with his gift for planning that he would have the situation in control.
>
> The night of the wedding, we were driving to Tahoe when Frank suddenly stopped at a group of rundown motel cottages. He had planned this but not told me about it.
>
> Our cottage, we weren't even given a key, sat in scrub-pine, the faded wood of the floor and walls still hot and dry, smelling of the resin although it was already turning cool. Inside something scurried away as we entered, and I stood rooted to one spot till Frank had lit the kerosene lamp. As the flame lit up the room came alive in all its shabbiness. A chipped bedframe held a sway backed mattress and on a small dresser a half filled carafe of water next to a metal bowl. No toilet! No sink! No privacy for god's sake. How was I going to manage the diaphragm? The jelly? Putting that damn thing in with Frank there right in the room.
>
> 'Frank, please, can't we just drive on to Tahoe, I'm not comfortable here.'
>
> 'It's a beautiful place,' he said, 'quiet and simple. What's wrong with it?'

Frank wrote home from their lodge at Connelly's Bijou to say that 'we are having a wonderful time ... both of us have commented that we have never felt better'.[55] The surviving records of Liljan's experience are only retrospective but suggest that she saw it rather differently. 'I remember nothing about our nights together, or where we ate our meals, or how we spent our time. It's all been obscured for reasons I don't understand. I only remember the night we returned to Pasadena and opening the door of the little house with a sigh of relief.'[56]

There was one detail, however, that Liljan did not forget. Some time after they settled into the house on Cordova Street, her new husband told her that, notwithstanding his employment on a classified military programme, he had decided to join the Communist Party.[57]

UNIT 122

LILJAN'S FIRST INKLING of Frank's deep political seriousness came with the many books he brought her as tokens of his love. Where some suitors might have arrived with flowers or candy or jewellery, Frank turned up with Marx, Freud and Kierkegaard. And not just at the beginning of their courtship but week after week: Lenin, Jung, Adler, Baudelaire, Woolf.[1] 'To the point of intellectual indigestion,' recalled Liljan, 'I read books of philosophy, art, history and politics, psychology ... I was still at an age when knowledge is absorbed rapidly – painlessly – I was impressed with my own ability at this kind of learning and, for the most part, enjoying it.'[2] It must have been apparent that these weren't gifts in the conventional sense (unless, for Frank, a humanities schooling was in itself a gift). Rather it 'had been in preparation for my joining him in the Party': the development of a common philosophical background for their shared political life. 'I began to get an education far beyond the one I had acquired in high school,' she later told an interviewer. 'This was the start of our problems'.[3]

Even during their engagement there were signs that Frank considered their relationship to be part of a bigger political project, though he always took care not to be too specific in correspondence. 'I ... think you are in harmony with my goal,' he told her, 'a goal [that] ... makes life worthwhile. To feel that one is making this world a happier place for fellow human beings, to be useful, is a bless-ing.'[4] While written references to the Communist Party were usually veiled, between them there was greater openness. It's certainly hard to see how Frank's membership came as a great conjugal revelation to Liljan. On the contrary, letters in the months before the wedding suggest she shared her fiancé's political sympathies, telling him how at Otis Art School she had 'indulged in a heated argument with a girl who would make an excellent fascist. She called the poor starv-ing people "white trash" and said they were no more good than to be servants. You can imagine how my blood boiled.'[5] Liljan was in no way intimidated: 'we soon had them cornered where they didn't have any more answers except that Roosevelt was a dictator and we were a bunch of Communists.'[6] If the extent of Frank's commitment was perhaps a gradual realisation, there also seems to have been a moment of specific clarity: after 'Frank had confided in me that he was a member of the Communist Party and attended meetings regu-larly ... of course I was to join him'.[7]

Liljan was soon absorbed into Frank's circles, which centred on the von Kármáns on the one hand and Caltech's communist cadres on the other. Though there was some overlap between these two groups, von Kármán himself kept his distance from the party. He seemed instinctively wary. Frank would often spend the evening at the von Kármáns', either working with the Boss in his study or social-ising with other visiting students and faculty. Liljan started to resent having to attend these events, stuck as she was with 'faculty wives I didn't like', sitting 'correctly in pillowed wicker chairs', pretending 'to listen to the men arguing theories and aeronautical concepts'. 'It grew to be a chore I could hardly bear.'[8] Worse still was that not only did she have to manage the von Kármáns' expectations as their adop-tive daughter-in-law, but also those of Frank, who looked to them as the quintessential model of the good life. 'When we were married [it became clear that he] ... needed someone to take care of him, in the same way that Theodore was taken care of by Pipö' [...] 'there had to be someone to entertain, to do the dishes'.[9]

When Frank was not talking rockets at the von Kármáns', he'd often have his colleagues over at their own rented house at 1288

Cordova Street, a vine-covered single-storey with a living room, bedroom and a small study. It was in this study that Parsons, Forman and Tsien would meet to talk about their work, but also to thrash out the politics of the day. Also joining them was another colleague, the physicist Martin Summerfield, a short, awkward and almost super-naturally brilliant New Yorker who was orphaned at an early age and dodged a career at his uncle's grocery store by being a child prodigy. His ascent to Caltech was a success that no one expected, though he tended not to talk about this, nor about the grandparents who raised him.[10] He quietly did two PhDs in the time that Frank undertook one. That was the thing about rocktery, it attracted oddballs and outsiders; the extent of its theoretical complexity made it an appealing home for Summerfield, who was Frank's roommate until he got married.

Overhearing their animated conversation, yet never included, Liljan was expected to be on hand to serve drinks or coffee and cake, even if she had to be up early for art school the next morning.[11] She found herself 'too tired after my own day in art classes and making dinner, to much care about the ... work they were doing'.[12] If they weren't talking about liquid propellants, their discussions were most animated about what was happening in Spain or in Russia, the events that were unfolding in Czechoslovakia, or about the violence and cor-ruption of the Los Angeles mayoralty. All of these topics seemed to raise bigger questions: capitalism, communism, socialism, the rise of fascism and what could be done about it. Malina thought it was the job of educated young people to take sides in the world, views that were shared, and most likely influenced, by Sidney Weinbaum, Malina's Russian teacher. Weinbaum was the only person apart from von Kármán to whom Malina afforded such singular intellectual respect.

SIDNEY WEINBAUM, born in 1898 in Kamianets-Podilskyi in what is now Ukraine, was slim of build, with brown eyes and what others have called 'aquiline features' – an oblique reference to his Jewishness. There is no doubt that he was subject to anti-Semitism, explicitly in Kamianets-Podilskyi (where the best schools capped the number of Jewish students), and implicitly in California. As a child he had been misdiagnosed with a heart complaint which, being forbidden to run or play games with other children, had left him incurably studious.[13] It was something of a miracle that after fleeing the Russian Revolution Sidney had reached Caltech, the kind of place where his aptitude for maths could be both noticed and nurtured. Perhaps the first person to recognise and develop this talent was the chemist Linus Pauling, with whom Sidney studied towards a PhD on the application of quantum mechanics to molecular problems. He then became a research fellow in Pauling's X-ray crystallography team, crunching the numbers to determine the spiral structure of polypeptide chains; eventually this research won Pauling the Nobel Prize in Chemistry, though he seldom granted his assistant co-authorship on the published material.[14] Weinbaum seemed happy enough in a secondary role. In any case, he had other things on the go, like being a chess champion, a concert-level pianist, speaking four languages and corresponding with an international network of stamp collectors.

Sidney and Lina Weinbaum, then in their late thirties with a six-year-old daughter, Selina, represented a kind of domestic ideal for Frank. They were razor-sharp, politically aware, hospitable and attracted other international bright young things to their house at 1136 Steuben Street.[15] Not unlike Frank's family life growing up, the Weinbaums would pass social evenings playing and listening to music, or discussing world affairs. Frank would often be the only US-born person present. He loved these evenings and increasingly brought Weinbaum to his meetings with the rocketeers, such that a group of friends informally chatting grew into a more structured discussion and reading group, led (Parsons and Forman thought hijacked) by Weinbaum. They talked about Lenin's *State and Revolution*, Marx's *Capital*, as well as books by contemporary British Marxists like John Strachey and R. Palme Dutt.[16] Ed Forman made it clear that he had no interest or sympathy with this kind of talk and withdrew (though not before being cajoled by Weinbaum into a *Daily Worker* subscription).[17] Parsons continued to attend for about a year, though it was evident that Weinbaum, with Frank's tacit approval, had essentially turned it into a recruiting front for Professional Unit 122 of the Communist Party.

Liljan couldn't help noticing that the Weinbaums lived 'on the edge of poverty'. She described Sidney as wearing 'the same dirty brown sweater and shiny black jacket, the sweater a dreadful earth colour, made of harsh serviceable wool by the staunch Lina'.[18] Liljan recalled their 'rich Russian accents', and that 'they smiled a lot, but there was something tearful about their smiles'. Life had indeed been a struggle: 'one only had to see their faces,' said Liljan, 'to understand something of hardships beyond anything we knew.'[19] The Weinbaums didn't dwell on their past. Few people at Caltech realised that Sidney's father, a middle-class businessman named Joachim Weinbaum, had been killed by the Bolsheviks, or that Sidney had witnessed his killing.[20] Yet, having fled the revolution, Weinbaum remained genuinely supportive of the communist project, putting the death of his father down to the 'excesses' that often accompany major changes.[21] Despite seeing the Red Terror first-hand, he held on to a belief in revolutionary change. In fact, even after finding secure work with Pauling, he remained active in circulating petitions around Caltech faculty: one campaigned for the recognition of Russia, which Roosevelt eventually did in 1933; another was against California's anti-red Criminal Syndicalism Act.[22] Many of Caltech's faculty signed these petitions, though naturally Weinbaum only approached those he felt were likely to be sympathetic. And if the signatories wondered if Sidney was a member of the Communist Party, well, most felt that that was his business. Von Kármán would sometimes introduce Sidney at parties as his 'friend dealing in chemistry and communism', an alliteration that didn't raise too many eyebrows in the mid-1930s.[23]

People generally admired Sidney. They were more circumspect about Lina. Liljan Malina said she was 'a little weird' – that was among the more positive assessments.[24] Some were intimidated by her unapologetic intellectual streak, particularly because she made little effort to hide it. But she was a lively, social figure and, like her husband, a brilliant pianist. Sidney was completely devoted to her. Frank Malina enjoyed her keen mind and, it is clear from their correspondence, her command of interesting gossip.

Sidney and Lina became the heart of the group, but they weren't actually the first to join. The Unit started with Frank and Jackie

Oppenheimer: Frank, a Caltech physicist trying to advance research on beta-ray spectroscopy in the shadow of his older brother J. Robert Oppenheimer; Jackie, an economics graduate from Berkeley, who had been active in the Young Communist League. They formally joined the party after they moved to Pasadena, clipping out an ad in a newspaper and sending it in ('kind of a casual thing,' said Frank, 'but then we became very active').[25] At first they were involved in an open street unit – open in the sense that no one concealed their membership – where they worked alongside mostly unemployed African Americans campaigning for desegregation. But the party soon realised that Frank and Jackie Oppenheimer could be more useful setting up a separate 'professional unit' of scientists and intellectuals. They opened up their home, just along Cordova Street from the Malinas, for meetings that would recruit from the ranks of Caltech's faculty, research staff and students.[26] Joining the Oppenheimers as hosts of Unit 122 were Jacob ('Jan') Dubnoff, a postdoctoral Caltech biologist, and his wife Belle, who had recently returned from two years living in the Soviet Union; they in turn brought in the Weinbaums in September 1937.[27] And it was Sidney who became 'county organiser' – de facto unit leader – with one member later testifying that he was 'the guiding spirit', the 'best informed on matters of CP policy' and 'a rabid adherent'.[28]

It's not incidental that these original members of the unit were Jewish, nor that most of its members had direct knowledge of the rise of fascism abroad, or allied discrimination at home. Frank Oppenheimer had studied at Florence's Istituto di Arcetri observatory and had encountered fascist sympathies in the workplace.[29] Jan and Belle Dubnoff had seen the deteriorating picture in Europe for themselves, after returning from a visit to see family in the Soviet Union. And then there was the America that they now inhabited: no, it wasn't fascist, but it *was* characterised by brutal, systemic racism, violence against organised labour, and the corrupt nexus of big business, city hall and police.

The young Caltech scientists saw how tear gas and clubs were frequently used against union meetings; that in Alabama, lynch mobs and all-white juries were part of the 'Scottsboro boys' stitch-up of nine black teenagers falsely accused of raping two white women; that in California's Imperial Valley the police and white vigilantes helped the fruit growers beat up activists for merely enquiring into immigrant working conditions.[30] That's how things were. Yet there was one clear force pushing back against it all. It was the Communist Party, led by the Oppenheimers, that campaigned against the segregation of the local Pasadena swimming pool that ran 'blacks only' sessions on Wednesday afternoons, after which the pool was drained and cleaned for whites to return on Thursday morning.[31] It was the Communist Party who mobilised unemployed workers to march to city hall demanding relief. And it was the party that helped evicted tenants, behind on their rent, by organising squads to immediately return their furniture from the street back into their houses.

The party sought the wholesale transformation of society at every scale. It was both practical and pragmatic, though it hadn't always been that way.[32] When the writer Upton Sinclair ran in the gubernatorial election of 1934 on a platform of End Poverty in California (EPIC), the party dismissed him as a 'social fascist'. But after 1935, when the Comintern opened up a Popular Front strategy, it eventually embraced Sinclair as the paradigmatic example of the 'fellow traveller', a sympathetic supporter who was more useful outside the party than in. The term, of course, comes from Trotsky's *Literature and Revolution*. 'As regards a "fellow-traveller",' remarked the author tersely, 'the question always comes up – how far will they go?'[33] In 1943, the question came up for Berkeley physicist J. Robert Oppenheimer, or 'Oppie', Frank's celebrated older brother. He told a congressional hearing that he himself had been a fellow traveller, providing the helpful definition as 'someone who accepted part of the public programme of the Communist Party, but who was not a member'.[34] Though Oppie gave money to the party and attended at least one meeting at Unit 122, he could still say he wasn't willing to go all the way. The question of membership, what the historian David Caute calls the 'fetish of the Party card', can be overstated.[35] It's not a reliable guide to anything much – neither in the 1930s nor in the McCarthy era. But that isn't how the FBI saw it, particularly after the war. And as a consequence, the matter of who was and was not a member of the Communist Party at Caltech does take on a different kind of historical significance.

FRANK MALINA FORMALLY JOINED in November 1938.[36] That's the same month that his 'facts and fancies' talk led him to Washington, where the US Army Air Corps agreed to fund his rocket research; the same month that he admitted to being not 'very enthusiastic over making rockets for murdering purposes'; the same month, too, that military support brought with it military secrecy.[37] It wasn't exactly an impulsive decision: he had been talking politics with Weinbaum for over two years, freely predicting to his parents back in 1936 that the 'good points' of the 'Russian method' would soon be examined in the US.[38] But it's notable that Malina became a member at the moment that the Communist Party's political orbit passed closest to the mainstream. The context was the gubernatorial election of 8 November 1938, where the Democrats were trying to unseat the Republican incumbent Frank Merriam. 'Old Baldy' Merriam was the most pro-business governor in the country, and had presided over the 1934 'Bloody Thursday' attack on striking longshoremen that left two workers dead.

Challenging Merriam was a Democratic candidate named Culbert Olson, a New Dealer who was campaigning against the statewide initiative 'Proposition 1'. Prop 1 would have restricted picketing and secondary boycotts, which is why the Communist Party in general, and Sidney Weinbaum in particular, fought hard to create a broad left alliance inside the Democratic Party.[39] The result: a stunning victory for Olson, labour and New Deal Democrats. Within months, the CP was active in drafting new bills for a compulsory state health insurance programme, for workers' right to picket, for a state minimum wage and for the prohibition of racial discrimination in public places.[40] This, in other words, was the platform of many who supported the Communist Party. Not all of their bills got through state legislature, but even so, their aims were far more ambitious than the New Deal. To put it another way, there was no shame about joining the Communist Party in 1938. It was for people serious about making change.

Like other members, Frank Malina completed a form and

paid his dues.[41] Unlike others, however, Malina's involvement soon extended to the administration of the group itself, becoming its secretary and frequently chairing its meetings.[42] He wasn't on the fringe; 'they are not the quiet type', said one member in reference to Frank and Liljan.[43] Jack Parsons, by contrast, was never much of a communist, though his attendance at meetings can be partly put down to Sidney Weinbaum's deft understanding of personal psychology. Sidney could see that Parsons' obvious vulnerability was his lack of scientific training (Malina never concealed his belief that Parsons was 'not mathematically talented' – a bad reputation to have around Caltech).[44] So Sidney offered him some tutoring in mathematics, in the course of which he 'solicited [Parsons] ... to join an "inner circle" of intellectuals which ... included some prominent people' including 'Hollywood personalities'.[45]

Parsons was well disposed to exclusive groups. He later told the FBI that Weinbaum showed him a card and said that if he signed one just like it he would be eligible for meetings and he'd then know more.[46] But the earnest, learned tenor of the meetings wasn't to his taste, and he was never going to shine among a group of high-minded Caltech scientists. Regarding Sidney Weinbaum, Parsons later told the FBI that 'our relations cooled considerably'.[47]

Martin Summerfield and Hsue-Shen Tsien were also regular attendees of the group, joining the party under the pseudonyms Fred Kane and John Decker, respectively.[48] Malina's party name was Frank Parma. Weinbaum was Sidney Empson, Oppenheimer was Frank Folsom, while Dubnoff was John Kelly. In each case, members paid their dues, and then either Dubnoff as treasurer or Malina as secretary passed on these forms – complete with details of their real name, party name, address, date joined, age, place of employment and so on – to eventually arrive on the desk of William Wallace, a party functionary who served as assistant to the LA County membership director. No one took the risk of storing such sensitive records in offices where they could be stolen. Indeed the California Communist Party had issued a directive in 1933 explicitly forbidding this: 'keep all membership lists in a safe place and in cipher ... no records of any kind [are] to be kept in headquarters, offices etc.'[49]

Some historians have argued that this kind of concealment marks out the Communist Party as a departure from democratic norms, given that such clandestine procedures were readily adaptable to Soviet intelligence.[50] There's some truth in this, though it ignores the fact that openness was practically impossible in the political climate. In any case, for Malina and his colleagues the secrecy proved to be futile: William Wallace was not William Wallace, he was William Ward Kimple – an undercover agent working for the LAPD Intelligence Bureau.[51]

THE INTELLIGENCE BUREAU, more popularly known as the 'red squad', was a cell of violent anti-communists led by Kimple's LAPD supervisor, Captain William F. Hynes. The group had deep roots in the history of labour suppression in California. Back in the 1920s, Hynes had moved from being a corporate spy to an undercover LAPD officer planted in the Industrial Workers of the World – the Wobblies, as they were known. He pioneered the use of vigilantes in strike-breaking, and acted as an *agent provocateur* in a dockside dispute, trying to persuade the Wobblies to blow up ships, all the better to justify the inevitable police crackdown.[52] As leader of the red squad, he routinely clubbed and tear-gassed his way through party events, public and private. At a licensed demonstration of the unemployed in 1932, Hynes and his detail waded into the crowd with knuckledusters flailing; men and women were beaten indiscriminately.[53] In the 1934 Imperial Valley growers strike, the red squad bombed the union headquarters and beat up the families of the strikers.[54]

This campaign of terror wasn't cheap to arrange, but Hynes helped by setting up the red squad office inside the Chamber of Commerce. That made it easier for him to dole out corporate cash to the hired goons who would attack workers, and for the surveillance of New Dealers and anti-corruption activists.[55] One observer noted that Hynes 'was permitted by the city's ruling powers to pillage at will in return for protecting their interests'.[56] Beatings were routine. And California's Criminal Syndicalism Act – the same legislation that Weinbaum petitioned against – gave plenty of scope to arrest communists.[57]

Like his boss and mentor, William Ward Kimple was a true believer: so deep was his commitment to maintaining cover among the hated reds that he had himself declared officially dead – even to the extent of deceiving his own mother (as his heir, she collected on his life insurance).[58] By the time he was finally suspected and ousted by the CP in September 1939, Kimple had been squirrelling secrets away for nine years.[59] And at no stage did any of them – Malina, Weinbaum, Summerfield, Dubnoff, Oppenheimer or Tsien – have the slightest clue that their details had been compromised. As it turned out, their names must have been among the last that Kimple recorded. When Culbert Olson was elected governor with communist help in November 1938, he soon disbanded the red squad and turned over their records to the National Labor Relations Board.[60]

The climate of secrecy surrounding the party might in retrospect seem unduly furtive; the very idea of party names certainly wouldn't look too good in the McCarthy era, even if they were never much used at Unit 122. But even aside from the threats from Hynes and his gang, admitting association with the party carried an obvious risk to the professional careers of these young Caltech scientists. On this matter, there was some division within group. The strict observance of secrecy became a point of disagreement between the Weinbaums, who needed its protection, and the Oppenheimers, who didn't.[61] Insulated by their inherited wealth, Frank and Jackie were quite open about their membership, and disdained those who feared losing their jobs. As an immigrant, Sidney's employment situation was more precarious. Even if his affiliation was something of an open secret on campus, he was all too aware of Robert Millikan's antipathy to liberals, and especially communists. 'Like all rich people,' said Weinbaum of Oppenheimer years later, 'they don't give a damn about anything'. He was irritated by the fact that Oppenheimer once left his party card in a suit he sent to the dry cleaners.[62]

Their consensus in the end was for strict secrecy. At meetings, there was a 'tacit understanding', recalled one member, 'that communism or [the] Communist Party was never mentioned'.[63] The same member described the group as a 'recruiting front' from which Weinbaum might then approach individuals privately.[64] The whole thing, said another member, was 'very cloak and daggerish'.[65] At first Malina didn't entirely hide it from his GALCIT colleagues ('he talked a little about it', said one); soon the habit of secrecy became strict and inviolable.[66]

IF THERE WAS ONE thing the members of Unit 122 shared, it was a desire to fight fascists. A few came to the group from the Hollywood Anti-Nazi League. Others joined because they read the newspapers and felt they had to do *something*. The engineer Andrew Fejer arrived to work with von Kármán in September 1938 at just the moment that his native Czechoslovakia fell to the Nazis. The network around Unit 122 became, for the Fejers, the main locus of anti-fascist activity on campus.[67]

Frank Malina knew of the horrors of what was happening in Europe from his own family. One letter from a cousin, Andělka, detailed how people were fleeing the Nazis, abandoning homes and possessions. 'It's enough to break one's heart to read of their sorrow,' said Frank's mother, and 'the Nazi persecution of the Jews is horrible to read about. I can't understand why things like this can be tolerated. England and France just look on and do nothing.'[68] The Dubnoffs had lived in Russia, but as Jews travelling back through Europe in the mid-1930s, it didn't feel safe. And then there was the fight against racism at home. Even Tsien, arguably the least politically motivated of the group, had his own experience of being asked to move seats in a cinema so that a white patron didn't have to sit next to 'a Chinese'.[69]

A particularly important motivation for joining was General Franco's war against the Second Spanish Republic. Take Richard N. Lewis, a graduate chemist whose father, Gilbert Lewis, discovered the covalent bond and coined the term 'photon'. Richard wanted to spread his wings from Berkeley, California, and so spent the academic year 1937/8 among the gothic spires and left-leaning scholars of Oxford, England. He didn't need to be a student radical to be appalled at the British government's drift towards appeasement. 'I was anti-fascist from the time I first heard of Fascism', he later wrote.[70] The complacency regarding Franco demanded nothing less than a radical response. On his voyage home to the US, Lewis met a contingent from the Communist Party's Abraham Lincoln Brigade returning from Spain, and was moved by the example of young people risking their lives for the cause. When Lewis eventually enrolled at Caltech, he 'felt it was time to become involved', so he sought out Frank Oppenheimer, a contact recommended by English friends, and joined Unit 122 at the same time as Frank Malina.[71]

People came to these meetings because they wanted to, because their shared desire for a just world was enlivening. It was also quite fun. The first meeting Liljan attended seemed 'as innocuous as a church social'.[72] To her it was not so much a question of climbing the barricades and wiping out the present government, she recalled, but 'an attempt to be informed and ready when capitalism disintegrated of its own weight'.[73] In the late 1930s, that felt imminent. Some of their concerns can be found in scribbled talking points, pencilled in Malina's handwriting on the back of an old Caltech calendar:

How harmful is smoking, drinking?
Are we a nation of hypocrites?
Are gambling, alcohol, & women primary instincts of man?
Does the word liberty mean anything to the average worker?
Is the police of any use to the masses?
Can the scientific method be applied in government?
Is the industrial system destroying individualism?
Are farmers lazy?[74]

The questions might now seem rather quaint. They certainly reflect the preoccupations as well as the prejudices of the time – an ethical concern with vices and appetites, collectivity and control. The party was a vehicle that picked up all manner of travellers, with little in the sense of a single destination.[75] Few of those in Unit 122 were determined revolutionaries; rather they were, said Liljan, 'sincerely dedicated to … equal opportunity, fair distribution of wealth, the abolishment of poverty and hunger, and the building of a moral and decent society'.[76] Most of them anyway. One conceded that he 'was not political' but joined the group because he 'was attracted to girls of loose morals that he found among its members'.[77] It was a diverse crowd.

One thing that they all shared was a love of playing and listening to music. Meetings at the Dubnoffs would make use of Jan's fine hi-fi system, unusual for the late 1930s, along with their huge collection of classical records.[78] Dubnoff was not given to chattiness, and often sat in silence smoking his pipe, listening to others talk and play. Weinbaum was an outstanding pianist and loved the challenge of contemporary Russian composers such as Rachmaninoff, whom he knew, and Prokofiev, with whom he shared a mutual friend.[79] Oppenheimer and Malina were both musically talented. On these evenings, Tsien would sometimes tuck himself away in a corner to play Vivaldi on the alto recorder.[80]

Unit 122 was less a singular group than an agglomeration of different groups: each met on a different weekday evening, and attendees who regularly met on, say, a Tuesday wouldn't necessarily know those who met on a Friday. The party called them 'squads' as if to emphasise the expectation of campaign work (though this kind of commitment wasn't much in evidence at Caltech).[81] All squads together totalled twenty-five to thirty members, with varying degrees of involvement.[82] The exact constituency of meetings could be porous, but outside the leadership no one could know for sure who was and wasn't a member. That tactical indeterminacy was important: it afforded them some protection if, say, the FBI were ever to come knocking on their laboratory doors.

There were some Caltech staff who heard about these meetings – the music, the chat, the politics – and bitterly resented them. Clark B. Millikan, in particular, didn't like the politics of these people; he didn't like how the politics leaked into the science and he especially didn't like how conversations in everyday GALCIT contexts could sometimes turn red. What was the point in making engineers for the aviation industry only for them to be radicalised before they arrived on the factory floor? 'We had nothing to do with Russia,' said Liljan Malina of Unit 122. 'But when you mentioned this word "Communism" to Clark Millikan, he would go into orbit.'[83] The basis for the growing ill feeling between Malina and Millikan is not altogether clear. Either way, a fissure started to develop at GALCIT that wasn't about rockets per se (even Clark Millikan was coming round to that idea) but about the politics of those who worked on them.

WAR GAMES

THE TROUBLE WITH BEING A COMMUNIST was that the belief in the transformative power of organised labour meant a lot of practical graft, from leafleting workplaces to selling the *Daily Worker*. 'Every Evening to Party Work' ran one party slogan (admittedly not used for recruitment).[1] When one Unit 122 member, Louis Berkus, baulked at the expectation of regular 'handbilling', Sidney Weinbaum reputedly told him to 'get active or resign'.[2] The weekly unit meetings and the other committees were time-consuming enough, but then there was the expected involvement in various fronts, from professional societies and other political organisations to all manner of civic groups. (One member, the Caltech physicist and dance enthusiast Sylvan Rubin, was part of an alleged communist infiltration of the Mojave Desert Folk Dancers.)[3] Malina had no time for this sort of thing. He could see that the success of the party depended in part on the hyperactive citizenship of its members – but it wasn't for him. Instead, he encouraged Liljan to do it.

A few weeks after their wedding, Malina wrote home saying that while he had been busy 'working on rockets research most of the time', Liljan had now joined the Young Democrats of Pasadena.[4] Along with her work at Otis Art Institute, where she had been elected secretary to the student board, the Young Dems were to become the public face of Liljan's activism.[5] 'She enjoys the political goings on very much and should acquire some excellent experience for some future day,' wrote Frank in his weekly missive.[6] The enjoyment was true; it was a welcome change of scene from Caltech's engineers and their spouses. At least here she could be her own person.

The Young Dems in Pasadena were less an alternative to the Communist Party than its clandestine extension. Yet it wasn't just a case of entryism: many of these Democratic groups had evolved from Upton Sinclair's EPIC clubs, which had always implicitly relied on party support.[7] Frank would occasionally accompany Liljan to Young Dem meetings just to show his support for her campaigning, but for the most part his energies were kept for research. This was how he felt he could make the most difference in the world, even if his alignment of science and politics was becoming harder to sustain now that the army-funded rocket work was unavoidably tailored to a war he opposed. That was awkward.

Rockets were just one part of Malina's research. A few months after his engagement to Liljan, he took on a part-time job as a 'co-operative agent' with the Department of Agriculture's Soil Conservation Service. This was part of Roosevelt's response to the Dust Bowl storms of 1933 that had destroyed millions of acres of crops, killed countless livestock and bankrupted thousands of farmers.[8] With technical research needed to inform preventative measures, Theodore von Kármán was contracted to study the aerodynamics of soil movement. For $125 a month, a welcome addition to household income, Malina came on board as the hands-on engineer.[9] It was science in the service of the people, exploring how mechanised agriculture and ecological ignorance had produced a double ruination – robbing soil from the Great Plains only for it to fall as choking black blizzards over East Coast cities. Frank conducted fieldwork in Texas in June 1939 and 'saw examples of the worst wind can do'.[10] 'What makes people live in these regions is beyond me,' he wrote to Liljan. 'They have more guts than the average or have no brains. It is really a crime to commit the land to be ruined the way it is. Good use could be made of all that "battleship" money.'[11]

Battleship money – or rocket money, perhaps? In the aftermath of Malina's party membership, it's clear that he was very unhappy with the war orientation of their GALCIT work. 'My enthusiasm vanishes when I am forced to develop better munitions,' he wrote to his parents.[12] By contrast, his work designing a Caltech wind tunnel to study the mechanics of wind erosion was plainly in support of poor tenant farmers. A letter home, just a couple of months after he joined the party, gives a glimpse of his conflict:

> I prefer to keep on with SCS [Soil Conservation Service] as it is further from warmongering. Still, as Tsien says, we must fight fascism with US capitalists so better armaments are needed. Of course, one cannot be certain that such a path will not boomerang.[13]

When Theodore von Kármán recruited Malina to work on the Soil Conservation Service contract it was, in part, because of his extensive experience testing airplane models in Caltech's ten-foot wind tunnel. The relationship between soil particles and aircraft may not seem immediately obvious, but what engineers had learned about airfoil design or propeller dynamics was largely down to their ability to model airflow in experimental conditions. 'The problems of the soil conservationist in fighting wind-erosion,' figured Malina, 'are similar to those encountered by the pioneers in the development of aircraft.'[14] That's why von Kármán and Malina wanted to study 'saltation' (the process of wind-blown soil movement) by developing a specialist soil-blowing tunnel at Caltech, along with a portable tunnel that could be used in the field.[15] But it felt as if there was something important behind this split between, on the one hand, those who wanted to model the aerodynamics of soil movement for social benefit and, on the other, those who wanted to test airfoil designs for the aviation industry.

Lots of von Kármán's students did wind tunnel work. But it was notably the left-leaning engineers who were attracted to soil erosion research. The Czech Andrew Fejer and the Japanese George Morikawa, both friends of Sidney Weinbaum, applied to Soil Conservation Service, only to find their appointment blocked by the nationalist prejudices of the time.[16] Frank was outraged: '[T]he dust storm problem is more important than little men and something should be done about it,' he fumed. 'The President has written an executive order forbidding discrimination. Such mockery!'[17] But behind his frustration lay a glint of steel: 'one of these days I will be in a position to do something.'[18] It was the sort of reaction that meant it wasn't just Clark Millikan who was wondering where Frank's decisiveness might lead him.

HOMER JOE STEWART was a pudgy-faced engineer and self-described conservative Republican from the rural Midwest. He belonged to von Kármán's close circle of graduate students, and would eventually join the faculty at GALCIT (unlike Frank, he got on well with Clark Millikan).[19] In the meantime, Stewart was partly employed on the GALCIT Rocket Research Project, to which Millikan too now had some advisory oversight. Like Malina, Stewart was juggling lots of jobs – airfoil work in the wind tunnel, some rocket research, his own PhD – but as a result of a staff shortage Kármán also asked him to teach a class on dynamic meteorology to a group of military students. It was an unusual opportunity given that he was still two years from finishing his doctorate, but Stewart was a talented engineer and a popular teacher.

In May 1940, with the academic year nearing its conclusion, the senior officer of the group, Lt Roy Lester, hosted a pre-graduation party for his fellow officers and faculty at his house. At some point in the course of this event, the host took Stewart aside. He told Stewart that their 'counter-intelligence people' had discovered that some classified information had been stolen from the Caltech wind tunnel.[20] Lester had in fact been authorised by the FBI to ask Stewart to 'watch for any irregularities in the aeronautical branch of the school'.[21] The officer wanted to talk in some detail about how the wind tunnel was organised, and who, exactly, had access. Nothing more was said, and no charges were made, but in private Stewart was horrified.

Because FBI agents had impressed on Lester 'the strict confidential treatment of all information and inquiries', he didn't tell Stewart that the FBI's prime suspect was the then Head of Physics, a Prussian-born Nazi sympathiser named Alexander Goetz, who was thought to be leaking material to Germany.[22] The FBI were at that time investigating all GALCIT staff with links to Germany, even von Kármán, a Hungarian Jew who vocally hated the Nazis. Goetz, by contrast, had barely concealed sympathies. He was later said to have given a toast to the German battleship *Bismarck* after it sank the Royal Navy's HMS *Hood*.[23] The investigation into Goetz and his links to Nazi 'un-American groups' was ordered by J. Edgar Hoover himself, but Stewart had no idea about the fascist nature of the concern in this case. His suspicions lay elsewhere. Much later in life, after he retired as Professor of Aeronautics, Stewart recorded an oral history with a Caltech historian. In the course of this interview, Stewart amplified his own contribution to the development of rocketry and the crucial leadership of Clark Millikan (Millikan, he said, was 'the administrative, academic leader who kept things going').[24] As soon as the historian brought up Frank Malina, however, Stewart's manner shifted. 'There's a whole background that I think was most unfortunate,' he said. 'There was some real honest-to-God poison around Caltech.' Of course, Stewart didn't explicitly name Malina as his own suspect. He didn't need to. 'As far as I know,' said Stewart pointedly, 'they never caught anybody that did it.'[25]

FRANK MALINA HAD BEEN in the Communist Party for less than a year when, almost overnight, the movement was dealt a blow from which it would never recover. The careful coalition-building of the Popular Front – with its united alliance of New Dealers, liberals, anti-Nazis and trade unionists – was summarily executed by Stalin's non-aggression treaty with Germany. The Molotov–Ribbentrop Pact on 23 August 1939 put in writing an agreement that neither government would assist an enemy of the other. Having been the principal advocates for collective security against fascism, Stalin's 'Munich' now withdrew support from those fighting the Nazis. People left the party in droves, particularly in Los Angeles, where the Jewish community in the needle trades made up a core of membership.[26] The fellow travellers and trade unionists whose support was so essential to local campaigns and strikes felt personally betrayed. They dropped, as the *New Republic* put it, 'like ripe plums in a hurricane'.[27] Unit 122 had its share of this crop, like the folk-dancing physicist Sylvan Rubin, another assistant on the GALCIT wind tunnel. But most of the group were determined to adhere to the party, favouring an old metaphor of Lenin's: 'when the locomotive of history takes a sharp turn only the steadfast cling to the train.'[28]

Weinbaum and Malina were among the steadfast. And if Unit 122 members weren't all left strewn along the railroad by this development, it probably reflects their academic distance from the conflict that was tearing apart the Los Angeles party. 'He came to this political thing almost as an intellectual game,' Liljan later observed of Frank.[29] She too stayed in the party, but 'quickly became bored with the meetings. There were constantly such erudite discussions that I couldn't help but wonder what connection all this had to do with political actions.'[30] For many of them, it was a bit like chess – 'theoretical acrobatics', as Liljan would call it.[31] The analogy is quite apt: after the business part of unit meetings, as people drank and chatted, often two or three set up a board in the corner, perhaps with a Stravinksy record playing on Jan Dubnoff's hi-fi.[32] But plain old chess was too straightforward for these Caltech scientists; they played *Kriegspiel* (literally 'war game' in German but also known as 'blind chess'), in which players could only see their own pieces but not those of their opponent, depending instead on a third party to update them on the progress of the game. The metaphor appears to have passed them by.

'So far it seems as though the non-aggression pact has only done good for democratic (?) countries,' Frank insisted in his letter home. 'I do not know of course what future events will show, however, I am hopeful that good leadership of the people in Europe will bring about a socialistic union of the various countries.'[33]

Malina's parents clearly took a different view. Their letters have not survived, so we can only infer their contents from Malina's replies. 'I did not mean that an Allied victory would be bad,' he told them, clarifying an earlier letter that has since been destroyed. 'I merely do not see much promise of anything good. The main thing is that the US keep out [of the war] and that means that Roosevelt should not be elected for a third term.'[34] With Young Dems like this, the president might have wondered, who needs enemies? Part of Frank's problem with his parents was having to distinguish between his anti-war position and that of the isolationist right, spearheaded by the aviator Charles Lindbergh, under the rubric of 'America First'. Having admired Lindbergh as a teenager – he had even drawn a portrait of him for his dorm-room wall – Malina eventually met him on a visit to see rocketry work at Caltech.[35] Politically, however, they had little in common. 'Lindbergh says many things that are true,' he told his parents, 'however, his convictions are no doubt fascistic – at least his writings show definite influences, so that one has to be careful not to let some of his true statements lead to false conclusions.'[36] Frank was highly sceptical of Allied political leadership. By the end of 1940, he said he 'wouldn't trust Mr Churchill as far as Hitler could throw a rock'; in that sense, Frank didn't deviate from the Communist Party line, embodied in slogans such as '"The Yanks Are Not Coming" and "Vote FDR is Vote for War"'.[37]

But then in June 1941 it was over. Frank and Liljan disembarked a plane at Los Angeles to learn that Germany had declared war on the USSR. As one Californian communist put it, it was 'switch-eroo, boom, there we were ... on the majority side for a change'.[38] No longer was this the Second Imperialist War. Finally, everyone had the same objective: fighting the fascists. 'Wonder if the communists will be called fifth columnists now?' asked Frank, needling his parents. At last, his rockets could be aligned with the struggles of the people. Elsewhere in Pasadena, one former member of Unit 122, Richard Rosanoff, ran into Sidney Weinbaum on campus. Rosanoff had by his own admission 'flunked out' of Caltech in 1939, joined the army, but then returned in June 1941 while passing through on furlough. If what Rosanoff later told the FBI is to be believed, 'poor old Weinbaum was a week behind the time' and hadn't been briefed about the new line. He 'was still calling it a dirty capitalistic war'.[39]

LILJAN WAS COMING TO REALISE that she would need a life of her own. Much later, she'd reflect on how 'great men, and also not so great men, look to their women still to free them of all the nasty little inevitable daily chores so that they can go on with their important work'.[40]

But she had her own important work, both at Otis Art School and, increasingly, for the party. She wasn't just going to await Frank's return in order to cook him some food. She also had to fend off his suggestion that they start a family.[41] At just 20 years old, she found that married life wasn't straightforward, and that while being Frank's wife was limiting, his encouragement of party activism gave her an expanded understanding of liberty and emancipation – including her own.

When Liljan was not studying at Otis, she'd sometimes borrow Frank's car to drive from Pasadena to East Los Angeles to meet up with a group of Young Dems. She'd go to meetings, print fliers, make banners and even worked to set up other Young Dem clubs across the city. 'We were so earnest and so sure of our ability to affect the world,' she recalled, 'that's something very specific that belongs to youth.'[42] Travelling up to an activist conference in San Francisco in June 1940, Liljan came to know that youth also held other potencies. 'The convention promises to be hot stuff,' Liljan wrote to Frank, 'plenty of fights and plenty of work'.[43] Her political sentiments were nothing if not passionate. 'Inevitably,' she confessed in her journal, 'there was a man'.[44]

Jim Burford worked for the Congress of Industrial Organisations (CIO) – a powerful federation of unions – for which he directed the establishment and communist inclination of the Young Dem clubs (he was 'the moving power behind all this frenetic activity,' as Liljan put it).[45] Burford had joined the party at Berkeley (Unit 5) at the beginning of the Popular Front era, another veteran of Upton Sinclair's EPIC defeat, and had taken the party name 'Ron Hillyer' (we know this because he, too, was a victim of William Ward Kimple's meticulous file-keeping).[46] So Liljan found herself with Jim and a few others staying in Sausalito, spending time with the actress and politician Helen Gahagan Douglas, who had found fame playing the title role (She-who-must-be-obeyed) in the 1935 film of H. Rider Haggard's novel *She: A History of Adventure*.[47] On the ferry across San Francisco Bay, Liljan stood close – *very* close – to Jim. Her hair touched his face.[48] 'I fell madly in love,' admitted Liljan, 'for a brief time anyway,' adding that 'after the first night I knew this had to be the dullest lover a woman could have'.[49]

Burford ended up 'hurt but not mortally'; they remained friends. She even introduced him to Frank, though there is no suggestion that her husband knew what had passed between them. As a professional organiser, Burford didn't warm to Malina. He didn't like how Frank used his research to excuse himself from campaign work, complaining to Liljan that 'there are a number of people, organisations, etc right at his elbow which he could tie into if he really meant business'.[50]

Frank was away a lot and Liljan was acutely sensitive to being left on her own. At the same time she was chaffing against the constraints of what she felt was practically an arranged marriage. She met Chuck Bernstein, another Young Dem, who was 'an Adonis, a marvel … [with] a build that commanded immediate attention'. He and Liljan 'just naturally fell into bed together one night half way through beer and sandwiches at his place, ostensibly working on copy for a flyer'.[51] And then there was another Jim – Jim Bowling – a draftsman and photographer with whom Liljan shared an art class.[52]

This last relationship was the most serious. It coincided with Frank's frequent research trips for soil erosion work and, latterly, for rocketry meetings in Washington. For the nine months before Bowling was conscripted into the army, Liljan's time with him was 'the best of a most exciting and satisfying folly' – 'just the most important thing in my life'.[53] Bowling was 'one of the handsomest men I have ever met', and 'seeing him leave in uniform was one of the most heart-wrenching moments I've known'.[54] Perhaps 'the Bowling episode', as Frank would later call it, grew in significance for Liljan because it represented a retreat, an outside, both from the alliance of Frank and her parents, and from the social intensities of the party.[55]

At 20 years old, Liljan was desperately unhappy. One weekend, when Frank and Liljan were at the beach, they sat down by the water watching the breakers in the cool night air. Liljan began to cry. She tried to tell Frank about her sadness, about how alone she felt with him so often away with work, though there were dimensions to her loneliness that couldn't readily be expressed. 'How can you be so miserable?' asked Frank, with genuine surprise. The question stung.

How could I be so miserable when you, Frank, were doing so well, and wasn't I in art school and spending as much time as I wanted with my silly friends. What more do I want? How could I come to you with all these dissatisfactions and criticisms when you were so involved in a monumental project of great importance to the country?

The remark was not lost on me. I realized that besides being downright difficult right now and making you squirm, I was also being unpatriotic![56]

She had no answers for him after that. How could she 'be so awful as to even think such thoughts'?

'I think it's time you had a baby to care for,' Frank said.[57]

GIVE THE FASCISTS HELL

THE GROWING MILITARY INVESTMENT in rocket research meant that Frank was often away at meetings in Washington DC or in Ohio. He wrote to Liljan saying how much he missed her, letters that were always heartfelt but also oblivious to the impact of his absence. Even when he and Liljan were both in Pasadena, he never really stopped working; their lives were bent to his own iron discipline. Late into the night he'd sit in a wicker chair wearing an old sweater that Baba, his Czech grandmother, had knitted for him when he first went to college. The collar would be turned up against the cold of the back porch, while a cigarette curled smoke from his left hand. He smoked a lot and slept very little.[1] Some evenings would be spent at the von Kármáns, but these days Frank was mostly unaccompanied by Liljan. She had little patience left for engineers and their spouses, nor for the grandstanding Caltech communists, though she still hosted Unit 122 meetings at their home on a Tuesday evening. Life in Cordova Street was punctuated by occasional arguments. After Liljan raised some complaint or other, Frank outlined his priorities: 'I am ... involved in something that is very important, and all of this is simply going to have to wait.'[2]

The funding from the National Academy of Sciences Air Corps – the $10,000 that von Kármán had secured in July 1939 to study Jet

Assisted Take-Off (JATO) – had in one sense diverted the GALCIT team from their original aim of creating a high-altitude rocket. Who needed space flight in a world consumed by war? But the build-up of the American military in the early 1940s meant that support for basic research in jet propulsion was directed to aeronautical, rather than astronautical, ends. Rockets couldn't be weaponised directly but they could be used to lift a military plane. And were it not for Malina's initial discomfort about the war, this actually suited him well. He knew that the outstanding problems were fundamental, that only a team of mathematically inclined engineers could hope to solve them, and that the object of rocketry was less important at this stage than its practical development.

As well as experimenting with liquid propellants, the GALCIT team had persevered with solid fuel, including a 'machine gun'-style design with a mechanism for reloading successive explosive cartridges so that the rocket could 'punch' its way through the air. True to form, Parsons and Forman courted the popular press, being photographed with the rocket, while Malina and Tsien quietly published the scientific analysis of how it might work.[3] As it turned out, however, this particular model didn't work. Solid fuel was easier to store and handle – ideal for military purposes – but it generally had a burn duration of not more than three seconds. Primitive rockets of this sort had been around for hundreds of years, but in order to boost a propeller-driven plane, a rocket motor would have to burn for ten to twenty seconds. So-called 'long-duration' solid propellants didn't exist, and many engineers intuitively thought they *couldn't* exist.

Their next solid fuel design involved a long charge of black powder mixed with potassium nitrate or sodium nitrate as oxidant, which was intended to burn like a cigarette from end to end over twelve seconds.[4] Once ignited, the solid fuel did what the scientific consensus said it would: burn like fury with pressure increasing in the chamber until it exploded – usually about a second later. Parsons tried numerous different solid propellant combinations but the experiments invariably ended the same way. Boom. It was particularly frustrating for him that none of these experiments involved the thrill of launching an actual rocket. They simply meant laborious preparations for a static test that might literally be over in a flash. The work was exasperating.

Even from behind the closed doors of his Caltech office, Theodore von Kármán couldn't help hearing the ear-splitting results of Parsons' experiments. Like a maestro irritated by a busker, the Boss could stand it no more, and finally asked him to stop.[5] Malina and von Kármán realised that the problem needed to be resolved theoretically: there was little merit in trying new propellant combinations until the principles of combustion dynamics could be better understood.[6] Was the problem mechanical? If, for instance, there was a crack in the compacted powder, it would mean that the ratio between surface area of the burning propellant and the area of the throat nozzle would be highly unstable. What if the problem was more fundamental? Von Kármán sought help from his first love – mathematics.[7]

One spring evening in 1940, he disappeared into his study on Marengo Avenue – a large room stuffed with exotic furniture and Japanese art – and returned some time later with four differential equations for Frank to solve.[8] This wasn't an easy undertaking, given that they were elaborate non-linear equations, the solutions to which depended on specific mathematical functions assigned to different derivatives.[9] Malina worked away integrating their experimental data into the equations, and found that stable pressure in the chamber could be maintained 'as long as the ratio of the area of the throat of the exhaust nozzle to the burning area of the propellant charge remained constant'. In other words, the problem *wasn't* fundamental. Parsons should continue. They just needed to renew their focus on the integrity of the cast, such that from now on there could be no cracks. It was their most important breakthrough to date.

A few weeks after this progress, Malina gave von Kármán a cartoon drawn to mark his 59th birthday. The father of Caltech aeronautics is shown in his natural habitat, surrounded by his GALCIT colleagues and students, with a note in the bottom-right-hand corner reading 'Happy Birthday Boss from Liljan and Frank'. Unsurprisingly, Malina made himself the primary recipient of the professor's attention. Tsien is a more liminal figure, toying with an egg on which is written USA/CHINA, while Homer Joe Stewart – pictured asleep, he was so smart he didn't have to be awake to do maths – is the most remote in this circle. William Bollay is there, and Frank's GALCIT friend Bill Sears, apparently addressing his wife Mabel, who looks down from the walls. The sacred embrace between mathematics and engineering, represented in the open book, is the sign under which they all work, watched over by their own engineering saints, professors Goldstein and Michal.[10] It's hard not to notice that the women in their lives – those shouldering the burden of practical living – are rendered as two-dimensional spectators on male engineering (Pipö, the Boss's fur-decked sister, is the only exception). And then there's Jack Parsons and Ed Forman, who are conspicuous by their absence; they didn't figure among von Kármán's disciples, and enjoyed only a marginal link to GALCIT (when they once got invited to a Caltech staff ball at the Atheneum, it turned out to be a mistake).[11]

Parsons and Forman, for whom calculus was a foreign language, were unmoved by Malina's success with the equations. They resented how Frank could come and go from rockets to soil erosion to political activities, yet drop in, as it were, in a supervisory capacity while they were putting in long hours at the test stand. The previous year he had written home admitting that 'rockets have been having another sleep', while 'Parsons is doing some experimenting with powder and is disgusted with me for not devoting more time to the research'.[12] The reality, of course, was that Malina's episodic theoretical work was an essential guide for Parsons' and Forman's experiments. Without the mathematical knowledge of von Kármán and Malina – not to mention the scientific credibility behind their funding applications – Parsons and Forman would still have been smoking roll-ups, watching their black powder rockets disappear into the Arroyo.

THE VON KÁRMÁN-MALINA THEORY of constant-thrust, long-duration engines was a leap forward from which other successes would soon follow.[13] The team now had a renewed confidence in the empirical work (admittedly, Parsons might have carried on propellant testing no matter what the maths said). Morale was also lifted by having a conceptual development to report to the National Academy of Sciences, progress that was duly rewarded with a doubling of their grant to $22,000 on 1 July 1940. When the news filtered back to Robert Goddard, he was furious – this was serious money for a 'student project', when he was struggling for funding of his own. One biographer even described von Kármán and Malina as 'Goddard's demons'.[14] The additional funding enabled the GALCIT team to lease six acres of land on the western bank of the Arroyo Seco to put up fixed test stands and some primitive buildings.[15] They hired Malina's Caltech friend and comrade Martin Summerfield, who had, like Frank, just finished his doctorate; his first task in professional rocketry was to clear the Arroyo's vegetation with a hand scythe.[16] After that, Summerfield set up office in his car while the other rocketeers, Malina included, lent a hand knocking together some basic structures from wood and corrugated iron.[17]

By the spring of 1941, the sound of Parsons' powder trials at the test stand was shifting from a boom to a disciplined roar. 'The rocket project is beginning to make a little headway at last,' wrote Frank. 'I told Liljan that I was afraid that the damn rocket would work … I know what that will mean.'[18] That was before Germany opened up an eastern line of attack in June that stretched from the Baltic to the Black Sea. Now his rocket research was on the same side as the Soviet Union. It was about fighting fascism – much like the work of another well-known local communist sympathiser, Woody Guthrie, who that year scrawled THIS MACHINE **KILLS** FASCISTS on the front of his guitar.

In August, with their grant increased again to $125,000, they were now preparing to test their first JATO – freshly filled with Parsons' latest solid fuel recipe. The code name GALCIT 27 belied the rather improvised 'office supplies' character of the new propellant, which consisted of amidic black powder and ammonium nitrate mixed in corn starch and Le Page's stationery glue, then pressed into a heavy steel case lined with ordinary blotting paper.[19] Fred Miller, an engineer who worked with Parsons on solid fuel, called it 'goop' – but it was pretty special goop given that they had tried and rejected at least forty other combinations.[20] On the Arroyo stand, GALCIT 27 burned for twelve seconds, producing twenty-eight pounds of thrust. The real test, however, was whether it could assist take-off for an actual aircraft. To trial the JATOs on a small Ercoupe monoplane, they drove west to March Field Air Force base. The pilot was to be Homer A. Boushey, a former student of von Kármán's who had since become a rocket enthusiast.

The Ercoupe was chosen for being 'spin-proof', so stable it was said that anyone who could drive a car could fly it.[21] A battery of tests were scheduled to start on 6 August, with eighteen new JATOs being driven across Los Angeles from Pasadena every other day. On a bright summer morning, the first trial started with just one unit bolted underneath each wing, with the stationary Ercoupe anchored to the runway. That went okay. Eventually the rocketeers worked their way up to testing the JATOs in flight, with Boushey taking the Ercoupe to 3,000 feet before igniting the rockets. Footage from another plane shows the Ercoupe sauntering along before suddenly soaring into the cobalt blue sky, etching the JATO's white contrail in its wake. On one trial the signature exhaust plumes stopped prematurely, and when Boushey landed it was apparent that one of the rockets had 'blown'. There wasn't any damage – Malina had designed shear bolts to prevent a bigger explosion if a critical pressure was breached – but clearly something had gone wrong with the charge.[22]

On 8 August, further stationary tests with four JATOS on the Ercoupe resulted in an explosion. When the smoke cleared, the team could see from the smattering of damage that one rocket nozzle had bounced off the runway and gouged a hole in the rear fuselage. 'Well, at least it isn't a big hole,' someone said. Homer Boushey inserted his entire hand and forearm into the punctured chassis.[23] Von Kármán didn't seem particularly perturbed; he was just happy 'that the basic problem had been licked'. Mere mechanical issues were a lower order difficulty than his big theory, even if they presented a challenge to Malina and Parsons, not to mention the fearless Homer Boushey.

The rocketeers observed that recently prepared charges more or less burned as expected, while those stored overnight – and subject to potential contraction and cracking from the cool night air – usually exploded. A new arrangement was made so that JATOs could be made and then immediately driven the one hour to March Field for loading onto the Ercoupe. Finally, on 12 August, they were ready to test true rocket-assisted take-off. Boushey let off the brakes, rolled down the runway and, after gaining a little momentum, ignited the JATOs. Smoke gushed from the nozzles and, as von Kármán described it, 'the plane shot off the ground as if released from a slingshot … none of us had ever seen a plane at such a steep angle'.[24] This was not aviation as anyone knew it – a spectacle seemingly in defiance of gravity.

As ever with experiments, however, it was the data that really mattered. To that end, two new human 'computers' were hired – a couple, Richard and Barbara Canright – who now carefully recorded and analysed the Ercoupe trajectory. Barbara Canright was the first of many women employed for this work, though in 1941 she was the sole exception to this male domain.[25] It's not just that women weren't hired for engineering jobs; Caltech wouldn't even let them enroll as students for another thirty years. Barby Canright's calculations were crucial, showing unequivocally that not only was this the first time in the United States that a plane had taken off assisted by rocket power, but that it had cut take-off distance by nearly half, from 580 feet to 300 feet.[26]

For an encore, the group tried out an unplanned test – as much stunt as experiment – to see if the Ercoupe could fly with *only* rocket power. This impulsive idea came from von Kármán, an example of the theoretician's breezy confidence even with the pilot's life on the line. The loyal Boushey was willing to give it a try, so the team attached some additional JATOs and unbolted the propeller – the plane's only available source of thrust should the rockets fail. The apparent recklessness of this move was acknowledged when someone pasted a health and safety poster onto the now bare nose of the Ercoupe that read BE ALERT! DON'T GET HURT![27]

As the group gathered around the plane for a team conference, Boushey grew nervous. He had doubts about setting off twelve JATOs (the maximum number possible on the Ercoupe) simultaneously; instead he wanted to try sequential firing of six followed by another six a few seconds later. Everyone agreed to this, but from a standing start the little Ercoupe couldn't even reach flight speed. A photograph captures yet another meeting around the Ercoupe, with von Kármán jotting down calculations on the wing, surrounded by Malina, Summerfield, Clark Millikan and a visibly attentive Boushey.

At this point, Boushey offered to hold a tow rope out of the cockpit window to get some initial momentum before igniting all of the JATOs simultaneously. 'It was surprising how much pull force was exerted on my arm,' he noted with admirable understatement; 'the point of release was somewhat below that wherein the arm bone would have been pulled out of the shoulder socket'.[28] That was the moment when he flicked the ignition. The JATOs' blast sent the plane lurching into the air in an almost unnatural sigmoid curve. No sooner had they finished their twelve-second burn than the Ercoupe went into a steep dive. The spent rockets were acting as an air brake. If not for Homer Boushey's deft handling, the trial would have ended in tragedy. He remained remarkably buoyant, saying that 'the first US flight of an airplane powered solely by rockets was brief but exciting', only later muttering about 'those birds who got me to fly an aeroplane without a prop!'[29]

Despite this success, rocket-powered aviation was still a long way from being a viable technology. GALCIT 27 was too volatile for military use. A new contract signed with the US Navy to develop an eight-second, 200-pound-thrust solid-fuel JATO increased the pressure on the team, yet the solution to stable combustion was far from obvious. Black powder had been Parsons' fuel ingredient of choice, but it tended leave a cast that was brittle and prone to failure. What they needed was an absolutely uniform mix of fuel and oxidant, one that didn't crack with the temperature gradients from Alaska to Africa, and a mix that could evenly fill the casing without trapping rogue pockets of gas. None of the amidic black powder propellants satisfied these criteria.

Exactly who worked out the ingenious solution to this problem, and how, is a matter of contention. The usual story is that Jack Parsons noticed some construction workers tarring a roof with liquid asphalt. He would surely have known about asphalt's provenance in the ancient world, from the 'Greek Fire' incendiaries of the Byzantine Empire to the pitched basket of the infant Moses. Although asphalt can occur naturally, its conventional form is the glutinous black liquid that remains after crude oil has been distilled to form gasoline and other volatile fractions. The claggy residue of this process makes a good binding agent for aggregates on a road surface but also, as Parsons found out, an ideal fuel matrix to evenly carry a granular oxidant. The result was a treacle-smooth goop that Parsons carefully poured into the cyclinder casing, initially just enough to 'prime', but then building up the charge until it hardened like Brighton rock.

On the day of the roof tar inspiration, Malina thought that something 'in the back of ... [Parsons'] mind jelled'; congealment, at any rate, was central to this breakthrough.[30] For the first time, Parsons had developed a *castable* solid propellant – not pressed or extruded or molded – but poured and set like a jelly or a child's plaster of Paris model. They called it GALCIT 53, and it was nothing less than a new paradigm in rocketry. Part of its brilliance was their decision to use potassium perchlorate as the oxidant – a salt previously deemed too dangerous by the Ordnance Department (Parsons was undeterred). In any case, their asphalt-potassium perchlorate mix threw out a regular, controlled exhaust flame, could be stored indefinitely and was stable within a vast temperature range. GALCIT 53 worked so well that it inaugurated an entirely new approach to rocket motor design. It became the antecedent of composite propellants used in everything from Trident nuclear missiles to space shuttle boosters and ejector seats.

Much of this innovation can be credited to Parsons and, in Malina's view, to 'his poetic spirits'.[31] Less spectral presences may also have had a hand in the design. Despite the fact that a patent was eventually credited to Parsons, his close GALCIT team members Mark Mills and Fred Miller both advanced their own claims.[32] Helen Parsons once told an interviewer that it was *her* idea.[33] Malina certainly thought that Parsons' unconventional knowledge was at work, even if it was founded on a theory that he scarcely understood.

The more successful the rocket research became, the more the engineers had to adjust to military secrecy. 'Everything now is confidential and we have to suspect spies under every piece of paper,' Malina told his parents.[34] He didn't exactly conceal his frustration with the restrictions of a classified programme. He was also quite open about his political sympathies, even handing out literature to fellow employees and privately soliciting them to join the party.[35] But as he moved into a management role – Chief Engineer, under the ultimate direction of von Kármán – Malina necessarily became a little more cautious. At only 29, he was younger than many of the growing workforce, and felt a need to be taken seriously. Now that Caltech had made him Assistant Professor he started to wear a necktie as a sartorial emblem of authority, even amid the sweat and grime of the Arroyo test pits.[36] It didn't wash: a mechanic armed with a pair of scissors walked up to him and simply cut it off just below the knot. He decided to dress less formally after that.[37]

The responsibility that came with wartime military funding meant that staff were expected to endure long hours in conditions so primitive that there was ample scope for petty resentments. In the interminable delays between JATO tests, an assistant to Martin Summerfield in the liquid propellants section, Walter Powell, had been playing with a model glider. When other team members were distracted by the plane, Frank got irritated: 'Put away the toy, Walt. It's not a playground.'[38] One account has it that Malina actually impounded the model.[39] Either way, the grievance smouldered until Powell snapped, grabbed a hatchet and started smashing his way through the closed wooden door to Frank's office.[40] Malina was terrified. Other workers managed to calm Powell down, and he walked away embarrassed, but it remained the talk of GALCIT for some time.

There was something wild about the scene at the Arroyo – it was frontier science in every sense: the facilities nestled in the chaparral scrub, where deer could be found in the parking lot, roadrunners would parade past office windows, and tarantulas lurked in office corridors.[41] Space for humans, however, was at a premium. In early 1942, there were only two buildings – a new one that had limited accommodation for engineers, secretaries and the new 'computers', and an old barn that housed the machinists.[42] If an engineer was lucky enough to have an office, then the periodic roar from static tests effectively suspended any conversation that might take place there. Malina shared his office – eight feet by ten feet with a view of the liquid propellant test pit – with his secretary, Dorothy Lewis, and an administrative assistant, Eugene Pierce.[43] In these conditions the Chief Engineer understood that, notwithstanding the occasional axe attack, a collegial atmosphere was important. He was concerned that new staff should be easy to work with and was personally involved in hiring, often through word of mouth.

In the absence of a cafeteria, most workers took their packed lunch outside at noon under a giant oak tree, where they could swap problems across the new specialised sections.[44] Good communications and free exchange of ideas were essential. One inevitable discomfort of Malina's management role was that his oversight of Parsons and Forman had become formalised in the structure, even if they, in turn, led other staff in the solid propellants team and the machine shop, respectively. The easy rapport that Malina and Parsons shared in their early rocket designs was shifting to a more professional contact. Some of their original warmth ebbed away as they gradually inhabited different circles.

MALINA'S CLOSEST WORKING RELATIONSHIP was with Martin Summerfield. Frank put his former roommate and Unit 122 comrade in charge of the liquid JATO programme that ran in parallel with Parsons' solid fuel. Liquid propellants were not new – Robert Goddard had been experimenting with them for decades, though with limited success. In theory, they promised greater thrust per weight of propellant, and could be controlled through advanced plumbing (unlike, say, Parsons' twelve-second cigarette, which couldn't be extinguished once ignited). But the theory hadn't quite been borne out. Goddard, for instance, had never come near their current goal of a thousand pounds of thrust for one minute. Other experts doubted that this would ever be possible; one told von Kármán that they'd need a combustion chamber the size of a small house.[45] Summerfield's approach was the opposite of Parsons. Inspired by a passage in an old British chemistry textbook that claimed that a liquid hydrocarbon could burn completely in a millisecond, he dived headlong into the theory, and calculated that a small burning chamber should provide enough thrust.[46] But Summerfield also built on earlier work by Parsons and the suicide squad, including the experiment that rusted Caltech's lab equipment, from which Malina and von Kármán had settled on using Red Fuming Nitric Acid (RFNA) as a potential oxidant.[47] It's a brutally corrosive substance but, unlike liquid oxygen, it didn't have to be kept at ultra-low temperatures.

Summerfield set to work with his axeman assistant, Walter Powell, to see if they could perfect a gasoline-RFNA motor. They quickly discovered that, though the propellants would obligingly combust in an open crucible test, the chamber of a rocket motor was a bit more complicated. In smaller motors such as, say, a three-inch-diameter chamber giving 200 pounds of thrust, the tests mostly worked. But in scaling up the design, the motors started to pulse dramatically, each pulse gathering intensity until, if it wasn't shut off, the chamber would explode. This combustion instability – yes, 'throbbing' – would turn out to be a persistent problem, even with a modest four-inch cylinder. It was, as the Boss put it, 'the knottiest problem in liquid propellants at Caltech'.[48]

The solution was soon at hand. Frank visited the Naval Engineering Experiment Station at Annapolis to talk to Lt Robert C. Truax, who was undertaking his own liquid propellant research, albeit with limited success. Malina and Truax may have been rivals, but they got on well and freely shared information. Truax, though initially inspired by Robert Goddard, knew very well the pitfalls of furtive science, so he was happy to share his colleague Ray C. Stiff's observation that aniline reacted spontaneously with nitric acid. Maybe if they added aniline, it might help with their throbbing problem. Malina turned this idea over in his head, thinking about the many benefits of a so-called 'hypergolic' reaction, one that combusted without ignition. But why add aniline to gasoline? Why not just use it to *replace* gasoline?[49] The next morning he wired Summerfield in Pasadena asking him to react aniline with RFNA. Summerfield and his assistants, Walter Powell and Edward Crofut, carefully poured the toxic RFNA into a small open crucible.[50] Separately, they filled a long-handled beaker with the aniline, trying not to inhale the signature smell of rotten fish. As soon as Summerfield tilted the beaker, spilling the aniline, a ball of flame mushroomed from the crucible.

'From then on,' recalled Powell, 'we were acid-aniline people'.[51] By the time Malina made it back to Pasadena, Summerfield had already perfected the combination in static tests.

The journey from GALCIT's first introduction to hypergolic propellants in early February 1942 to their first flight test took just two months. March passed in a blur. 'Not written for some time,' apologised Malina in his letter home, 'mainly because of the pressure of work during the past hectic two weeks. One does not live through many similar periods, at least I hope not.'[52] By April, they were ready to apply liquid rocket power to a Douglas A-20A at Muroc Field in the Mojave Desert. 'Keep your fingers crossed for us,' Malina asked his mother, testing the boundaries of his security restrictions. 'I have hopes that we will repeat what we did last summer so nicely. Can you make sense of these riddles?'[53]

The Douglas A-20A, however, was no Ercoupe: it was a twin-engine attack bomber, a rugged workhorse of the air war in which liquid JATO units were permanently installed. With each unit offering a thousand pounds of thrust, this was a real-world military application of liquid propellants. Another former student of von Kármán, Major Paul H. Dane, was brought in as the test pilot, while Clark Millikan came to give his input on the theoretical side. Despite his initial scepticism, Millikan had become increasingly involved in the rocket work.

After the first static ground tests, the plane was readied for true rocket-assisted take-off. The engines rumbled into life, the parking break was released and the rockets were ignited. The bomber rose into the air with a roar. 'Almost straight up,' remembered von Kármán, 'the A-20 continued to rise as though scooped up by a sudden draft'.[54] It wasn't a one-off: they did forty-four successive tests without JATO failure, in each case dramatically shortening the take-off distance and duration. Malina was jubilant. 'The most important factor is that this weekend the outlook for the future is the brightest it has ever been,' he wrote home. 'We now have something that really works and we should be able to help give the fascists hell!'[55]

JATOs became operational in 1943 with a patent registered in the name of Malina and Parsons (though, oddly, not Summerfield).

By the end of World War II, their use in rescuing military personnel from isolated sites with short runways is estimated to have saved 4,500 lives.[56] And as von Kármán put it, their success marked 'the beginning of practical rocketry in the United States'.[57]

THE WAY OF THE BEAST

AT 1746 WINONA BOULEVARD, a gong signalled the start of the Gnostic Mass. Jack Parsons followed as the congregants made their way up three flights of stairs and filed into the Temple − a dimly lit attic of a Hollywood mansion that was now given over to the Ordo Templi Orientis (OTO). On the east side of the room, on a chequerboard dais, the High Altar was bedecked with candles and flanked at each end by one black and one white obelisk. The east side was chosen because it was the closest to Boleskine House, a stately hunting lodge five thousand miles away on the shores of Loch Ness, Scotland. None of the congregants had ever been to Boleskine, an obscure and distant Mecca for the followers of its one-time owner: the English occultist, magician and poet Aleister Crowley, aka The Beast.

Aleister Crowley didn't start out as The Beast, nor even as 'Aleister'. He was plain old 'Edward Alexander', born in 1875 to devoutly Brethren parents, whose religious and orthographic conventions he defied, in part, by coining a new name for himself − the first of many. The inventive spelling of Aleister may have hinted at a unique young man, but it would soon be apparent that no mere Christian name could ever capture his anti-Christian multiplicity. On the whole, Crowley preferred titles to names. He had the peculiar English burden of being born into wealth but not into aristocracy (his family's brewing fortune was considered déclassé), leaving him acutely sensitive to subtle gradations of the British class system. When he became sole prophet of his new occult religion, Thelema, he seized the opportunity to enoble himself. He would now be known as 'The Great Beast', '666', 'To Mega Therion' and 'BAPHOMET'. Social class, for Crowley, was like the degree rites of his magick: hard-won status in a fixed hierarchy. For this reason, among many others, he hated both Christianity and the 'vermin socialists'.[1]

As far as the British tabloids were concerned, Crowley was 'the wickedest man in the world'. Even in California, which had less orthodox ideas about religion, the popular press referred to his acolytes as the 'purple cult'. Not that this would have discouraged Parsons when his friend John Baxter invited him to attend a mass at the OTO.[2] Nothing attracted Jack like a bad reputation. For Baxter, the libertarian spirit of the OTO made it an accommodating home in the days when homosexuality was illegal; an ethos of permissiveness was not merely a lifestyle at the OTO but a part of their religious practice. Parsons, who oscillated between same-sex attraction and revulsion, found the scene intriguing.[3] As with his experiments on rocket propellants, there was a technical knowledge to be acquired and secrets to be kept, but there was also a sense that such knowledge might yield infinite possibility. This excited him.

At the top of the Temple's stage, Jack could just make out a coffin or sarcophagus, partially obscured by a gauze curtain. A white-robed Priest entered to the sound of organ music, passing the fifteen or so worshippers as he made his way to the altar. Underneath the robe was Wilfred Talbot Smith, by day a clerk in a gas company, though his astral life was taken up with OTO 'Agape Lodge' No. 1, which he had founded. The Priest kissed *The Book of the Law* before reading aloud its primary injunction: 'Do what thou wilt shall be the whole of the Law.' This central tenet of Thelema, the OTO believed, was not authored by Aleister Crowley but rather was the revelation of a 'praeterhuman' presence that in 1904 had interrupted The Beast's Egyptian honeymoon and asked him to take dictation. Crowley duly recorded the text that would inaugurate an Æon of Horus – later given ritual expression when he wrote an accompanying Gnostic Mass.

Even among those with esoteric interests, the Gnostic Mass was little known and even less performed; together with an LA dance teacher named Regina Kahl – the Priestess – Smith became its first public practitioner. By the time Jack first attended in 1939, it was being staged once a week.

The Priestess had assumed a supplicant position and was now stroking the Priest's lance. He declared that he would now 'take thee, Virgin pure without spot'. At one stage, the Priest fell, as the liturgy demanded, in 'adoration' of the Priestess, kissing her knees; later he kissed between her breasts. It was heady stuff. One Crowley biographer claimed that Smith made cunnilingus the central act of the Gnostic Mass. That's probably not accurate, but a degree of timely disrobing was still, on occasion, part of the performance.[4]

The whole ceremony was long and earnest, but if the Thelemites were eager to impress The Beast, he was having none of it. 'Most of them look loopy,' Crowley remarked on seeing photos of the Mass; 'none of them look as if they had any birth, breeding or education whatsoever.'[5] He hadn't met most of these disciples in person but managed to direct the group via transatlantic correspondence. Given his reduced circumstances in a small London boarding house and his gargantuan heroin habit, he was hardly going to turn down their financial support. It was still easier than selling 'elixir pills' that contained his own semen as the active ingredient.[6]

Jack and Helen felt themselves increasingly drawn to the OTO, if for different reasons. Helen was wary after their initial contact in January 1939. She baulked at Crowley's treatment of women as moral inferiors, yet found in his writings something serious and purposeful.[7] She was a more attentive reader than Jack, whose interest, at once mystical and poetic, was significantly founded on the lure of the transgressive. Helen, by contrast, saw in the initiates of the OTO a more mature and disciplined community than the social circles that they currently inhabited. There was a closeness about the OTO community and their 'profess house' at Winona Boulevard, which, Helen hoped, might be a corrective to Parsons' wayward tendencies.[8]

It had by now become painfully apparent to her that Jack's free-spirited approach to life also extended to their marriage. On a number of occasions Helen had arrived home to their house on Terrace Drive in Pasadena only to find another woman slipping out the back door. While Jack might have had a habit of signing off letters to Helen with *Semper Fidelis* (always faithful), it wasn't a valediction he observed in practice.[9] In this respect, Helen's hopes of the OTO were ill-founded. At the time of their initiation into the Agape Lodge, it was a hothouse of sexual adventure, jealousy, gossip and recrimination. The OTO didn't so much mitigate Parsons' waywardness as inject it with an amphetamine rush of occult energy.

Jack and Helen had their initiation on 15 February 1941. It was a solemn affair. Wilfred Smith was dressed, as the ceremony required, in the guise of the Muslim warrior Saladin, with vertical striped robes and a white turban (Crowley saying the costume made him look 'silly').[10] Like all candidates for the Minerval 0°, the couple were asked to demonstrate their understanding of *The Book of the Law*. Saladin first asked them to explain the nature of the law.

'Do what thou wilt shall be the whole of the law,' they replied. Were they willing to stake their life to defend these principles?

'I am,' they replied in unison.

At this point, Saladin expounded on the paradox that freedom to follow their will was founded on their submission to the discipline of the OTO. Even the Communist Party wasn't quite this demanding, nor did the Comintern specify 'mutilation' as the inevitable consequence of breaking rank.[11]

At 56, the English-born Smith was an earnest if at times ineffectual Priest, who ruled over social and ceremonial activities with a bumbling air. He dished out instructions and prohibitions to members but was often beset by insecurities, inflamed by The Beast himself. Crowley's correspondence with Agape members, Smith in particular, is characterised by endless complaint and disappointment: they didn't send 666 enough money, they couldn't attract wealthy members, they lacked talent or class or bearing. 'He has no presence,' despaired Crowley of Smith. 'He can't wear robes without looking ridiculous, and' – here was the real sin – 'he simply exudes lower ... middle class'.[12] The only member of Agape who actually knew Crowley well was Jane Wolfe, a former star of the silver screen who had given up Hollywood to live with The Beast at his Abbey of Thelema in Cefalù, Sicily, in the early 1920s. She and Smith had been lovers until the arrival of Regina Kahl eventually displaced Wolfe as Priestess. Kahl became a stalwart of the lodge, as well as its most effective evangelist, drawing in many of her own drama students, including Phyllis and Paul Seckler, who would in turn play notable roles in the OTO.[13] But there was no pleasing The Beast. In correspondence, he referred to Regina as 'Vagina', dismissing her as 'probably a good fat old whore'.[14]

This was the world to which Jack returned, exhausted from the trial and error of solid propellant JATOs. It was the community that Helen hoped would nurture a more mature relationship with Jack. They were certainly given a warm welcome. Jane Wolfe described Parsons as 'Crowleyesque in attainment' … 'vital, potential bi-sexual at the very least … now engaged in Cal Tech chemical laboratories developing "bigger and better" explosives for Uncle Sam'. Jane thought that Jack could well be their next leader, concluding that he was 'the natural successor of Therion'.[15] In March, Parsons became a Probationer under Smith, taking the name Frater T.O.P.A.N., which, he said, stood for *Thelemum Obtentum Procedero Amoris Nuptaie*, or 'the attainment of Thelema through the nuptials of love' (ever quick with a dry word of discouragement, Crowley wrote that Parsons' mangled Latin was a 'language beyond my powers of understanding'.)[16] The new disciple, at any rate, took the motto seriously – like his master, he would do his level best to flout conventional morality, and expected Helen to do the same.

IN THE SUMMER OF 1941, when Frank Malina was at meetings in Washington, Jack and Helen invited Liljan to a party at Winona Boulevard, an event that was aimed in part to attract new members to Agape Lodge. Just as Frank had been known to encourage his colleagues to attend Unit 122, Jack too recruited from among his GALCIT friends. The loyal Ed Forman and his wife Phyllis signed up without demur. From among the rocket team, Jack also brought in the 'human computers', Richard and Barbara Canright, along with his solid propellant colleague, Fred Miller.[17] Parsons even persuaded

Helen's teenage half-sister, Sara Northrup – everyone called her Betty – to join. Liljan, with a demanding social life in the Communist Party and in art school, wasn't the least bit tempted. She had no interest in Thelema, but she hadn't forgotten the Parsons' friendliness when she and Frank had first started dating. She had no reason to feel inhibited from turning up at the huge house – until the moment she entered the front door.

'It was like walking into a Fellini movie,' she remembered. 'Women were walking around in diaphanous togas and weird make-up. Some dressed up like animals, like a costume party. Something was going on, but it was a little strange. Eventually there was this big thing about going up three flights of stairs and ending up in an attic for what I thought was to be entertainment, but appeared to be a rite.'[18] 'The woman in the see-through thing [was] doing a very sinuous dance around a pot of something glowing in the middle of the floor.' Even Jack's famous Parsons Poison Punch didn't take the edge off Liljan's anxiety.[19] 'All I could think of was if this place catches fire, we are all goners. It was too much.' She noticed that 'couples were disappearing … it was very strange'. She told Frank everything when he got back from Washington, but he simply shrugged: 'don't pay too much attention to it'.[20] He didn't want to know.

Within months of the Parsons joining the OTO, the lodge was aquiver with low-level intrigue, by no means all of it occasioned by their arrival. Regina Kahl was upset that Wilfred Smith had been paying nightly visits to the room of her former dance pupil, Phyllis Seckler, a 24-year-old bank clerk and OTO Probationer. Phyllis, meanwhile, was initially happy to see the unexpected arrival of her husband, Paul Seckler, whose girlfriend had recently ditched him. This, in turn, displeased Wilfred Smith, who, as lodge autocrat, tried to forbid Paul from being upstairs or in Phyllis's room.[21] Not that Paul's interests had exclusively returned to Phyllis – he had also taken a shine to Liljan Malina, having met her at the Winona Boulevard party ('it was just getting too embarrassing,' said Liljan, though she noticed that he was 'a very handsome young man').[22]

The arrival of Helen's half-sister, Betty (Sara) Northrup, caused a particular stir. At 17, she was tall, blonde and willowy, with a sensual frankness that papered over a broken childhood. She had come to the lodge as an escapee from the sexual abuse of her father, Burton North-rup, a debt collector and convicted fraudster.[23] Helen understood this because she, too, had once been subject to the same violence, but their shared trauma ended up being a point of difference between the sisters. When Helen tried to talk about it, Betty shut the conversation down as if it was of no concern.[24]

In June 1941, Helen returned from a desert holiday to find, as she put it, 'my own sister in my own house with my own husband'.[25] Betty was wearing Helen's handmade clothes and claiming to be Jack's new wife. Apparently, the involvement had been going on for some time. Jack was not only unrepentant at this development, he matter-of-factly told Helen that it was because he preferred Betty sexually.[26] Were this not cruelty enough, his response to Helen's hurt was to arrange for Wilfred Smith, thirty years her senior, as some kind of consolation prize. Parsons told the Priest that he should now consider Helen as his own, though neither of them thought to ask her consent to this backroom deal. The first she knew of it was a bewildering and unwanted advance from Wilfred.[27]

This wasn't the beginning of the abuse, nor its end. Shortly before her death in 1998, Betty Northrup told her daughter Alexis that Jack initiated their sexual relationship when she was just 13 – two years before the Parsons were involved in the OTO.[28] Judging by his memoir, Jack thrived on the sense of transgression:

> Betty served to effect a transference from Helen at a critical period. Had this not occurred your repressed homosexual com-ponent could have caused a serious disorder. Your passion for Betty also gave you the magical force needed at the time, and the act of adultery tinged with incest, served as your magical confirmation in the Law of Thelema.[29]

In Crowley's schema, the violation of social and sexual taboos was a potent source of magickal power, all the greater with a person who, by virtue of their youth, was deemed pure and sanctified.[30] Even if the coercion of Betty Northrup preceded their life in the OTO, abuse wasn't exactly aberrant behaviour at Agape Lodge. Parsons' biographer, John Carter, details how a 16-year-old boy told the police about being 'forcibly sodomised' by 'three of Parsons' followers', but that the professional respectability of some OTO members reassured the police that there was nothing to investigate.[31] All of this suggests that the aggressive sexuality of Crowley's *Hymn to Pan* was, for Jack, more than figurative. Parsons liked to recite the hymn as a kind of consecration of JATO tests, igniting the charge at its violent conclusion:

> I am Pan! Io Pan! Io Pan Pan! Pan!
> I am thy mate, I am thy man,
> Goat of thy flock, I am gold, I am god,
> Flesh to thy bone, flower to thy rod.
> And I rave, and I rape and I rip and I rend
> Everlasting, world without end,
> Mannikin, maiden, Maenad, man,
> In the might of Pan.
> Io Pan! Io Pan Pan! Pan! Io Pan! [32]

Barbara and Richard Canright, the rocket 'computers', were drawn into the scene for a year or two. Barby Canright is rightly celebrated as one of the pioneering women making technical contributions to the rocket programme.[33] It's notable, however, that she never once spoke of GALCIT. Why not? We know nothing about her life at this time other than that years later she had a nervous breakdown from which she did not recover.[34] One possibility is that this whole period was mixed up in the trauma of her time in Agape Lodge. Through her domineering husband, Richard Canright, as well as Jack Parsons, she became part of a group in which coercive sexual behaviour was to a large degree normalised.[35] Given the truly historic role that Barby Canright occupied, the silence in her testimony feels ominous.

For her part, Helen Parsons seemed relieved to have split from Jack, though it took some time for the rows to subside. Jane Wolfe noticed 'cracked doors' as 'mute witnesses of combats with Jack', suggesting that he remained physically as well as sexually abusive.[36]

Helen and Jack didn't formally divorce at this stage – that would have been to concede too much to the 'detestable institution' of marriage – rather, they simply came to inhabit the same house with their different partners.[37] For Helen, spending intimate time with Smith made her forget 'the sore spot where my heart used to be'.[38] They all ate together, though Jane still saw 'the politics played, the scheming, the subtleties used to gain and keep power'.[39] Though Jack and Helen irrevocably lost their commitment to each other, they both remained devoted to Thelema, and each started to exercise greater influence on the direction of the lodge. Jack pushed the group to move to a new house, one that would not only be closer to his work but might also enable an expansion of their membership. Helen gradually assumed the role of Priestess to Smith's Priest, a transition that pained the ousted Regina Kahl. 'It was a pretty turbulent time,' admitted Phyllis Seckler, and it wouldn't be long before the vortices from the Ordo Templis Orientis collided with those of the rocket project.

FOR ALL THAT THELEMA taught 'Love is the law, love under will', a culture of episodic violence prevailed. Even aside from what was happening between Jack and Helen, relations between Helen and Betty also descended into fist fighting.[40] Paul Seckler, who, despite Smith's initial decree eventually came to live in the lodge, unleashed further aggression. He and Phyllis had patched things up; in May 1941 she gave birth to their second daughter.[41] Paul found odd jobs, initially in a restaurant, until he was later hired – presumably on Jack's recommendation – as a night watchman on the new GALCIT premises at the Arroyo.[42] The following month, Paul Seckler played the role of Emir at the initiation of three new members: a soldier, Grady McMurtry, whom Jack had met at the Los Angeles Science Fiction League; McMurtry's fiancée, Claire Palmer; and Betty Northrup. Phyllis had been staying with Paul's relatives in Perris when, together with their toddler and newborn, she returned to the lodge in the midst of the post-initiation celebrations. Initiation into the OTO's First Degree, in which Paul Seckler had officiated, requires the candidate to foreswear any undue indulgence in alcohol. No one took that promise too seriously, least of all Seckler, who was very drunk when Phyllis arrived with their children. She recorded only that he 'nearly tried to kill me', concluding that 'there was nothing for it but to leave'.[43]

A similar scene unfolded in November. Frank was at the GALCIT office when he got a phone call from the police saying that one of his employees was in jail. It wasn't exactly clear what had happened, but Paul Seckler had been drinking with Parsons and Forman when the evening took a turn. Malina, who always took a studied disinterest in the details of Parsons' magickal life, thought they had conducted a séance.[44] It's more likely to have been an OTO ritual that somehow disturbed Seckler, who, being drunk and possibly high, now grabbed Parsons' Colt pistol and ran out into the street. 'There was a couple necking in a car,' Frank recalled. 'He forced them out at the point of a gun, took the car, drove to Hollywood evidently not quite knowing what he was going to do, and then after a certain amount of time drove back to Pasadena.'[45]

The police were waiting for Seckler at the Colorado Street Bridge as he returned in the early hours of the morning. Frank left GALCIT and went down to the jail to see 'what exactly made him do a stupid thing like that'. He found Seckler 'very vague'.[46] More worrying for Malina, he 'couldn't get anything out of Parsons or out of Forman as to why this happened'.[47]

Whatever mutual respect had once supported their rocketry collaboration, Parsons and Forman now openly strained against Malina's authority. They gave no reassurance or explanation. It was left to Frank to placate the chief of police, persuading him to arrange things 'so that the newspaper reporters did not get our project involved'.[48] Seckler was eventually convicted of grand theft auto and spent two years in San Quentin State Prison.[49] For the rocket project, however, the outcome of this incident was a lasting breakdown in trust. 'I always sort of smiled at Jack,' said Frank of his friend's 'mystical approach'.[50] 'I knew he was involved in some of this magic kind of thing but he [had] never made any effort to pull me in.' With little knowledge of life at Agape Lodge, Frank tended to think the best of Jack. 'I don't know,' Frank said later, 'maybe I used to josh him a bit when he would pull out some of these exotic books. I just couldn't take it seriously.'[51] The Seckler incident changed things. It 'then became quite evident to me that whatever it was that Parsons and Forman (much less) were playing with had certain worrisome aspects'.[52] At the time, however, Frank felt disinclined to investigate further. Every time Liljan tried to talk about the OTO, her concerns were rebuffed. 'I can still see myself standing by his chair where he was, as usual, working,' she wrote in her journal. 'He never looked up. He kept his eyes on the papers spread out before him and told me I was being ridiculous.'[53]

ENTERPRISE

EVER SINCE THE SUCCESS of the Ercoupe, Malina had been toying with the idea of setting up a company to manufacture JATOs. He had seen how Caltech had done well out of military research; along with Harvard, MIT and Columbia, the pursuit of weapons science was part of a university's wartime duty, one for which Caltech alone was awarded a staggering $83 million during World War II.[1] But neither Malina nor von Kármán felt that the liberal ideal of the university should extend to manufacturing armaments. They had seen how the Caltech physicist Charles Lauritsen had started a different rocket programme for the Navy and had been drawn into production contracts – a 'great mistake', thought Malina. He felt strongly that Caltech was in the business of minds not missiles, which is why he first suggested the idea for a rocket company to von Kármán in September 1941.[2] Surely if JATOs were going to work, and were likely to be made, why shouldn't they be the ones to profit from their own invention? Parsons and Forman certainly felt that way. For all his communist inclination, Malina felt that 'if a human being can get financial independence, then he's much freer to do what he wants'.[3]

The first step was for von Kármán to sound out aircraft manufacturers who might have an interest in building JATOs under licence. Von Kármán went to see his friend Jack Northrop, the founder of

Northrop Corporation, who had started his aircraft factory in a small garage. He wasn't encouraging.[4] Others all said the same thing: rockets are not a viable business if the only possible customer is the US government.[5] Von Kármán was experienced enough to know that Caltech itself might not be too pleased if two of its faculty started a private company to profit from GALCIT research. So he diplomatically consulted Robert A. Millikan, the Caltech chair of council, and an arch supporter of American enterprise. Millikan was unsure, less about the principle of a spin-off company than about whether mere faculty members could successfully manage a business.[6] In the end, Millikan raised no major objection, but all these discussions were still unresolved when the Seckler incident blew up. The last thing any of them needed was bad publicity, particularly when Frank was investing all efforts into smoothing the first potential contracts, not to mention making sure the JATOs actually worked.

Crashing through these tentative negotiations came the blunt force of history. On 7 December 1941, Frank woke as usual and turned on the radio to listen to the symphony concert, only for the broadcast to be interrupted by news of the attack on Pearl Harbor. 'The whole state of affairs seems rather unreal,' he wrote to his parents.[7] The following week, Frank and Liljan went outside to the street to join their neighbours in watching the lights go out across the city. 'We do not know if the blackout was just for practice or if bombers were around.' Yet he could still find room for optimism: 'the Nazis for once are going backwards and the Russians are showing the rest of the world how to do it'.[8] But with the United States now drawn into the war, the prospects for a company in the business of aerial supremacy took a marked turn for the better. Malina's practical anti-fascism might even turn a profit.

In the meantime, other issues occupied his attention. Who, for instance, should be invited to join the company as stockholders? The chief investor and advisor would be von Kármán's attorney, Andrew G. Haley, a well-dressed businessman and sharp lawyer with the jowly appearance of Alfred Hitchcock.[9] Andy Haley – no one ever called him Andrew – could see serious business potential if the Navy and Air Force could be kept onside, so he volunteered as company secretary and sunk the lion's share of capital, $2,000, into the venture.[10] Von

Kármán would be president, with Malina as treasurer, and Parsons, Forman and Summerfield as vice presidents – all of whom put in $200 and assigned any patents they held. But it's notable who *wasn't* on the Board of Directors. Clark Millikan heard about the venture, either from von Kármán or from his father, Robert Millikan, and smarted at not being asked to join. Tsien was a surprising omission, given his theoretical assistance at various junctures; Homer Stewart too was at the margins of conceptual contribution but not invited to the party. Haley aside, a stake in the company was reserved for the absolute core staff.

Another question for the new business was what to call it. Von Kármán suggested 'Superpower' – it's not clear why – but this was rejected in favour of the modern portmanteau 'Aerojet'. The company incorporated on 19 March 1942, setting up offices in an old juice cannery on East Colorado Street in Pasadena.[11] At first, things went smoothly. Then in September, von Kármán got a letter from the Air Corps to say that they wouldn't be renewing their contract. This was the hazard of having only one client. With no explanation forthcoming, von Kármán and Malina flew to Washington to talk to General Ben Chidlaw of the Air Corps. 'We like you very much, Doctor,' said the General, 'but only in cap and gown to advise us what to do in science. The derby hat of the businessman does not befit you.'[12] Malina and the Boss schmoozed high and low in Washington (one military advisor, with whom Frank was negotiating, asked to send his regards to Liljan. 'You made a big impression on him, my Mata Hari,' wrote Frank in a letter.)[13] A new arrangement was finally worked out whereby Andy Haley – a hard-drinking dealmaker with an affinity for detail – was made the new President of Aerojet Engineering Corporation. Somehow, this reopened the tap of military contracts. Haley's glad-handing would be a large part of Aerojet's success, and Malina for one was relieved. 'None of us have the time to devote to that job,' he told his parents. Haley 'is a lawyer and one almost has to be lawyer to understand all the government forms that other lawyers have cooked up'.[14]

THE NEW CONTRACTS REPRESENTED a significant financial confidence in rocketry from the US military at a time of war.[15] The problem was the interminable workload for Aerojet staff. 'Every-thing is going as fast around me that it's bewildering,' Frank told his parents. 'The company will either be a wonderful success or a glorious flop, there won't be any in-between.'[16] All of 'the old gang' were now routinely putting in twelve-hour days, though they were mostly doing so at Aerojet rather than at GALCIT.[17] Parsons, Forman and Summerfield all moved full-time to supervise production of solid and liquid fuel JATOs, while Malina stayed on the rocket project at GALCIT, with his Aerojet consulting role and directorship on the side.[18] A short-term consequence for Frank and Liljan was that they actually had less money in the bank. Plans to buy a house were shelved in order to sink capital into the company; they decided instead to rent a bigger place on South Hudson Street, finally gaining entry on 10 June 1942.[19]

Three months after they moved in, Frank and Liljan held a house-warming party for about forty friends and colleagues. It was 'sorta a morale uplifter,' he told his parents, 'with dancing and punch drinking etc'.[20] He didn't detail who was invited but it included work colleagues, political friends and Liljan's art school contemporaries. Frank wasn't partisan about his friendships: if a colleague was a conservative Republican, that needn't get in the way of cordial relations. He always felt friendly towards Homer J. Stewart, even if this wasn't always reciprocated. Frank's differences with Clark Millikan were bigger than both rocketry and politics. And then there was Louis Gerhardus Dunn, another talented if conservative engineer. 'I soon realised that Louis Dunn's attitude to questions that I thought were important like racial equality were quite different,' Frank once told an interviewer. 'Louis Dunn and I were great pals, great friends but there were territories that we just left alone.'[21]

It's not clear that Dunn felt the same way. He had arrived from South Africa, the youngest son from a religious family who turned away from his expected career as a church leader after the sight of an airplane flying over the family ranch captivated him. He decided to become a pilot and enrolled in a flight school in Los Angeles, only for the school to close down. Somehow he ended up at Caltech, as yet another disciple of Theodore von Kármán.[22] Dunn may have been happy to leave South Africa, but he held on to his country's racial politics. He didn't like liberals, far less communists. He was one of those at the Malinas' house-warming who could not just set aside his politics for the sake of a pleasant evening. He didn't forget how that night, as the cocktails flowed, the conversation at the Malinas' party went 'from red to redder'.[23] For the most part, it was a cheerful event that lasted until about midnight. That, at least, is what 'Confidential Informant T-7' told the FBI.[24]

It's a tiny detail, of no real significance, but the duration of the Malinas' house-warming is the very first entry in what would become a system of total surveillance, a system that hoovered up information on their domestic lives at the request of none other than the FBI Director himself. The ordinary comings and goings of Frank and Liljan – visitors, journeys, car registrations – would all be sucked into the database that J. Edgar Hoover called the 'Custodial Detention Index', or CDI – a vast register of potential subversives.[25] Malina wasn't formally placed on the index in 1942, but its machinery became a shadow, at first unseen, that followed his every movement and conversation, extending to friends and colleagues, from whose ranks other informants could be recruited. Informant T-7 was most likely a neighbour, given that they provided other information about visitors. But then there was T-6, who was even closer to home, an unidentified GALCIT colleague who told an agent that he'd heard rumours that Frank was a communist. Not only that, but that he'd invited other colleagues to attend communist meetings.[26]

IT WASN'T EASY to balance precision manufacturing at Aerojet with the social turmoil of the OTO. Still, the difficulties presented by these rival demands had one obvious panacea: drugs. Jack Parsons loved drugs.[27] He loved how they heightened and extended his psychonautic adventures. Besides, it had been a constituent part of Crowley's magickal metier, as well as of his literary output. So it's hardly surprising if Parsons thought that imitative behaviour could be a tribute to the author of *Diary of a Drug Fiend*. In early 1943, Jack and Wilfred Smith decided to start an OTO magazine called *Oriflamme*, in which Jack contributed a poem that opened:

I height Don Quixote, I live on Peyote
    marihuana, morphine and cocaine.
I never knew a sadness but only a madness
    that burns at the heart and the brain,
I see each charwoman ecstatic, inhuman,
    angelic, demonic, divine,
Each wagon a dragon, each beer mug a flagon
    that brims with ambrosial wine.

Crowley responded with his usual asperity, declaiming the magazine as 'a ragbag of imbecility' and 'the most amateurish production I have ever set eyes on'.[28] Others had reservations about the drugs, if not the verse. The new OTO initiate Grady McMurtry complained that Wilfred Smith had Parsons 'coked up like a snowbird', while his rocketry colleagues noticed that he now wore a perpetual sweat – a common sign of amphetamine use.[29]

It was about this time that the pregnancies began. Parsons arranged for an abortion for McMurtry's wife, Claire, then another one a week or two later for Betty, but he didn't miss the opportunity to taunt Helen about how she'd feel about the idea of her sister bearing his child.[30] Helen had long wanted a baby, which Jack had always refused, though by this stage she and Smith were now expecting one of their own. Crowley decided that the entire mess – from the lack of ritual seriousness to the endless sexual jealousies – was Wilfred Smith's fault. For years, Crowley had lambasted and belittled Smith, 'a cheeky Cockney, devoid of atmosphere, lacking presence and personality, who doesn't know quite know what to do with his hands'.[31] Yet Parsons somehow emerged with his reputation intact. Crowley was coming round to the idea that Jack could be the next leader. And it doubtless helped that he now brought in the largest income in the OTO. Crowley certainly noticed the founding of Aerojet; he told Jack that $2,000 would be required to bring The Beast to the United States (alas, the State Department eventually denied him a visa) and to publish his new essay on the philosophy of tarot cards.[32]

Jack agreed to find the money for his master, but there were other pressures on his wallet.[33] In 1942, he persuaded Smith and the Agape

Lodge to relocate to a new, larger 'profess house' in Pasadena rather than Hollywood – to allow him easier access to Aerojet. Though Smith was still Priest and leader, it was clear that the group increasingly organised itself around Jack's life and work. The new house at 1003 South Orange Grove Avenue represented a personal return to his wealthy childhood, to the same street on which his grandfather, Walter Whiteside, had bought his grand Italianate mansion. Granted, the millionaire's row had never quite recovered its sheen after the stock market crash; the lawns were shabbier and the houses lacked the impressive maintenance staffs of the roaring twenties, but it was still a homecoming to a desirable street. For Jack, it was the restoration of status lost in the Depression. A friend of his mother's had given them the lease for a discounted price on condition that they assumed liability for upkeep. Smith sunk his retirement savings from his job as a gas clerk into the house fund, but it was Parsons on which the house financially depended.[34]

Agape took possession of 'Grim Gables', as it came to be known by its inhabitants, at exactly 11 a.m. on 9 June 1942 – Wilfred Smith's 57th birthday.[35] As the senior member, Jane Wolfe entered first, carrying over the threshold a signed portrait of Crowley. Helen took up the rear, bearing a wooden ark into which Smith deposited *The Book of the Law*.[36] A large living room to the left of the front door was designated as the Temple, though the ritual observance of the Gnostic Mass had all but ceased amid recent upheavals. All told, there were sixteen rooms, including a library with hand-tooled leather murals, a music room with a piano and Jack's phonograph, plus five bathrooms, two wine cellars, an immense garage and laundry, not to mention extensive grounds.[37] It soon filled up. Jack and Betty took a suite of rooms at the back of the house, while the master bedroom went to Wilfred and Helen. Other members occupied the remaining bedrooms, followed by a swarm of bees, which Jane took to be a good sign. The discovery of bats in the attic was deemed less magickally propitious; they were exterminated.[38]

In September 1942, the Agape Lodge held an equinox gathering – a house-warming of sorts, the same month as Frank and Liljan's party. Jack swept the house for copies of Crowley's recent one-page libertarian manifesto, *Liber Oz*, lest it be found by what Jane Wolfe called his 'army associates'. She didn't approve of this cleaning up, and wrote to Crowley to tell him about it; predictably, The Beast worked himself into a lather about 'that white-livered lunatic Jack Parsons'.[39] But Parsons could see well enough that Crowley's teachings had the potential to give him a security headache, especially since *Liber Oz* concluded with the breezy declaration that 'Man has the right to kill those who would thwart these rights'.[40] In fact, Jack's instinct for caution in the face of FBI scrutiny was warranted. A couple of agents arrived at the house on the evening of 16 January 1943. Jack told them that the Church of Thelema was 'dedicated to the freedom and liberty of the individual … that they were anti-communistic and anti-fascist'.[41] The agents found 'nothing of a subversive nature', but recorded that there had been complaints of 'strange goings on' in what was rumoured to be 'a gathering place of perverts'.[42] The fact that the equinox party had ended with everyone stripping naked and parading around the fountain probably didn't dispel this impression.

Had any FBI agents caught a glimpse of this scene, it might have suggested a carefree, Dionysian atmosphere at Grim Gables. The reality was different. Relations in the house had not been improved by the relocation from Winona Drive and, by 1943, the residents were miserable. Regina Kahl was in poor health and unhappy at no longer being Smith's Priestess; she quit Agape to move to Texas. Phyllis Seckler had returned to the lodge after her violent husband was jailed, only to find her brethren deeply irritated by her two young children. She finally packed her bags for the last time after an argument about her daughter Stella's tantrums (the root of the problem lay in undiag-nosed deafness, but everyone bar Phyllis thought spanking would fix it).[43] Jack didn't care; he wrote to Crowley that she was just an 'indi-gent cook', a remark that Crowley relayed straight back to Phyllis.[44]

And then there was Smith, whom The Beast had decided was the root of all difficulties within the lodge. Crowley's solution to the problem of Smith was typical: he drew on his numerological knowledge and cast a 'natal chart', concluding that the particular planetary aspect at Smith's birth made him uniquely favoured. The Beast had never seen such an 'astonishingly fortunate' horoscope, nor comparable evidence of 'greatness' in any other human, despite – and here was the kicker – 'no breeding, no education; physically … meagre … with every vice, every defect, conceivable'.[45] Crowley's conclusion was stark: '*Wilfred T. Smith, Fra.·. 132, is not a man at all; he is the Incarnation of some God*'.[46] And the duty of this unknown god was now to discover his true identity, alone, in the desert. (Perhaps only The Beast could excommunicate a disciple by declaring them divine.) Smith, to his credit, worried in his diary that Crowley's elaborate number-crunching 'could be balls', but it was plain that Frater 132 had lost the support of his master.[47] Crowley informed the others in the lodge that 'Smith was ignominiously kicked out … for malfeasance in office and larceny, but I wished to get rid of him without disruption'.[48] When Jack complained about Crowley's treatment of Smith, The Beast retorted that Jack was 'a bit of a marshmallow Sundae' … 'he does what the last person to talk to him tells him'.[49]

After two years in the OTO, everything that Parsons had worked for was unravelling. At the very least, Agape Lodge and the Gnostic Mass that first attracted Jack and Helen were no longer what they had been; nor, for that matter, were Jack and Helen.

BY 1943, THEODORE VON KÁRMÁN had become a remote figure at Caltech. He was less and less involved in the day-to-day operations at both Aerojet and the GALCIT rocket project, depending more on Malina to manage the growing workforce. He spent much of his time in Washington and worked at the very heart of the American military establishment. His professional reputation there was without equal, which was useful for Aerojet. One day in late July, Army Ordnance approached the Boss with three top-secret aerial photographs taken by the British RAF on the French Atlantic coast. Could he try to identify the structures? As he peered through the stereoscope, the images worried him. He asked Malina and Tsien for a second opinion.[50] Von Kármán told the Army that they appeared to be 'pictures of a launch platform and storage center for missiles', adding with some trepidation that he 'had never seen anything that big before'.[51]

Their interpretation, despatched on 2 August, confirmed numerous other Allied reports that the Nazis had some kind of long-range missile. Two weeks after the Caltech team looked at the photographs, the RAF launched Operation Hydra, a strategic bombing campaign against the German missile centre at Peenemünde on the Baltic Sea. On 17 August, as British bombs rained on the rival rocket programme, Malina once again sat down to write to his parents about how busy he was. 'Progress is a hard master and at the present time it's a whip cracking over our heads without letup.'[52] It was about to get much worse.

The first response to the German missile discovery was that the Army Air Forces' liaison officer at Caltech, Colonel W. H. Joiner, asked Malina to immediately prepare a report on the potential for long-range projectiles. Malina turned to H. S. Tsien for help.[53] His old friend had since been granted a faculty position at Caltech, and was now dividing his time between the rocket project and teaching at GALCIT. They were still close, and had remained so since Unit 122 days, even if there was a slight fraternal rivalry for von Kármán's attention. Together they completed their report by November 1943, arguing that with existing technology, rockets wouldn't have a range over 100 miles, but bigger rockets, and indeed bigger payloads, could be achieved with further investment.

Von Kármán heartily endorsed this report, adding his own detailed memorandum asking for the investment to be made without delay.[54] The Boss suggested that the report be operationalised into a four-stage programme: a small, 350-pound solid propellant rocket; a 2,000-pound liquid propellant rocket; theoretical studies of a ramjet engine; and, finally, a 10,000-pound missile with a range of seventy-five miles.[55]

As if in anticipation of Army support, the report by Malina and Tsien had a new designation. It was report number JPL-1, from an institution that for the first time was being called the 'Jet Propulsion Laboratory'. Since 1936, rocket research at GALCIT had gone from being an out-of-hours experiment, to a doctoral candidature, to the Rocket Research Project, to the US Army Air Corps Jet Propulsion Research Project, to the Jet Propulsion Laboratory, which now employed about a hundred people. The R-word had become big business, but nobody wanted to cause a fuss by spelling it out.

When JPL-1 report was forwarded to the Army Air Forces, as well as to the Army Ordnance, it initially foundered on an inter-service rivalry over categorisation. Were rockets essentially a big bullet for Army Ordnance, or a pilotless aircraft for the Air Force?[56] To everyone's surprise, it was Army Ordnance who not only picked up the report but asked for a new proposal to deliver a more extensive programme of research, which they wanted to fund in full. Army Ordnance now requested a detailed chronology of experimental studies and models to be built, cautioning that the funding would only be for one year in the first instance, and that it shouldn't exceed $3 million.

Three *million* dollars. It was just over seven years previous that Malina and Parsons couldn't find $60 for a few high-pressure tanks. The sums offered on 15 January 1944 'threw us into a proper dither!' admitted Frank. This kind of money came with its attendant bureaucracy of lawyers, negotiations, memoranda, letters of intent and, eventually, a contract. After all the diverted applications of rocket motors in aircraft, here, at last, was a commitment to build 'a long-range rocket missile', delivering a high-explosive payload of a thousand pounds over a range of 150 miles.[57] Malina persuaded himself that this was wartime, that they were fighting fascists and that this new weapon might be important for Allied victory.

The ORDCIT Project (Ordnance-California Institute of Technology), as it became known, was so immense that the Jet Propulsion Laboratory needed to invent a different administrative structure. JPL was a lot more than just a fresh name. It was a fundamentally different kind of institution: from the advanced degree of specialisation required and the massive new infrastructure, to its expected growth and its reconfigured status as a US Army facility operated by Caltech. In 1944, it comprised four major projects: JPL-1 on JATO devices (not to be confused with the 1943 JPL-1 memo); JPL-2 on a 'hydrobomb' underwater torpedo, overseen by Louis Dunn; JPL-3 on ramjet engines; and, most significantly, JPL-4, which was the ORDCIT contract. Unlike an industrial laboratory, however, JPL was organised along academic lines, with eleven different sections analogous to university departments, each of which had their own head reporting to Malina as Chief Engineer.[58] The scale of JPL, not to mention its dizzying budget, presented serious challenges of governance.

Throughout the whole thing, the rock at the heart of all these changes was Theodore von Kármán, who continued as Director. But before work could officially start on 1 July 1944, the Boss was taken seriously ill. He was diagnosed with a carcinoma in late June and was sent to be operated on by a well-known German surgeon in New York. The surgery was successful, but the convalescence took months (the Boss was unhappy to be left with a hernia, complaining to the surgeon that 'if an aircraft mechanic had left a weld ... similar to the junction ... in my intestines he would be fired').[59] Then in September, before von Kármán had even recovered, General Hap Arnold asked him to join an Air Force scientific advisory board at the Pentagon. So if von Kármán was instrumental in founding the Jet Propulsion Laboratory, he was scarcely present as Director before Frank, at just 31, had to take over.

Malina was named Acting Director, but in some sense that was to simply formalise the job he'd always had as chair of the weekly research conferences; it was less a case of being promoted into this position than of the new bureaucracy recognising the management role that came with being the *co-founder*, along with von Kármán, of the Jet Propulsion Laboratory. Tsien had important co-authorship of the first JPL report, but his involvement with JPL research was always peripheral to his GALCIT faculty position. The initiator of the formal research programme was Malina, albeit assisted by the fortuitous meeting with Parsons and Forman in 1935; yes, von Kármán had been an early supporter and mathematical mentor, but at every turn it was Malina driving the work, both theoretically and institutionally. And of the original team, now it was only Malina who was left – Parsons, Forman and Summerfield having more or less moved to Aerojet two years before.[60] By the late twentieth century, when the particularity of Parsons' life and work had attracted a new interest, JPL would sometimes be jokingly referred to as the 'Jack Parsons Laboratory'. But the truth is that while Parsons was a founder of Aerojet, he had no discernible role in the institutional emergence of the Jet Propulsion Laboratory.[61] There were good reasons why not.

THE RAPID SUCCESS of Aerojet was largely down to the confident leadership of Andy Haley, an astute, larger-than-life figure who kept a watchful eye on the wage bill. One observer compared him to Delos D. Harriman, the protagonist in Robert Heinlein's *The Man Who Sold the Moon*, who made space flight possible yet lived with the sorrow of being left behind on Earth.[62] Haley was tough, occasionally given to grandiosity and more than able to hold his own with the military top brass, especially with a few drinks under his belt. While travelling to secure the next contract, he would send cables to Aerojet staff urging them to 'drive night and day with indomitable purpose and inspired leadership'.[63]

Sometimes an appeal to greatness was easier than simply meeting the payroll. Cashflow was often perilous as the workforce expanded tenfold from October 1942 to early 1943 and the company opened up a new plant in Azusa.[64] Malina's position as the first secretary-treasurer made him acutely aware of how poorly capitalised Aerojet was – it had the worst capital position of any military contractor in the country, a fact that greatly troubled the Navy when they placed large JATO orders for carrier-based aircraft.[65] Labour, too, was in short supply, from ordinary mechanics to cutting-edge theoretical engineers (even Liljan's father, René Darcourt, got a job as a machinist on the Aerojet factory floor).[66] Robert Truax and Ray Stiff – Malina's friendly rivals from the Navy station at Annapolis – found themselves seconded to work on the same hypergolic propellant that they had inspired the year before.[67] So, too, did Amo Smith, returning to become Aerojet's Chief Engineer many years after his debut with the suicide squad.[68]

Such was the demand for scarce expertise in wartime that the company even sucked in Caltech professors. The dark matter physicist Fritz Zwicky, a student of Einstein's, arrived as chief of research and development (ironic given that Zwicky had once told Malina that he was 'a bloody fool' for believing that rockets could work in the vacuum of space). And then there was Clark Millikan, who so resented not being included in the original start-up that he lobbied von Kármán to let him buy into the company. The Boss didn't mind letting him have some shares, but he guessed that Malina wouldn't be so keen. He guessed right. Millikan put his engineering expertise at Aerojet's disposal and in turn was finally granted some stock in 1943, though only after what Frank called 'considerable hesitation on the part of the original founders'.[69] For Malina, this was doubly vexatious. Not only did Millikan have a profit-sharing stake in the very work he had initially discouraged, but after von Kármán's departure, Caltech appointed him chair of JPL's executive board.

It was soon apparent that the tightness of the monthly Aerojet payroll pointed to a bigger structural problem: the company could never meet the escalating JATO production schedule without additional capital.[70] But because all their work was highly classified, they couldn't just ask a bank. As ever, it was Andy Haley who worked the revolving doors and smoke-filled offices of Washington DC to bring home a murky triumph. With permission from the military, he approached a client of his own law firm, William O'Neil, the President of General Tire and Rubber Company, who was interested in acquisitions.[71] A full order book and a weight of expertise meant that General Tire looked favourably on Aerojet. They immediately offered a $500,000 line of credit and assumed responsibility for the performance of the Navy contract. That was good. But in exchange, they wanted a 'consideration' on future stock: in other words, control of the company.[72]

Exactly what happened at Aerojet in 1943/4 isn't clear beyond evidence of hard-fought internal struggles. Parsons' behaviour had become noticeably erratic. He indicated to Crowley that he was fearful of losing his job, and Jane Wolfe complained that his 'lack of stability ruins much'.[73] There was a particular tension between Parsons and Fritz Zwicky, who, like Clark Millikan, was now an Aerojet shareholder. As Director of Research, Zwicky was effectively Jack's manager. But Parsons so passionately disagreed with the professor's curiosity about using nitromethane as a monopropellant that he and Ed Forman unilaterally destroyed – exploded – an expensive consignment of the compound that Zwicky had ordered for Aerojet. Malina recalled 'a rather bitter conference' at which Zwicky was 'very angry at me' for backing Parsons' view on nitromethane (as much on technical grounds as out of personal loyalty to Jack).[74]

Then there was the OTO problem. Late in December 1943, Liljan caught up with an old school friend, LaVerne George, along with other mutual friends from the world of broadcasting, for a night of beer and chat, 'cover[ing] the world situation from start to finish'.[75] By this stage, Liljan's political activism had extended to the Office for War Information, where she helped make radio programmes for transmission in France.[76] She told Frank in a letter about how this scene was populated by 'honest liberals who are using their place in the field of radio to very good advantage'.[77] But then this:

```
Most disconcerting thing I have heard this week is a
story from LaVerne concerning friend Parsons.  It seems that
certain people in the field of radio are not taking his cult
as lightly as you are.  They are well informed as to the ideas
held therin and are being called the Intellectual Fascists of
Cal-Tech.  They are very much aware of Aerojet in regards this
whole nauseating mess, and it seems you and VonKarman are the
only two people they fear.  This story was given to L. V. by
someone who doesn't know us at all ***OOps, paper slipping
```

Liljan fed a fresh sheet of paper into the typewriter platen and carried on.

> The story goes further, in as much as the FBI is on their trail because they are sure there is some honest to god leak in the whole organisation! By these people, Parsons and the other leaders are considered clever people, and from what they know, people who will stop at nothing.

It's true that the FBI *were* on the trail of the OTO. Who and what were leaking is less clear. It was well known that 'friend Parsons' – is that sarcasm? – had close contacts in radio and had boasted to Wilfred Smith that he should be able to get the Priest some airtime for OTO purposes.[78] This wasn't the first time Liljan had tried to get Frank's attention about what was going on at Agape Lodge.

So again I come to the thing I have said so many times that I think you and anyone else in Aerojet who is not of their ilk is in serious danger. You know Frank that I am not a person to jump at such conclusions without a reason, but remember I saw some of their goings-on too, and it is no joke. It would give me great pleasure if you and Haley and Kármán would begin thinking a little more seriously about the whole thing. [79]

The danger to which Liljan refers isn't clear. Perhaps the most surprising titbit from LaVerne was that '[Clark] Millikan has been advised of the existence of this group and that he won't do anything about it because [his wife] Helen Millikan is also involved!, plus the fact that he is trying to save face as far as Caltech is concerned'.[80] It seemed as if the OTO's influence in the world of rocketry was extending, with the Parsons, the Formans, the Canrights, Fred Miller and, most unlikely of all, Helen Millikan all involved.

None of these fragments quite add up to a coherent picture, but a few things are evident. Some kind of serious tussle took place over Aerojet. Much later, Liljan recalled in her journal that the contest over the company was 'a dirty story filled with less than appealing characters. Somewhere in there I remember that Jack P. was in cahoots with some ex-con who was trying to rest [sic] control of Aerojet out of the hands of von Kármán … and Frank.'[81] At the very least, it's apparent that Parsons and Forman were mistrusted by their own colleagues and were consequently edged out of the company they had helped found. In these circumstances, it's unlikely that Malina and von Kármán would have involved Parsons in the ORDCIT contract and the founding of JPL, even if he was for a short time used as a consultant on solid fuels. JPL's primary developer of solid fuels, Charles E. Bartley, said that from the start of the ORDCIT contract he 'didn't remember having much association with Jack' as 'he was over to Aerojet most of the time'.[82]

The offer from General Tire, when it came, was markedly less generous than anyone had expected. They offered $75,000 for 51 per cent of Aerojet stock. Not a great deal, but in the absence of any other option, von Kármán recommended that they accept. Before proceeding, however, the Caltech stockholders (von Kármán, Malina, Summerfield, Millikan and Zwicky) all agreed that Parsons and Forman should not be included in the refigured company (General Tire may have made the same condition). It's hard to see this decision being made lightly or without reason. Malina certainly had the capacity to harbour resentments – Clark Millikan being the obvious example – but for all Liljan's concern, Frank never seemed bitter or even especially cautious about Jack. On the contrary, later in life he took particular care to chronicle Jack's solid fuel contribution. But something must have triggered the ousting of Parsons and Forman, with the result that Andy Haley had to get them to sell their stock to General Tire.[83] Estimates of what they received vary wildly – Forman is said to have pocketed $11,000 with Parsons presumably getting a similar amount.[84] Malina and all the other shareholders sold their required share to General Tire in January 1945 in return for stock in the parent company.[85]

Jack and Ed now had a decent chunk of capital that would sustain their reputations in the OTO for being something of a success. There's no record of complaint on Jack's part about what happened at Aerojet – with the end of the war in sight, perhaps he thought it a good opportunity to get his capital out. Charles Bartley reportedly heard them boasting that 'this is the end of rockets', and that they were now planning on going into the laundromat business.[86] For a while at least, they lived the unpressured lives of people who suddenly come into a lot of money. Ed Forman decided to buy a plane. With his share of the windfall, Jack bought out the lease for 1003 South Orange Grove Avenue, and then eventually purchased the entire property, in which he and Betty flitted around reading and drinking. Grim Gables gave way to a looser, rooming house of tenants that coalesced around Jack rather than around the OTO per se (which is when the house became known as 'the Parsonage').

But it didn't take long for Jack to find himself in financial difficulty after his services were no longer needed at either Aerojet or at JPL. The success of both enterprises, to which he had contributed a great deal, paradoxically created the conditions that favoured Caltech-style specialisation and professionalism over Jack's quirky inventiveness. With the GALCIT 53 propellant, Parsons had made a formative breakthrough in practical rocketry, but in the end it was theoretical understanding that triumphed. As the head of solid propellants at the new JPL, Bartley soon superseded Parsons' asphalt propellant with a polysulfide polymer, while also swapping potassium perchlorate for ammonium perchlorate to create APCP (ammonium perchlorate composite propellant).[87] APCP has since become a standard propellant in myriad applications, from ejector seats to the descent stages of JPL's Martian rovers (a jotting in one of Frank's notebooks from 1942 or 1943 says 'Have Parsons check ... on Ammonium Perchlorate', which suggests he may have anticipated this breakthrough).[88] Bartley's lasting innovations did not end there: he also redesigned the cast to a star shape that increased the burning area of the propellant.

Parsons had more than made his mark on rocketry; Forman too. But though they eventually continued some corporate rocketry work for other companies, the era of their technical contribution in Pasadena was over. The same cannot be said, however, for their influence over the fates of their former colleagues. Jack and Ed departed the scene, still the same duo as when they had first arrived on campus – before their optimism as pioneers of the unknown had turned to pure resentment at those who would make that dream a reality.

THE QUEST FOR SPACE

IN THE SUMMER OF 1944, a few weeks before the formal birth of the Jet Propulsion Laboratory, a different organisation to which Frank Malina was committed came to an apparent end. At the twelfth convention of the US Communist Party, General Secretary Earl Browder shocked the assembled delegates by saying, 'I hereby move that the Communist Party of America be, and hereby is, dissolved.' Almost immediately, the delegation reorganised itself as the Communist Political Association (CPA). This move was in the service of the 'Tehran line' – a new bid for wartime unity exemplified by the meeting of Roosevelt, Churchill and Stalin in Iran at the end of 1943. Browder's logic was that wartime required its own discipline, and that ordinary forms of labour struggle could reasonably be suspended in order to maximise productivity and thus ensure victory over fascism.

In fact, his thinking went even further: Browder anticipated an extended period of international unity in the post-war period, epitomised by his remarkable offer to shake the hand of banker J. P. Morgan, if the arch-capitalist could support a more social democratic settlement.[1] In these circumstances, thought Browder, a Communist *Party* was no longer the best way of advancing proletarian interests; instead, the movement should adapt to the two-party tradition of American democracy by extending its influence over the Democratic

Party and supporting Roosevelt's common cause with the Allied powers. It was unfortunate that members hated its don't-rock-the-boat-ism (the Communist Party of the Soviet Union likewise), that Churchill was a zealous anti-communist, and that Roosevelt soon died.

The Tehran line was yet another botch of the sort that had seen a leak of members since the non-aggression pact of 1939. Policy had invariably been a top-down affair, which irritated many of those in Unit 122. Of the party's purported 'democratic centralism' – a notional reciprocity between the leadership and local groups – Frank Oppenheimer observed that, while it was obviously 'centralism', it was in no way 'democratic'.[2] But even before Browder's dissolution, something of his ecumenical spirit struck a chord with the Caltech communists. They were arguably ahead of him, having effectively dissolved Unit 122 in the summer of 1941, when the Axis powers opened up the Eastern Front. From that moment on, they were all entirely dedicated to the war effort.

Frank Oppenheimer moved to the University of California's Radiation Laboratory, where he worked on the separation of uranium isotopes – he felt that helping to develop the Bomb was an 'obligation' in the face of fascism.[3] Sidney Weinbaum's wartime commitment saw him move east to the Curtiss Wright Research Laboratory, where he found himself at odds with military management (they fussed about his lab not being 'as clean as the boudoir of a kept woman').[4] But many others stayed in Pasadena: Frank's friends Jacob and Belle Dubnoff and Andrew and Edith Fejer remained politically active, though most of the real work was shouldered by Belle and Edith. Liljan Malina was involved with the Yugoslavian War Relief Council, served on the Board of the Czech Red Cross and painted posters in aid of Russian War Relief.[5] She even started learning Russian, but sometimes the work just got too much.[6] 'I get a bit angry,' she complained to Frank after making some posters, that 'people like Edith and Belle give me these things to do overnight, I break my neck getting them out, big ones in color too, and then do you think they would even call up to thank me? Not on your life! They expect it, take it and no more is said.'[7] Everyone kept busy, even if Unit 122 was no longer formally meeting under the auspices of the party; some memberships likely lapsed in this period, but there's no sense that the Caltech communists went quiet or tempered their activities in case the FBI were watching. They still socialised together. Frank and Liljan even called their new kitten 'Tito'.[8]

Whatever the nature of Frank and Liljan's troubles at this stage, they weren't political, though they did seem careworn in different ways. For nearly a year, Frank had invested everything in the nego-tiations at Aerojet and, simultaneously, in the delicate diplomacy around the ORDCIT contract and JPL. And there was still the every-day engineering challenges of practical rocketry – almost all of this in the absence of von Kármán. It had been exhausting. Frank confessed to his parents that he felt 'somewhat burned out', by which he meant that he 'didn't have very much ambition'.[9] He wondered if 'perhaps everyone is getting war weary', noting that friends in the military 'want either to go home or take a year's vacation in the mountains. I have similar feelings at times.'[10]

Liljan, too, was tired – both of Frank's presence and of his absence. Something important happened between them in or around April 1944. Frank's letters from Washington DC have an unusual intensity, addressing his wife as 'Dear Guardian' or 'Dear Philosopher of the Heart'. The picture isn't entirely clear, as some letters have since been destroyed. But in one especially heartfelt note, Frank wrote that the 'thundering jolt you have given me has made me think of other things than elusive rockets for the first time in months, perhaps years'.[11] There are suggestions that Liljan wanted a divorce, to leave him and seek a new life in New York; clues can be found in letters that she continued to exchange with Jim Burford – the CIO organiser and 'dullest lover' – long after their affair ended.[12] The letters suggest that while Liljan had made the case to Frank that their married life should encompass her plans as well as his, he had contrasted his important rocket work with her 'whims'.[13] Burford was typically unsympathetic to Frank:

> I was always struck by the fact that for a man of progressive political ideas he almost completely followed the pattern of the

typical old-fashioned male in his attitude towards women and his wife in particular. The pattern of life which he seems to find agreeable would appear to be fantastically impossible for you.

Burford advised Liljan to turn to Lenin for insight ('I think that every woman should read and understand the historical and theoretical aspects of what Vladimir L. once wrote about as the "Woman Question"'), before eventually concluding that maybe 'divorce is the only course'.[14]

Not long after that, Liljan got on a train and left for Mexico. They decided to live apart for six months, though it wasn't a public separation. As far as everyone else was concerned, they were just pursuing individual projects until the end of the year. Frank asked that they delay a final decision on their relationship until after the war – that would give him more time to plan a future that he hoped would suit them both. He also wanted a change of work scene, and in September 1944 he took the opportunity of becoming a military attaché to the Royal Army Ordnance Department in London.

A British attaché, Colonel F. F. 'Froggy' Reed, had visited JPL that summer, and Malina decided to reciprocate so that he could study the state of rocket research in the UK and visit the same V-1 launching sites that he had helped von Kármán identify the previous year. Liljan meanwhile was in Mexico, finding the space she needed, travelling for the sake of travelling and occasionally painting along the way.[15] That their ordinary relationship rules were in abeyance, perhaps by mutual agreement between them, perhaps not, is evident in the manner in which they would sometimes tease each other about rival suitors.

Liljan had a wild time. With an art school friend called Mary Tracy she 'found ... three millionaires on the train down, and they served us very well'.[16] That was just the outward journey. 'We've been night-clubbing every single night with private cars and chauffeurs ... it's like something out of a novel.'[17] Everything about the trip spoke to unfulfilled needs. 'I've never in my life had so much attention paid me from all sides. The kind of attention that would make any woman look in a mirror and admit she is beautiful, charming, lovely, intelligent, bewitching, graceful, and attractive.'[18]

At 23, Liljan was learning to shed the identity of 'Mrs Frank Malina'. She hung out with painters and poets, radio producers and Hollywood people, and stayed long after Mary had returned to California. 'I haven't the vaguest idea where you are – I hope everything is under control,' she wrote to her husband. 'I can't say I'm homesick, it would be a lie. If you were me, you'd know what I mean.'[19] The trip was pure escape right to the very end. On the night before her flight home, she didn't bother going to bed – 'my American friends had a bang-up party for me, and we ended up in jail, and fianlly [sic] at the Tampico Club at six in the morning for oyster soup.'[20] It was by way of an afterthought that she asked him not to 'worry about the jail incident, it's the easiest thing in the world for Americans to land in a Mexican jail ... it added the last touch of excitement that we needed'.[21]

Frank's life, by contrast, was all work. When Liljan was partying, he was trying to direct JPL in the vacuum of von Kármán's leadership. He didn't find it easy. 'There is almost a complete lack of decision and as a result not even a muddled policy to proceed on,' he wrote to Liljan. 'One reason for this is a lack of conviction on the part of some that our goal is worth striving for. You know that as far as I am concerned I am willing to put my shoulder to the wheel for one more sizeable effort in this business.'[22] After the best part of a decade on rocketry work, Malina found that he now had limited energy to expend on space exploration. He was willing to give it another heave, but he also longed for something different. The attaché post would give him a change of scene as well as a more critical wartime role, but it would also place him directly in the path of Nazi rockets; misgivings about his technology's potential for human suffering would hardly be allayed by inspecting the suburban impact craters of the rumoured V-2.

At the beginning of September, Frank was called to Washington DC to await orders to fly, but problems with paperwork kept him grounded for three weeks at the Hotel Raleigh, a grand Beaux Arts building on Pennsylvania Avenue; it was, as he told Liljan, 'an enforced vacation ... a change from the continual grind of the past few years'.[23] They still exchanged regular letters, mostly warm in tone.

An unhurried letter to Liljan was Frank's preferred space for reflec-
tion, for musing on his feelings, and on the problem of feeling itself.
For an engineer obsessed with reason and rationality, Frank often
found himself confounded by his own emotional responses, and this
both troubled and intrigued him. He held an image of himself as a
supreme logician, someone who should be able to rise above petty
sentiments. It surprised him that he couldn't just think his way out
of the marriage crisis.

One night at the Raleigh, he went to bed early but woke with a
start at midnight. He reached for the light and pulled out the pad of
hotel notepaper in order to share his thoughts with Liljan.[24] A vivid
dream had disturbed him. He had borrowed his father's car to go to
a picnic in Texas, but when he returned, it was missing. '[M]y father
controlled himself very well,' Frank wrote, yet his own response was
pure panic and distress. 'I am conscious of a great fear of my father
... [he] ... belittled my intelligence many times ... as you know,
he was at one time a butcher and had a meat market and I shied
away from that business.' The prospect of seeing the consequences
of rocket attacks terrified Frank. He confessed to Liljan that he had
'an unmanageable fear of physical pain, not so much for myself, but
when it happens to others'. He detailed other instances of his debili-
tating fear, like 'when I hear lectures on matters I do not quickly
grasp' or 'in movies or operas if there was physical harm done to any
of the actors'. Frank and Liljan had both been reading about psychoanalysis that summer, and Malina was curious about the challenge
of analysing his dream. 'Now what I wonder is am I transferring my
suppressed fears of my father into present occurrences in my life. I
don't think it's impossible.'[25]

THE FIRST V-2 ARRIVED in Britain a couple of days after Frank's
dream. It was just after teatime on 8 September 1944 when the rocket
took off from Holland. Within five minutes it found its way to
Staveley Road in Chiswick, an ordinary West London street, killing
three people and injuring another twenty-two. Rosemary Clarke was
among the first victims; she lay undisturbed in her cot at number 1
but her lungs collapsed from the air vacuum created by the amatol

warhead. She was three years old. The public explanation was that a gas main had exploded. The number of officials who visited the crater suggested something more unusual. Despite the strict secrecy surrounding the Staveley Road blast, the news eventually filtered back to the V-2 engineers, who cracked open the champagne to celebrate their accuracy.[26]

A few weeks later, Frank flew into Prestwick, Scotland, before taking the train to London and settling into his new position.[27] 'Psychologically I'm gradually becoming hardened by the many firsthand reports I have heard and some of the sights I have seen,' he told Liljan.[28] But as the weeks wore on and the rockets rained, the full horror of the V-2 became apparent. The silence of its arrival, unlike the 'buzz bomb' V-1, made it a weapon like no other. Some landed harmlessly in parks; others destroyed houses, factories and offices. For a city that had already felt the heat of the Blitz, life could only continue as normal – or as normal as wartime would allow.

Late in November, when a new consignment of saucepans had arrived at Woolworths on New Cross Road, a queue of hopeful shoppers snaked their way outside.[29] At 12.26 p.m., a flash of light. 'Woolies' bulged outwards and then collapsed, sending up a shower of masonry, plaster, body parts and homewares. The crowded Co-op next door imploded; a double-decker bus was picked up and spun like a toy, its passengers found dead in their seats like dusty white statues. With 168 dead, it was one of Britain's worst civilian disasters in World War II. Others would follow: 110 dead at Smithfield market; 68 in the Prince of Wales pub; and hundreds of lesser tragedies, as rockets fell into gardens and on roads, killing a few people here and a few people there.

Amid these incidents, work provided a focus and a distraction for Frank. In less than three months, he spoke with eighty-two different officials, most of whom were quite open about the state of British rocketry. He also inspected parts of a relatively intact V-2 that had fallen in Sweden. Size aside, it gave no technical revelation that made him rethink JPL's own rocket design.[30] In many respects, JPL was clearly more advanced. Nevertheless, at a conference at the British Projectile Development Establishment at Fort Halstead, Kent, the meeting was 'shook-up' by an incoming rocket that 'left me rather disturbed, but my colleagues continued … as though nothing had happened'.[31]

In a strange way he felt exhilarated by some of this work. 'I have been feeling wonderfully well,' he wrote to his parents at the beginning of November, almost as if he had discovered a new vigour. With Liljan's colourful accounts from Mexico still trickling in among batches of late mail forwarded from Washington, Frank felt compelled to reciprocate. He wrote about a date to the cinema with an unnamed 'girl I have met'; they saw *Marriage is a Private Affair*, an MGM Studios war comedy about the dilemmas of a party girl married to a straight-edged Air Corps lieutenant.[32] That was an interesting choice of movie, not least because the female lead was played by Liljan's Hollywood High School contemporary Lana Turner. 'I wondered if you had seen it?' he asked Liljan.[33] Later he would talk about this time as being the first occasion that he had 'express[ed] my sexual ego with any woman beside my wife', but it was a transgression that didn't sit comfortably.[34] The life he wanted was with Liljan.

When he finally made it back to the US on a lumbering B-17 bomber, Frank knew that his marriage was far from certain. JPL and Aerojet were more stable fixtures, but he was less sure about whether their objectives would be aligned with his own after the war. Of course, much had been achieved in the decade since he moved to Caltech, but it had come at the cost of his relationship; he had also let his vehicle for civilian science be taken over by the Army. Crossing the Atlantic to the drone of the B-17 engines, his mind turned to the core of his life's project. In 1936, Frank had sketched a chart showing all the different component parts necessary for their goal of a sounding rocket. He realised that what had once been a puzzle for his office wall was now well worked out – that the rocket he had dreamed of a decade before was actually within reach.[35] Goddard had spent his entire life attempting this, without success; how odd it would be for Malina to succeed in the technical design only to stop short of its practical demonstration at extreme altitudes. Having experienced first-hand the rocket as weapon, he wanted more than ever to redeem the civilian rocket ideal, to make an instrument of scientific exploration that would kill no one. All Frank had to do was persuade von Kármán and Army Ordnance.

BEFORE THEIR TEMPORARY SEPARATION in 1944, Frank had told Liljan that he wanted to make 'one more sizeable effort in this [rocket] business'.[36] That was probably the last thing she wanted to hear. Yet having returned from London and bargained for a stay of execution on the divorce, Frank still made rocketry his main priority. The basic ingredients were all now in place; when Frank had been in London, JPL staff had already made progress on ORDCIT's proposed long-range missile programme. It was to be developed in stages, starting with a small-scale unguided solid propellant rocket moving on to a larger and liquid-fuelled guided missile with a range of over one hundred miles. The sequential development of these missiles was to be represented in their naming: they'd start with Private, move on to Corporal and then Sergeant, eventually working their way up the army ranks. When Major General Gladeon Barnes, Chief of Army Ordnance, asked von Kármán how far up the ranks the missiles would go, von Kármán deadpanned 'certainly not over colonel', explaining that 'this is the highest rank that works'.[37]

What Frank now suggested was a slight departure from this programme. The first rocket, Private A – eight feet long with a solid asphalt-based propellant – had already been successfully tested in the Mojave Desert when Malina was in London, making it the first ballistic missile of the US military. It was never built to reach altitude: its purpose was to provide experimental data on using fixed fins to stabilise flight, and on a booster stage to launch the rocket out of guide rails.[38] Rather than moving from the Private straight to the liquid-fuelled Corporal, they instead attempted a smaller version that would carry twenty-five pounds of scientific instrumentation to a height of over a hundred thousand feet. It no doubt helped his case that the Signal Corps was looking for a cheap instrument carrier for meteorological purposes. But the more persuasive justification was that, while JPL knew how to build a rocket, they had next to no experience of launching one. Flight testing and instrumentation were severely lacking, and would need investment before the Corporal was viable.[39]

Frank reasoned that an interim vehicle could usefully test the complex launch systems, potentially heading off expensive mistakes on a scaled-up design. Because the US Army held grander ambitions for the Corporal, they asked Malina and his GALCIT colleague, Homer J. Stewart, to work up a fuller proposal, which was hammered out over a weekend in January 1945 and rapidly approved.[40] The new rocket was to be called the WAC Corporal: technically a small Corporal 'without attitude control' – in other words, without an inertial guidance system – though joking references to the Women's Auxiliary Corps and 'Corporal's little sister' proved hard to resist for some.[41] For Malina, this was finally a vehicle for science rather than war, a project that, as long as he didn't think about it too hard, seemed free from political contradiction.

It was now five years since Homer J. Stewart first developed his suspicions about Frank. Nothing in the interim had made him more comfortable with Malina's politics, though they hadn't had to work together since the days of the Ercoupe. Stewart had been out of rocketry for a couple of years, teaching meteorology. The advent of the ORDCIT contract saw him drafted in to H. S. Tsien's Research Analysis Section of JPL.[42] Having once been his GALCIT contemporary, Stewart was now working under Malina's authority on plans for the WAC Corporal. That was no problem for Malina. Stewart, however, couldn't help notice certain things that revived his suspicions.[43] When they went to a meeting at the Applied Physics Laboratory, it was compromised by the fact that the Navy hadn't forwarded Frank's security clearance; that meant no discussion of classified material.[44] On another occasion, Malina and others from JPL were due to observe some military tests in California when Malina's clearance was simply refused.[45] Stewart worried that these were more than mere bureaucratic errors. Why did they always happen to Frank?

When it came to designing the WAC Corporal, Malina, Stewart and the rest of the JPL team faced two main decisions: what kind of rocket motor and what kind of guidance? After all their JATO work, they had to decide which of their reliable solid and liquid JATOs would be most adaptable. Solid motors were fine for the Private but would be too heavy, relative to their thrust, to reach altitude. That meant using a modified version of the liquid RFNA and aniline motor that had already proven its worth as a JATO on the Douglas A-20. Aerojet – naturally – would make the engine that would deliver 1,500 pounds of thrust for forty-five seconds.[46]

The matter of stability took JPL on to less familiar territory, but they knew enough to avoid Goddard's signature mistake of having an elaborate gyroscopic guidance without the propulsive force to lift it. The conclusion at this stage was that if altitude was the primary object, then internal guidance was actually an encumbrance for a small sounding rocket. On the other hand, not having guidance wasn't an option: as soon as the rocket left its launcher, the combination of friction, gravitation and the eastward rotation of the Earth would pull the rocket into a curved flight.[47] But then they realised that the stability of the rocket in flight could, like the Private, be supplied externally to the rocket in the form of guide rails on the launching tower. It was a bit like a bullet fired from the long barrel of a rifle. Malina and Stewart worked out that if the rocket's launch speed was fast enough – over 400 feet per second when leaving the tower – then deviation from the vertical would be minimal. They also knew that, although their liquid motor couldn't provide that kind of acceleration, a solid propellant booster could be used to clear the launch tower. They now had a workable plan, but the pressure of the ORDCIT contract gave them less than a year to make it happen.

For a few months, life was almost normal. Frank and Liljan were back in South Hudson Street, thinking about a future, and living as if they had one together. As before, he worked all the time, while she immersed herself in activism. They even bought a plot of land in the west of Pasadena, with panoramic views south over Los Angeles and north to the San Gabriel mountains. The sale of some Aerojet stock to General Tire meant that they had the $3,000 cash up front and the promise of a fat dividend from remaining shares might be enough for the down payment on a new house.[48] The fact that this communist had done rather nicely out of a capitalist takeover was a point of family amusement ('the whole affair gives me a lot of laughs,' he admitted to his parents, 'for reasons that you know').[49]

By April 1945, Frank was once again away for weeks on end, leaving Liljan on her own. It was not a happy state. She felt vulnerable without someone else in the house. And as the Allies limped towards victory in Europe, she was also nervous about the imminent return of her battle-worn lover, Jim Bowling, the photographer with whom she had maintained airmail intimacy ever since his draft in 1942.[50] She thought about getting a dog but instead acquired a gun, a .32 that was 'guaranteed to kill anything within a range of 100 yards'. It was hard to imagine herself actually using it, but she thought it best to warn Frank just in case he had any ideas of making an unscheduled return.[51]

That wasn't likely. Frank was at Fort Bliss, Texas, entirely occupied with the range of his own projectile, wondering whether the addition of wings to the Private might make it go further. Several rounds of what was called the 'Private F' were launched, and all took off normally, but in every case the rocket quickly went into a spin, spiralling down to the desert floor and leaving a corkscrew contrail hanging in the air. After that fiasco, Frank went to oversee arrangements at the White Sands Proving Ground, which was getting ready for WAC Corporal testing later in the year. It was just before dinner in the White Sands Army mess hall when they heard about the death of Franklin D. Roosevelt – 'news so distressing that for a long time we refused to believe it'.[52] The upset was genuine but it marked a change from those days when Frank had taken the party line that FDR was a double-crosser who used war as the solution to unemployment.[53]

After White Sands, Frank was in Washington for Aerojet meetings, then back down to Texas to see his family. On the long train journey from Washington to New Orleans, Frank got chatting to another passenger, a young man from Dallas who turned out to be an FBI agent. It was, as he described it to Liljan, 'my first face-to-face encounter with a member of that organisation'.[54] Frank seemed mostly untroubled, other than that the conservative, anti-Roosevelt, anti-Soviet values of the agent were rather disheartening.

Frank got off the train in Brenham to find his family well, marvelling at how unchanged everything was. The view from his old bedroom window was exactly 'as it has been for as long as I can remember', so peaceful and quiet even though old school friends were 'scattered all over the world fighting wars'. He chatted to his grandmother Babi, now 85, about the San Francisco Conference that was about to take place, and from which the United Nations Charter would soon emerge.[55] Frank not only wanted to defeat fascism; he hoped that this historic meeting would end warfare as a strategy in international affairs.[56] His optimism, however, exceeded what was happening almost under his nose. By the time he made it back to Pasadena, the leaders of Germany's V-2 programme – actual fascists – had started their takeover of American rocketry.

As the war drew to a close in April 1945, the United States and Russia scrambled to capture the documents, materials and personnel of Germany's V-2. In his capacity as head of the US Army Air Force Scientific Advisory Group, Theodore von Kármán was asked to visit Germany and assess the state of German rocket research. General Hap Arnold wanted to steal a march on the Soviets when it came to post-war missile development, and Nazi engineering was now part of the war bounty. The Boss was happy to help, adding Tsien's expertise to the delegation. Having been forced to flee his own academic post at Aachen in 1929, von Kármán was curious to see the Nazification of German engineering. He didn't hold back his contempt for some former colleagues who had willingly collaborated with the Reich while others became its victims. In particular he sought out a former Caltech student, the French physicist Charles Sadron, who had been worked almost to death in V-2 production at the Mittelwork factory. The Boss found him at the ancillary camp of Nordhausen, a living skeleton. The visit was 'one of the most ghastly experiences I ever had'.[57] Tsien, meanwhile, was called to the village of Kochel, where he interrogated none other than Wernher von Braun himself. There are no surviving records of this meeting, but it appears to have left its mark. Years later, Tsien would refuse to attend a lunch with Nazi scientists. 'I'll learn from the Germans,' he remarked, 'but I won't eat with them.'[58]

IF THE EARLY HISTORY of American space exploration is a three-act movie plot, the flight of the WAC Corporal is a midpoint, between Goddard's experiments and the moon landings of Apollo 11. To its principal protagonist, however, it felt like an ending. Frank Malina was the only member of the original GALCIT rocket group to witness the culmination of a decade's work. Parsons and Forman were out of the game. Von Kármán and Tsien were in Washington advising on the future of warfare. Martin Summerfield had returned from Aerojet to JPL but was at his desk working out the mathematics of orbital access.[59] It was Malina who donned the desert boots to oversee the trials at White Sands Proving Ground.

White Sands was a newly opened facility built to accommodate the wide reach of modern rockets across 3,000 square miles of Chihuahuan Desert. It was still pretty primitive. 'The food is fine and we should have a pleasant outing,' he wrote to Liljan, 'providing we do not have too many technical difficulties with our baby.'[60] He didn't exactly expect difficulties, but there was still some tension. 'Tonight all of us are rather keyed up over the behaviour of our "little lady" tomorrow ... we have high hopes that she will not let us down [...] I will enjoy telling you about her someday. She is very pretty and up to now we love her dearly.'[61]

Frank could afford to have such confidence in the WAC Corporal. Every part of the system had been rigorously tested – it wasn't just a case of putting it all together and hoping it would fly. On 4 July they launched a 1:5 scale version, the so-called 'Baby WAC Corporal', to see whether the rocket could be stabilised with three fins rather than four, like feathers on an Apache arrow (Homer Joe Stewart's idea). On 26 September they tested the solid fuel booster, 'Tiny Tim'. Then on 1 October, they tried a partial charge of propellant to test the separation of the WAC from the Tiny Tim, followed by nose cone separation. The launch of the fully charged WAC Corporal was a test not just of the rocket but of their entire ecology of instruments and personnel; that, in essence, was the difference between the delegated expertise of Malina's systems approach and the concentrated authority of Robert Goddard. Had the older rocketeer not died just six weeks earlier, he would surely have been vexed to learn of JPL's progress so close to his own Roswell ranch.

The morning of 11 October started at 5 a.m. with a touch of patriarchal discipline. Frank picked up a bugle and blew out the

Reveille that he had learned from his father.[62] A few high clouds lingered at altitude but they didn't last long. The WAC and Tiny Tim were loaded side by side on to an adapted trailer which a Dodge half-ton truck pulled out to the launch tower. Once raised to the vertical, a quick photographic portrait was taken of Frank and his 'little lady'. The rocketeer rested a proprietorial arm on the WAC's yellow chassis. The WAC was then fuelled, hoisted into place and mated to its booster, the crew straining and pulling with wrenches and winches. Once the rocket was primed and in position, it seemed almost imprisoned by the hundred-foot steel gantry that sat stark against the western skyline of the Organ Mountains.

At the appointed hour, the crew managed a countdown of sorts – mostly hand signals – as the rituals of rocket launching were still in their infancy.[63] Five, four, three, two, one. Inside the improvised blockhouse, someone pushed a black button.

Ignition.

The movement was seamless. Tiny Tim burst up the gantry and separated cleanly, just as the WAC's hypergolic cocktail surged into life. The combustion gases ripped through the graphite nozzle, sending the rocket up the Y-axis in perfect co-measure to the motor's output, a motor that had been crafted and refined through the rigours of scientific method and a brush with the black arts. In just three seconds, the rocket was a distant object traced by a thread of dark smoke. It punched through the stratosphere at 2,113 miles per hour, nearly three times the speed of sound.[64] At 80,000 feet, when the propellants burned out, the rocket was well beyond the point at which the curvature of the Earth would be obvious. Above this hung an inky darkness that no human had ever seen.

Frank, who was always allergic to flag-waving and the language of conquest, had no thought of this voyage as a national achievement. He sought little glory for the USA, far less for himself. What quickened his heart was the common property of new knowledge: the promise of a better understanding of Earth's atmospheric envelope, and what it would mean for humanity to slip outside its protection.

He craned his neck as the rocket continued its ascent even after the burn had finished. One minute after launch it was no longer clear to the crew what they were looking at; the rocket certainly wasn't visible, and the radar data only made it so retrospectively. The upper contrail, curled and stretched by the stratospheric winds, looked like the last wisp from an extinguished cigarette. The WAC Corporal surged on through the ozone layer, the shield that absorbs the Sun's ultraviolet radiation, and up into the mesosphere: the coldest place on Earth. It was in this nether region, an unchartered ocean of darkness buffeted by strong atmospheric tides, that the rocket slowed. At the top of its trajectory there was just pure black space above and the bright carpet of Earth below. These days we call it 'near space'.

Frank stood at the empty launch tower letting his mind drift with the exhaust vapours and the desert dust. Later he'd say that in these minutes he thought of his 'other goals', of the world's 'desperate social problems'; it's clear from his letters that it was Liljan that most occupied his thoughts.[65] He had assured her that 'this ambition … at White Sands would be the last one for a long while', after which 'we would live only because of the joy of living'. Given that he was still sending her his dirty laundry, this may not have been the most persuasive vision – but then Frank was always an optimist.[66] 'I love you sweetheart. I want to ride a ship with you, see France and Czecho with you, live anywhere with you, see you happy and gay and watch over you, see you forget or understand those things that had to be.' 'Those things' were perhaps Liljan's other involvements.

Exactly 450 seconds after the launch, the WAC Corporal gouged the desert floor, just short of a kilometer from where it took off. The ascent was near vertical – in other words, perfect. The parachute recovery mechanism in the nose cone failed to deploy, a problem that wouldn't prove easy to fix, but that was a detail.[67] Radar tracking data showed that the rocket had reached nearly 73 kilometres, well over double their target of 30 kilometres, making it the first American rocket to reach an extreme altitude and the world's first sounding rocket. No, it wasn't space as we now know it, but then again, the limits of Earth and the boundaries of space were part of what this technology was yet to determine (a later launch reached 80 kilometres, making it the first American rocket to reach space as then defined).[68] Frank whooped with delight as the crew shook hands and slapped backs. They had built and flown a rocket into the upper atmosphere in about nine months, launched with only thirty-seven people. The V-2 that by today's definition is credited with being the first object to leave Earth's atmosphere had vastly greater funding and staffing, but its space flight was a secondary achievement to another goal: civilian terror.

The glow of Frank's fulfilled ambition was short-lived, however. Naturally Frank wrote to Liljan to tell her the news, or at least as well as he could under security restrictions: 'today was <u>THE DAY</u> and I don't think I will ever forget it. Our "little lady" performed beautifully on her maiden flight'.[69] But he was now beyond the point of self-deception about what it all meant, especially after President Truman authorised the atomic bombing of Hiroshima and Nagasaki. On a flight from White Sands, Frank asked the pilot to fly over the Trinity test site at Alamagardo on the other side of the Tularosa Basin from the WAC launch pad.[70] From the air it just looked like a dark, dry inkspot, hundreds of metres across. Still, the reality of mass destruction was plain enough to see. Frank had some inkling of the Bomb, in part because he had met a few of its personnel – Frank Oppenheimer, Nathan Eisen and, most notably, J. Robert Oppenheimer – through Communist Party networks. But only when the Manhattan Project was brought to terrible fruition did Malina glimpse that the post-war world would live under the shadow of a new totem, an alliance of the rocket and the Bomb.

It's no coincidence that one week after the atomic bombing of Japan, Frank finally asked von Kármán if he could take a year of absence from Caltech. He didn't know what he was going to do but he had a good idea what he wanted to avoid. 'As you know,' he wrote to his parents, 'I have never been convinced that what I was doing the past 10 years was right – at least in my own mind. Technical developments are now so far ahead of other human arrangements that it appears nonsensical for one who sees the gap to make it even wider.'[71] If he stayed as Director at JPL, he'd be overseeing the next rank in the ORDCIT programme: the Corporal missile, which the US Army intended as the first American missile to carry a nuclear warhead. His legacy would then become the targeted delivery of mass destruction. He would prefer to follow Liljan, wherever that took him.

FRANK SAW THE DAWN of the atomic age, but he had no idea of the impending detonation in his personal life. Jim Bowling's return from the war in Europe was the catalyst; the fallout was lasting. After years of intimate correspondence, Liljan found that she and Jim had little to say to each other in person.[72] At some point there was a revelation for Frank, as he realised that the depth of Liljan and Jim's relationship had been different from what she later called her other 'half-loves, bed partners and admirers'.[73] A scene ensued, perhaps more than one. Frank felt crushed, though he still fought for his vision of them continuing as a couple. But the strain of the crisis was not, as in the previous year, something that could just be kept between them. It naturally didn't help matters when Liljan's parents, Rene and Georgette, weighed in to remind their daughter that her place was with Frank, having his children, and keeping his home. The blowout finally ended with Georgette telling Liljan that she was no longer welcome in her house.[74] Desperate to escape all of them – Frank, Jim and especially her parents, Liljan booked a train ticket for New York.

It was at this time, when she was still at home in South Hudson Avenue, that Liljan got the phone call from Katja Liepmann, the wife of Frank's colleague, Hans Liepmann, to say that the FBI were on their way – that they'd be there within the hour. How exactly Katja knew this and why she was motivated to warn Liljan isn't clear. Hans and Frank weren't close; they had socialised a little in the late 1930s but since the war Hans hadn't entirely approved of Frank.[75] He was one of the more conservative members of GALCIT, and was friends with the other Republican faculty – Homer J. Stewart, Clark Millikan and Louis Dunn. Katja, however, was an independent-minded spouse who didn't share her husband's politics.[76] The most plausible explanation is that Hans had somehow got word of the raid, or had maybe seen agents arrive at GALCIT to interview Frank, and had told Katja, who may have acted on her own volition in making the call. The fact that Frank had been a little careful with political papers, destroying some while storing others in the attic, shows that they observed a general caution that their lives might be under surveillance. But the tone of Frank's description of meeting the FBI agent on the train to New Orleans suggests that they didn't take it too seriously. And, in fact, he was right. The FBI were at this point not much interested in Frank Malina. This wasn't the FBI.[77]

Frank never talked about this episode, and while Liljan testified to her own experience, she was wrong about the identity of the agents, as Frank must surely have known. His interrogation at JPL and the raid on his home was most likely by the US Army's Counter Intelligence Corps (CIC), the same security service that had worked with the FBI on the leak of classified data from the Caltech wind tunnel in 1940.[78] Now, though, the CIC had a much more serious concern.

Late in 1945 they had intercepted a Russian courier in Paris, in whose possession they found a cache of highly classified documents including 'blueprints' for the WAC Corporal.[79] Precisely which documents is not specified but, at the very least, it seems to have included Malina and Stewart's original proposal for the WAC – 'Considerations of the Feasibility of Developing a 100,000ft Altitude Rocket (The WAC Corporal)'. Naturally, CIC interviewed Stewart as one of the memo's authors; he was shocked but had nothing to say about how such documents could have ended up in Soviet hands.[80] Privately, of course, he nursed suspicions.

The CIC enquiry could not have been concealed from the others in JPL management – Louis Dunn as Assistant Director, Clark Millikan as chair of the council and, of course, Homer Joe Stewart – all of them fervent, vocal anti-communists. Had Frank ever mentioned his desire for a leave of absence? Had he talked about the importance of open, universal science? Had he expressed concern about the concentration of technology in the bi-polar world? Any one of these matters might have added to their suspicions. Yet there is no evidence to suggest that Malina had anything to do with the leak.

But 'leak' wasn't the word that agents were using to describe the loss of the WAC Corporal plans. They called it espionage.

THE SHRINKS

'D-DAY' ARRIVED. Frank took Liljan to the station, but as she boarded the train to New York he couldn't face up to what departure really meant. By this stage, she was desperate to get out. 'I don't think either Frank or my parents ... understood that I really meant to do this.' But when she wrote to him asking for details of a divorce lawyer, that was a harder signal to ignore (Frank asked his own attorney, the communist Jack Frankel, who recommended Carol Weiss King, famous for having defended the Scottsboro boys, longshoreman Harry Bridges and other *causes célèbres* of the Communist Party).[1] After seven years with Frank, Liljan described it as 'a bit like playing out a bad poker hand', where 'everything had to be sacrificed to the pot in hope of winning something better' ... 'I needed ... time to be alone, time to find out who I was and what I wanted.'[2] She was 24 years old.

The realisation that this was the end of their marriage left Frank adrift. The day after Liljan's departure, he wanted to go and find Jim Bowling, her uniformed ex-lover; he just had this feeling, 'an almost uncontrollable drive to do something'.[3] As it turned out, the 'something' wasn't particularly violent. He just wanted to tell Bowling to 'go see a psychologist'. He reassured Liljan that he wouldn't do anything rash, then drew in the letter a small picture of a broken heart.[4]

In the end, Frank left Bowling alone but took up psychoanalysis himself.

Frank and Liljan had been reading about the 'talking cure' for some time, and were curious about how it might help them. Many Caltech scientists were at the time looking to psychoanalysis as the primary theory of the self, most notably the physicist Paul Sophus Epstein, a pioneer of quantum mechanics who became depressed while working in Switzerland in the 1920s and was drawn into Carl Jung's circle. After a 'very intensive' psychoanalysis and a meeting with Freud, Epstein brought his new passion to Caltech, founding the Los Angeles Psychoanalytic Study Group, which in turn attracted eminent analysts like the Marxist Otto Fenichel and physicist colleagues like Richard C. Tolman, as well as Frank's Unit 122 friend, Sidney Weinbaum.[5] Even von Kármán was a sympathetic supporter. It was almost as if it was in the Caltech water. 'What the past has done to each of us is now to be discussed with a competent psychologist,' Frank told Liljan, which 'will help to understand our actions, and show us how to become better adjusted and more personally secure ... at the present stage of history'.[6] They settled on a kind of hiatus, albeit one that came with a precondition. 'I propose,' said Frank as part of a numbered list of stipulations, 'that during this period we abandon the "Do What Thou Wilt" philosophy in our physical lives.'[7] Jack Parsons was such an easy shorthand for a *laissez-faire* libido.

Therapy might have been the done thing at Caltech, but in the American Communist Party it was more controversial. Many of Frank and Liljan's communist friends saw analysts, though they tended to gravitate to those who were politically approved.[8] Given decades of repression, the party was understandably wary about members spilling their hearts to any old bourgeois therapist. What if they couldn't be trusted? It would be hard to conceive of a more effective surveillance apparatus than some kind of rat-shrink serving up the inner lives of party members to the FBI. Besides, the solution to the neurological stresses and sexual hypocrisies of capitalism, many communists felt, was not to be found on the couch but through revolutionary change.[9] Debate raged in the communist magazine *New Masses* until, in 1949, the party leadership proscribed the couch entirely.[10] Until then, the safest option was for members to be directed to analysts deemed sympathetic, mostly inside the fold. For all the official caution about psychoanalysis, it was through party networks and their lawyer Jack Frankel that Liljan was introduced to the prominent analyst Joseph Furst in New York, while Frank was lined up with Ernest Philip Cohen in Los Angeles.[11]

THE ANALYTIC HOUR STARTED when the double doors opened and the doctor gestured for the patient to come inside. Phil Cohen was short and stocky, with spectacles, a tidy moustache, a neat seersucker shirt and bow tie – a bit like a 'mild-mannered dentist'.[12] The patient would hang up their coat and settle down on a leather couch that was still warm from the preceding session, while Cohen assumed his usual seat and readied his notebook. One of those who took to the couch, albeit a few years after Frank, was the actor and party member Sterling Hayden, who wanted to better understand his mother, his marriage and his feelings about co-star Marilyn Monroe.[13] Phil Cohen listened as the analysand spoke – sometimes offering interpretations, sometimes not.[14] There were days when Hayden would bare his soul and yet feel that Cohen was 'maliciously detached'. There were other days when the words just wouldn't come. 'You must make an effort, Mr Hayden,' insisted Cohen, 'just say whatever comes into your mind.' Hayden lay still. 'Fuck you, Doc,' he thought. 'If you were a good analyst, you'd be working with prisoners or in Harlem instead of here in Beverly Hills. How about that?'[15] The session continued in silence.

Cohen was 'charming,' said one patient, 'shy, but ... nice,' said another. One of his professional peers was less complimentary: 'I always thought of him as a psychopath.'[16] Strictly speaking, Cohen wasn't a psychoanalyst but practised under the woolier term 'therapist', which lacked the same rigorous regimen of training and accreditation. He had still spent three years researching, if not quite completing, a psychology PhD at the University of Chicago before doing further study at the University of Washington.[17] In fact, it was in Seattle that Cohen joined the Communist Party, at the height of the Popular Front era, largely motivated – as Frank had been – by his support for practical anti-fascism. Unlike most communist professionals, however, Cohen gained business by becoming one of the handful of trusted shrinks to whom members could be reliably referred, and trade was nowhere more brisk than in Hollywood, where a cast of neurotic leftists were suffering from stage fright, writer's block and other Tinseltown maladies.

Like Hayden, Frank Malina wanted help with his marriage. He also wanted to get over his dread of witnessing the physical pain of others. He had always struggled with the horrors of rocketry – that lurching fear he had felt in London at the human consequences of Space Age warfare. As Freud put it in the closing sentences of *Civilization and Its Discontents*, 'Human beings have made such strides in controlling the forces of nature that, with the help of these forces, they will have no difficulty in exterminating one another, down to the last man. They know this, and it is this knowledge that accounts for much of their present disquiet, unhappiness and anxiety.'[18] Was this the source of Frank's unhappiness? Ever since he'd returned to work on what was to become the world's first nuclear missile, his phobia had assumed a different dimension. Frank would find himself in meetings on the future of warfare such as, say, the War Department Equipment Board, chaired by the caustic conservative Joseph 'Vinegar Joe' Stilwell, in which missile mass destruction was a topic of calm consideration.[19]

Sitting in a boardroom listening to Republican cold warriors talk about the future use of his rocket induced a kind of bodily shock. 'I found in these meetings, I was getting more and more disturbed,' recalled Frank.[20] His propensity to break into a cold sweat was, he felt, a neurotic symptom for which he needed help – if only he could find a window in Cohen's crowded diary. Initially, Frank was dismayed to find that the therapist 'didn't seem to feel that my problem was too critical or urgent'.[21] It came as a relief then that Cohen decided during a lunch in February 1946 that Frank had 'a number of insecurity problems' that could be helped by 'intensive work' that could be carried out before his leave of absence from the institute.[22] As a test of the treatment's scientific efficacy, analyst and patient together agreed that Frank would visit a hospital at the end of the work – 'if I do not break into a cold sweat I will know that it has been successful'.[23]

Phil Cohen wasn't an orthodox analyst. He'd sometimes pour a brandy to loosen the tongue of a more reticent analysand, and thought nothing of socialising with his patients – as he did with Frank – inviting them to dinner and even giving one a trip in his private plane.[24] And for a party shrink he didn't seem like much of a communist. Sterling Hayden paid a whopping $500 a month, $25 an hour, while others, like the communist screenwriter Bernard Gordon, were asked to pay whatever they could (naturally their generosity or parsimony would be subject to rigorous interpretation). The most notable aspect of Cohen's practice, however, was his keenness to receive material from the analysand in written as well as in verbal form. Bernard Gordon handed over sheaves of scribbled foolscap every week – dreams, hopes, ambitions, memories – though Cohen never quite got round to commenting on them. Gordon eventually accused him of never even looking at his accounts, and when the exasperated screenwriter eventually asked for some of this material back, none of it could be found. That was a little odd.[25]

In Frank's case, Cohen requested an 'emotional biography' as a prelude to their sessions. So Frank filled up a black notebook with accounts of his childhood, his meeting Liljan on Bastille Day and their subsequent life together.[26] Cohen discouraged him from 'reading books on psychology and related subjects', lest they 'confuse' him, but the confusion was there regardless.[27] 'We are a hopeless and confused couple of human beings at the present time – aren't we?' Frank wrote to Liljan. Nevertheless, their period of self-study was, he felt, 'one of the most important events in our lives … without our crises we would certainly not have made the attempt to straighten out our internal conflicts'.[28] Liljan felt similarly. 'I give you all my good wishes with Cohen,' she replied. 'They are helping us, but the realization comes slowly that the hard work is left for us to do, as it should.'[29]

Unlike Frank, Liljan found a rigorously trained analyst who taught psychiatry at the party's Jefferson School of Social Science.[30] Joseph Furst was a protégé of the influential German analyst Karen Horney. He was one of the most vociferous defenders of Freud in *New Masses*, and had founded the Benjamin Rush Society for Marxist psychiatry. In contrast to Cohen, he was politically energetic, and would have been sympathetic to Liljan's own continuing activism, though exactly how much Frank and Liljan told their shrinks about this part of their lives is not known. At any rate, the couch didn't seem to distract them from political involvement. Frank was still attending party meetings in Los Angeles, while the move to New York gave Liljan renewed vigour, as she worked as a textile designer and immersed herself in a campaign against the newly minted House Un-American Activities Committee (HUAC) that was spearheaded by the racist Democratic congressmen and KKK supporters John E. Rankin and John S. Wood.[31]

Years before HUAC eventually turned its attention to communist scientists in California, it opened investigations into alleged communist influence in Hollywood, the media and the State Department – stirring a new public hostility towards socialist ideas. To fight this, Liljan and other party friends in New York founded the Citizens United to Abolish the Wood-Rankin Committee, which took out a full page ad in the *New York Times* headlined 'You can't talk … It's Un-American', accusing HUAC of wanting to 'supervise' free speech.[32] This, of course, was precisely the kind of thing that could get somebody hauled before HUAC, as Liljan well knew ('the old bastard [Rankin] … is setting a couple of bear-traps for us' … 'he's really after us and we ain't hiding!!').[33] But given Frank's line of work, she worried a little about also being involved in the 'anti-atomic' movement.[34]

She needn't have worried. Frank heartily approved of all this activity, and had himself joined another group in LA, the Independent Citizens Committee for the Arts, Sciences and Professions (ICCASP) – a broadly pro-Soviet 'peace-orientated' organisation whose members included Orson Welles and the Caltech biochemist Linus Pauling.[35] Despite the CIC raid at the end of the previous year, Frank seemed surprisingly incautious about ramping up his political involvement. At the point at which he started his work with Cohen, he was on the cusp of a considerable change. Joining the ICCASP while chairing meetings on the design of the Corporal missile was one example, but Frank's aspirations went further. He started to wonder whether he might try to visit the Soviet Union, perhaps even to speak about rocketry. Given the unfolding concern about the leaked WAC Corporal plans, the idea was either obstinate or reckless, as he must have realised. 'The trip to the USSR seems very dubious,' he wrote to Liljan; 'some think that the US will not let me go there, much less give some lectures on rockets. I know too much!!'[36] One week later, on 5 March 1946, Winston Churchill gave a speech in Missouri, famously observing that 'from Stettin in the Baltic to Trieste in the Adriatic an "Iron Curtain" has descended across the continent'.[37] More than just a vivid metaphor, the phrase 'Iron Curtain' worked to fortify the very border it described, a border on which the Corporal missile would one day stand sentinel. In such circumstances, did Frank really share his USSR travel ambitions with his colleagues and with Cohen?

JACK PARSONS WAS NOT into psychoanalysis.[38] He knew about it, and he knew that some Thelemites saw merit in it, but when it came to dealing with his own emotional disturbances, his preferred methods were magick and masturbation.[39] For other residents of Grim Gables, these methods could be disturbing. The long-standing and long-suffering OTO member Jane Wolfe noted that one unnamed couple 'had been traumatised by the astral developments from Jack's work', the man often having to get up in the night to conduct a banishing ritual 'to free himself from inimical forces', while the woman ended up 'quite off balance'.[40] To alleviate their distress the couple started psychoanalysis with Israel Regardie, a one-time secretary to Aleister Crowley who had later trained as an analyst in the United States. Even within the OTO, psychoanalysis, psychology and psychiatry ended up in the metaphysical punchbowl along with elements of Crowleyian magick, Rabelaisian humanism and Freemasonry. The mix is notable in part because the new guest to arrive at South Orange Grove Avenue would strain this miscellaneous brew into the elixir of a new religion.

One afternoon in August 1945, 'Ron' turned up in the company of Lou Goldstone, a science fiction illustrator and a regular visitor to Grim Gables.[41] L. Ron Hubbard, a sci-fi writer, was dressed in full naval uniform – a suave and assertive personality who treated

Grim Gables as a community awaiting his own messianic arrival. As a navy lieutenant on shore leave, Hubbard felt no compulsion to return to his wife and children, so when Jack suggested that he join the household, the sailor didn't hesitate. They hadn't met before, but each knew the other by reputation and from their shared appearance in Anthony Boucher's 1942 roman à clef *Rocket to the Morgue*, based on the LA sci-fi scene (Hubbard as 'D. Vance Wimpole' and Jack as 'Hugo Chantrelle').[42] It didn't take Hubbard long to dominate the life of the house, spinning tall stories of fighting polar bears off the Aleutian Islands and showing off the scars of arrows thrown by hostile tribes in the rainforests of South America.[43] He had been everywhere, done everything, from studying atomic physics to enjoying a decorated naval career; Hubbard had collected insights from Blackfoot shamans in Montana, Indian holy men, Buddhist priests and the last remaining magician from the court of Kublai Khan; or, as Judge Paul Breckenridge put it in a 1984 Los Angeles Superior Court trial, he was 'virtually a pathological liar when it [came] … to his history, background, and achievements'.[44]

Jack, however, was sold – and he felt sure that Crowley would approve of his new friend. How often had they been berated by The Beast for taking 'no steps at all to attract the right kind of people'? When Crowley saw photos of Agape members, he despaired that his disciples could look 'like hoboes! … sluttish, slattern, no trace of birth or breeding'.[45] But here was Ron, in uniform and sunglasses, flashing his smile, the sort of man who, as one resident put it, 'could charm the shit out of anybody'.[46] 'Although he has no formal training in Magick,' wrote Jack to The Beast, 'he has an extraordinary amount of experience and understanding in the field … he is the most Thelemic person I have ever met and is in complete accord with our own principles.'[47]

Jack's own Thelemic commitment, however, had started to waver. He had grown tired of the OTO, and was making plans to sell Grim Gables. For many months he had been pursuing his own ritual experiments with his ever-willing sidekick, Ed Forman. Amid the boredom of their post-Aerojet life, Jack and Ed resumed their tinkering, albeit orientated towards a different form of transcendence. Idle hands are the devil's workshop: they now approached magick much like the search for a solid propellant, combining drugs and music while modulating different invocations, consecrations and divinations – just to see what would happen. Naturally this was a rebellion against the plodding degree work of Crowley's magick, just as they had rebelled against the rigours of Malina's mathematical theory. As ever, they wanted a shortcut, to 'go shoot', and it didn't much matter why or how, as long as they got a result. After several trials, they did. What it was, no one could say for sure, but soon Forman became terrified of banshees screaming outside his bedroom window; Jane Wolfe found 'troublesome spirits' haunting the third floor of Grim Gables.[48] Jack, however, was on a roll, and ploughed headlong into new preoccupations with OTO's advanced VIII° and IX° sex magick rituals, to which there was little prospect of him being formally admitted.

L. Ron Hubbard could have been the inspiration for all these diabolical experiments. A new energy possessed Jack, mostly out of his distress that Ron and Betty were conducting a very public affair (one luckless lodger stumbled across the pair tangled like 'a starfish on a clam').[49] Jack had never discouraged Betty from taking other lovers, but it was different now that he found himself nursing the kind of jealousy that he considered unworthy of a Thelemite. He hated this feeling. It was sublimated in bouts of frenzied fencing that he and Ron staged in the 1003 living room.[50] Of course, Jack made a great show of fraternal support to the new lovers – 'with brotherly and sisterly embraces,' said Jane – but it pained him that other embraces with Betty were now off the table.[51]

Jack needed to escape from this crisis. The solution he settled on was desperate. He now believed that magick could help him conjure a woman who would be more amenable to his own special needs. His attitude to women had long been similar to Crowley's: in real life they were dismissed, yet on the astral plane they were venerated as idealised, attentive, sexual beings awaiting contact from the thrusting magi back on Earth. Crowley had his Guardian Angel, 'Aiwass', who dictated *The Book of the Law*. And then there was 'Babalon' – a classic Crowley re-spelling – also known as the 'Scarlet Woman', the maternal embodiment of sexuality, lust and other worldly pleasures.

But whereas Crowley seems to have thought of the Scarlet Woman as an 'office' rather than a body in the flesh, Jack wanted something or someone tangibly carnal to take the place of Betty.[52] To invoke this elemental figure, Jack turned to 'Enochian magic' and the sixteenth-century practices of the Elizabethan polymath Dr John Dee, who, with his ceremonial partner Edward Kelley, had attempted to commune with angels (they apparently instructed Kelley to sleep with Dee's wife, an echo of Ron and Betty that was not lost on Jack).

What Jack had once learned from his Caltech colleagues about scientific recording, he now applied to his astral experiments – making detailed notes about his Enochian preparations, consecrations and invocations that involved drawing pentagrams in the air and then un-drawing them in banishing rites, all to the sound of Prokofiev's Second Violin Concerto. Masturbation was inevitably involved; indeed the sight or sound of Jack 'invoking his elemental' was one of the hazards for other residents of Grim Gables. Each ritual was time-consuming and emotionally exhausting and, starting in January 1946, he repeated it for weeks on end, all the time making observations about unusual occurrences. In one instance, a table lamp was 'thrown violently to the floor', while on other occasions Jack noted 'windstorms' ('very interesting but not what I asked for,' he reported to Crowley).[53]

L. Ron Hubbard was invariably in the background of all this activity. Parsons referred to him as 'the Scribe', though how much Ron merely 'recorded' magical phenomena and how much he actively choreographed them is a matter of speculation. On 15 January, for instance, Parsons 'heard the raps again, and a buzzing, metallic voice crying "let me go free." I felt a great pressure and tension in the house that night.'[54] Some pressure is perhaps understandable given that Ron had earlier that day got Jack to sign paperwork on a business partnership – Allied Enterprises – in which the trio pooled their financial resources into a company (Parsons $20,000, Ron, $1,200 and Betty, nothing). A few days later, when Jack and Ron were conducting rituals in the Mojave Desert, Jack felt the tension snap, and said 'it is done'. Feeling sure that the operation was now complete, he 'returned home, and found a young woman answering the requirements waiting for me'.[55]

Standing at the door was Marjorie Cameron – everyone called her Candy – a 26-year-old artist who had come to visit a friend only to find herself playing the lead in Jack Parsons' Enochian fantasy. This didn't seem to faze her in the least. 'I have my elemental!' wrote a jubilant Jack to Crowley.[56] It wasn't actually the first time Candy had been to Grim Gables, nor her first sight of its locally infamous owner. She had been curious about the scene for some time, and was unmistakably attracted to Jack, even tolerating the overlay of magick that covered every aspect of their budding relationship. He described her as 'an air of fire type with bronze red hair, fiery and subtle, determined and obstinate, sincere and perverse, with extraordinary personality, talent, and intelligence'.[57] Candy wasn't Babalon, the Mother of Abominations, but Jack considered her to be a potential instrument towards her ultimate incarnation. And on that basis, the couple began working through the sex magick rituals of the OTO's advanced IX°.[58]

The first round of Parsons' 'Babalon Working' rituals had been conducted with L. Ron Hubbard, and the second was with Candy, and now the Scribe returned to assist Jack in the final bid to bring about the incarnation of the magickal child, the one that Crowley had described as inaugurating the new Æon of Horus.[59] As Ron and Jack resumed their ritual labours, the Scribe 'recorded' the ceremonial instructions given to Jack. '[C]onsecrate each woman thou hast raped,' wrote Hubbard. 'Remember her, think upon her, move her into BABALON, bring her into BABALON … one by one until the flame of lust is high.'[60] The cumulative weight of past abuse, years of dismissed consent within Agape Lodge, was now for Jack a magickal resource to augment his own personal power.[61] He felt sure that the operation had worked, and that somehow, somewhere, Babalon was becoming incarnate. Crowley was less certain: 'Apparently Parsons or Hubbard or somebody is producing a Moonchild,' he wrote to Karl Germer, his appointed OTO representative in the United States, who himself had a dim view of Parsons. 'I get fairly frantic when I contemplate the idiocy of these goats.'[62] But that was to misjudge how much they had already learned from his own example.

FRANK'S PSYCHOANALYSIS was circumscribed from the outset. He told Cohen that it would have to be all over before his absence from Caltech, leaving sessions to be squeezed around Frank's JPL commitments during March and April 1946. There wasn't a regular weekly slot, but he fitted them in when work and Cohen's diary allowed – sometimes one hour, sometimes two (another example of Cohen's unorthodox practice). The 'very enlightening' sessions had only been going on for a few weeks when Frank wrote to Liljan that 'I can hardly believe that my problem has had such a simple origin. The more I think about it, the more certain I am that ... [Cohen] has uncovered the basic pressure that has triggered many of my actions.'[63] Yet he didn't say what this was. Liljan was unconvinced, replying that these 'last few remarks are really too vague for me to understand what has taken place'.[64] In correspondence, at least, it never became any clearer beyond that Frank seemed happy with the help he received. But if Frank had hoped that psychoanalysis might help resolve their differences as a couple, he was left disappointed.

In Frank's mind, the beginning of April 1946 was the deadline they had agreed by which they would reconcile their relationship.[65] The candour and intimacy of the letters exchanged in the meantime led him to be hopeful. Throughout March, however, it became clear that this wasn't going to happen. Instead, she wanted 'to continue my work with Furst until he is done with me'.[66] Her commitment to the analytical process was deeper than Frank's, but she also offered it as a reason why she couldn't now return to him. In fact, she had already met someone else, Lester Wunderman, an advertising executive who would go on to pioneer the technologies of direct marketing, inventing both the toll-free telephone number and the loyalty card. For all Frank's new spirit of self-examination and emotional growth he was reluctant to admit that the end had finally come. He expressed his bitterness by weaponising various points of contention over the years – particularly the question of children – into a volley of oblique provocations.

When news of the WAC Corporal launch was finally made public in March, he sent her the press clippings with the line that 'the newspapers make me a father at long last', adding that Liljan was 'at least ... the midwife'.[67] He later sent her a baby book, which confused and upset her, and then weirdly boasted about his own fertility.[68] She responded to the rocket news with a hypergolic blend of grace and anger:

> I can only tell you that I sweated it thru in more ways than you realise. I was always in the background being sacrificed to this achievement. That is of no importance in itself, for I say truthfully that the reports I read of your success made me very happy and make me know that it was all to some very good purpose. [69]

Frank was now left to fall back on the same plan he had in 1944, arranging a second mission as a military attaché to Europe, this time with a wider remit to report not just on rocket propulsion but on science and technology more generally.[70] In preparation for a departure in July 1946, he put his work with Cohen to the test by submitting himself to a battery of injections and blood samples, even if it wasn't quite the full hospital trial that they had intended.[71] Frank's analysis hadn't been long but he felt to some degree sustained by the experience, as many of Cohen's patients did. Endings are never easy; 'all analyses end badly,' as the writer Janet Malcolm once put it; they all leave 'the participants with the taste of ashes in their mouths'.[72] Yet Frank seemed buoyant, remarking that work with Cohen had given him 'a number of victories under my belt over situations that before gave me hell'.[73] Frank's problems, however, were not always the ones he knew about. For one thing, trusted Phil Cohen was working with both the FBI and HUAC.

Exactly when this started, and to what extent it affected Frank Malina, is something of a mystery. At some stage in the early 1940s, Cohen dropped out of the party after being found insufficiently Marxist in his teaching of Marxism, as well as for questioning the Soviet influence over the CPUSA.[74] But if he drifted away from communist ideas, he certainly didn't drift away from others who held them, remaining close enough to the party to benefit from their many referrals. Then something odd happened. While still working as a therapist, the 'mild-mannered' shrink decided to swap his seersucker shirt and bow tie for a captain's uniform of the Inyo County Sheriff, later becoming a special agent in California's Department of Justice.[75]

Law enforcement is, to put it mildly, an unusual sideline for a practicing psychotherapist, and Cohen's conflicts of interest went beyond the obvious. In the late 1940s, with a casebook full of party members in the darkening climate of anti-communism, distressed patients would often come to him for counsel after being asked to testify before HUAC. What should they do? Should they become a 'stool pigeon' and name names under oath, or 'take the fifth' (the fifth amendment to the US constitution that allows a citizen to avoid self-incrimination) and face being blacklisted? To this dilemma, Cohen often gave his patients a subtle steer towards full candour and cooperation. If the patient was agreeable, he would even take the initiative to set up a meeting with a friendly FBI agent, Mark Bright, and HUAC's investigator, Bill Wheeler (with whom Cohen went to football games).[76] Wheeler matter-of-factly told the journalist Victor Navasky that Cohen helped 'condition' his patients to testify to HUAC.[77]

The screenwriter Bernard Gordon, who was in therapy with Cohen at around the same time as Frank, only found out about Cohen's collaboration with the FBI and HUAC when his production line of stool pigeons became a matter of anxious party gossip.[78] Decades later, when the blacklisted Gordon finally got a copy of his own FBI file, he realised the extent to which much of the testimony against him consisted of intimate information that only Cohen could have known. The names of the informants in the Gordon file are redacted – blacked out – as is true for the majority of FBI files of this era. But Gordon was in little doubt that Cohen was a key informer, and that all his missing written material had ended up with the FBI.

Even if Cohen's efforts funnelling communists to HUAC mostly happened after Frank's season on the couch, there are still plenty of questions about the integrity of their therapeutic relationship. Did Cohen specifically inform on Malina? It's possible – probable, even – but with the original (unredacted) version of Frank Malina's FBI file now destroyed, it's unlikely we can ever know. Still, the intrigue does not end there. In her memoir of her life in the California Communist Party, the activist Dorothy Ray Healey recorded that one of her 'distasteful tasks' as organisation secretary in the 1940s was to have a weekly meeting with Cohen to get a surveillance update on their own members ('were they likely to have a breakdown? Should we regard them as security risks?').[79] In other words, while Cohen was playing the attentive therapist, helping his patients reconcile their inner conflicts, he resolved his own ambivalence about communism by betraying his patients to both the party *and* the FBI. When Freud wrote about 'countertransference', he probably didn't imagine anything quite so dramatic.

ALEISTER CROWLEY WAS NOW in poor health and reduced circumstances, but he still maintained his correspondence with his followers across the pond. Everything he heard about Hubbard left him wary. This was especially true when he learned about Allied Enterprises, the company that Hubbard, Parsons and Betty Northrup had founded with Parsons' Aerojet capital. By the spring of 1946, Hubbard had run out of money. He had exhausted his credit with the residents of 1003 and needed some cash quick. His solution to the problem was to propose a scheme in which Allied Enterprises would buy yachts in Florida and Ron and Betty would sail them to California, where they'd then sell them for a higher price. Jack readily consented to this arrangement, not knowing that Ron and Betty had also written to the Chief of Naval Personnel requesting permission to leave the United States for South America and China.[80] Rather than following through with their yacht-trading enterprise, they were evidently planning a world cruise, living the high life with Jack's money in Florida.

The suspicions of OTO members were raised in April, and by the end of May, Crowley sent a cable to Karl Germer: 'Suspect Ron playing confidence trick. Jack evidently weak fool. Obvious victim prowling swindlers.'[81] When Jack finally realised that the Scribe, to whom he had entrusted so much magickal knowledge, might be nothing more than a con merchant, he packed a case and took a train to Florida to recover what he could, from the business if not the friendship. In Miami he discovered that Allied Enterprises had in fact purchased two schooners, the *Harpoon* and the *Blue Water II*, as well as a yacht, the *Diane*, for which Ron had paid $12,000.[82] The pair were nowhere to be found. Eventually Jack got a phone call from a harbour on the County Causeway to say that the *Harpoon* had set sail that very afternoon – with Ron, Betty and a hired crew.

Jack wasn't in a position to give chase, so at eight o'clock in the evening he put on his robes, traced a circle on the carpet of his Miami hotel room, and stepped into the ring. He began by conducting the Banishing Ritual of the Pentagram, before proceeding to a full invocation of Bartzabel, the spirit of Mars, whose energy he now summoned against his errant partners. A windstorm hadn't always been a useful consequence of Jack's magickal work, but this time it was just what was needed. 'At the same time, so far as I can check,' he wrote to Crowley, 'his ship was struck by a sudden squall off the coast, which ripped off his sails and forced him back to port, where I took the boat in custody.'[83] Jack's methods now shifted from Bartzabel to a Bill of Complaint in the Dade County Circuit Court, in which his lawyer managed to claw back a few thousand dollars, though much of the money had already been spent.[84] Parsons returned to Pasadena with an additional promissory note for $2,900 for his stake in the *Harpoon*, which he left with Ron and Betty – a deal allegedly made in exchange for Betty not pressing statutory rape charges against Jack for 'unlawful sexual intercourse with a minor'.[85]

Jack never saw either of them again. He returned to Candy in Pasadena and they married later in the year, as did Betty and Ron. 'Betty was ... designed to tear you away from the now unneeded Oedipus complex, the overvaluation of women and romantic love,' Jack wrote to himself, dressing his usual misogyny in a piece of borrowed Freudianism. He decided that 'the final experience with Hubbard and Betty, and the OTO, was necessary to overcome your false and infantile reliance on others, although this was only partially accomplished at the time'.[86] He would now turn inward. The subsequently turbulent relationship with Candy further discouraged him from emotional entanglements, even when it came to magick. 'My ... partners [for ritual sex magick] consist of those that can be purchased or otherwise easily picked up and disposed of,' he later told Karl Germer.[87] Some of the women in the OTO might not have noticed this as much of a change.

Not long after Jack parted company with Ron and Betty, Hubbard developed his sci-fi writing and occult experience into a new science of the mind which he called 'Dianetics'. It was, he said, nothing less than the biggest breakthrough since the discovery of fire (although he also saw it as a contribution to the markedly more recent 'field of psychology and psychotherapy').[88] Where Freud's patients had neuroses, Hubbard's had 'blocked engrams'. Only a trained 'auditor' (therapist) could help the troubled 'pre-clear' (patient) to eventually 'go clear'. Unsurprisingly the professional psychology and psychiatry associations took a dim view of Dianetics, not helped by the fact that it was first published in the *Astounding Science Fiction* magazine, a criticism that in turn triggered Hubbard's lifelong and fanatical opposition to the profession.[89]

When he and Betty eventually split – she alleged that he'd been beating her since the Florida summer of 1946 – Hubbard was convinced that Betty was 'in league with' his arch enemies, the dreaded 'psychiatrist-psychologist-psychoanalyst clique'.[90] Perhaps if The Beast had not died in 1947 but had lived to pour scorn on Scientology, then Hubbard may have turned his fire on Crowley and the OTO instead of professional psychiatry. As it turned out, Hubbard's success in founding Scientology greatly eclipsed that of The Beast. The emergence of Scientology out of Dianetics showed Hubbard at his most inventive, the *bricoleur* of a new religion who variously drew on Crowley's magick, Freud's science of the self and elements of Hinduism and Buddhism – not to mention some purloined Aerojet wealth.[91]

The story of L. Ron Hubbard's entanglement with Jack Parsons is now relatively well known, ever since the publication in 1969 of an article by Alex Mitchell in the British *Sunday Times*.[92] But by far the oddest turn of events in this saga is that, although the Church of Scientology threatened legal action against the newspaper and won an out-of-court settlement, they also released a statement essentially confirming the Scribe's role in Jack's ritual workings; it's just that, they said, Hubbard was operating under cover for the navy, having been sent 'to break up black magic in America'.[93] There isn't any available evidence to support this claim. On the other hand, it's by no means improbable that the Office of Naval Intelligence (ONI) took an interest in Parsons and other residents of Grim Gables, especially since the FBI and CIC had Agape Lodge under surveillance. After all, Aerojet had navy contracts; Martin Summerfield, for instance, has an extensive ONI file. But for all it gets right, the Church's statement ends up presenting a very strangely slanted story about 'Dr [sic] Jack Parsons of Pasadena, California ... America's Number One solid fuel rocket expert', who conducted 'savage and bestial rites' with a 'black magic group [that] was dispersed and destroyed', from whom 'Hubbard rescued a girl they were using'.[94]

IF FRANK MALINA EVER FOUND out about Phil Cohen's other life, this discovery is nowhere mentioned in correspondence – hardly surprising, since he avoided referring to sensitive matters in print. But it's possible that he heard on the grapevine that his therapist was an informant. If so, he must have realised that anything he uttered on Cohen's couch might have ended up in a memo to J. Edgar Hoover. Still, at least Frank's experience was better than that of many of the other patients. The actor Sterling Hayden was one of those who, with Cohen's encouragement, cooperated with the FBI and HUAC, admitting his party membership in a bid to stay off the blacklist. 'I don't know why I got out of the Party any more than why I joined,' Hayden complained to his lawyer. 'I could say a lot of things about those people I knew in the Party ... it would all be good.'[95]

But in the end Hayden buckled to pressure and hated himself for it. Years later, the director Stanley Kubrick gave the actor the opportunity to express the absurdities of his accusers in his career-defining role as paranoid anti-communist Brigadier General Jack D. Ripper in the 1964 film *Dr. Strangelove*. 'I can no longer sit back,' says Ripper, 'and allow Communist infiltration, Communist indoctrination, Communist subversion, and the international Communist conspiracy to sap and impurify all of our precious bodily fluids.'[96] It wouldn't have been too far-fetched to hear that line from a congressman at HUAC.

Some time after confessing to the FBI, Hayden was back on the couch with Cohen. This time the analysand found no difficulty in speaking his mind. 'If it hadn't been for you,' Hayden accused his shrink, 'I wouldn't have turned into a stoolie for J. Edgar Hoover. I don't think you have the foggiest notion of the contempt I have had for myself since the day I did that thing.'[97] Hayden had testified in part because he wanted 'to remain employable in this town long enough to finish this … analysis'. But not for the first time in psychotherapy, the ending wasn't happy.[98]

'Fuck it!' he said to Cohen, taking his leave from the couch. 'And fuck you too.'[99]

THE PROBLEM OF ESCAPE

IN THE AUTUMN OF 1945, Frank was testing the WAC Corporal at White Sands when he learned that the army was keen to put the lab on a permanent footing. To that end, General Gladeon M. Barnes asked Caltech to operate JPL as a facility that would advance the army's strategic interest in basic rocket research.[1] Initially, that sounded like good news. But it was still research tailored to military objectives. And Frank was conscious that the Manhattan Project, which had also arisen from basic research, had recently led to the destruction of two Japanese cities. 'The way the world looks now too many people are getting ready for World War III,' he complained to his parents.[2] 'The problem [of JPL's future] is very complicated, with neither the government or Caltech really clear on what is to be done.'[3] It was partly a question of where this kind of science rightfully belonged. Should it be in a university to benefit from the open exchange of information? Or should a few researchers with security clearance be tucked away in a restricted military establishment? Or perhaps some hybrid of the two – like classified research inside a university lab?

Frank believed in the universality of science. He felt that ideas should transcend the petty politics of sovereign states, that they should be freely expressed and freely shared. That was not only desirable in itself, it also produced better work; he had learned from Robert

Goddard's difficulties that a collaborative culture of engineering was much more likely to succeed. This cooperative vision wasn't, however, shared by Malina's conservative colleagues. When it came to negotiating the army offer, it was Clark Millikan as Acting Chair of the JPL Executive Board, rather than Malina as Director, who really made the running. A seasoned anti-communist, Millikan thought that the values of a university were secondary to Caltech's moral duty to help fight the Soviet threat.[4] He strongly backed the Army's approach, persuading the board that JPL should be 'a special national defence laboratory supported by government funds'.[5] Frank hated this idea. 'I am so disgusted with the whole situation connected with this particular contract,' he wrote to Liljan, 'that at the least provocation I may blow my top'.[6] At one point Frank stormed out of a meeting and threatened to quit. 'If you feel that something is not right, and feel strongly about it, there's no point hiding it,' he told an interviewer, 'it's better to have a showdown.'[7]

Malina then countered with his own plan: that JPL should continue to be operated by Caltech for the Army, but that declassification of research 'should be continually pressed'. At the same time, Caltech faculty could operate a parallel rocket lab, a 'Department for Jet Propulsion Engineering', for 'completely unfettered research'.[8] The two labs would work closely together but the arrangement would put the onus on civilian rather than military rocket applications. The reality, of course, was that the Army held the purse strings, and were not much interested in Malina's idealistic vision. Nor were JPL's Republicans. 'Things ... that make ... [universities] more pure, may have been fine for some of the professors,' said Homer J. Stewart later, 'but it was very bad for the national interest.'[9] In any case, who would fund a civilian rocket lab?

By March 1946, Frank was telling Liljan that 'the future of the JP lab has looked very dark', but that they were now beginning 'to see the resolution of a number of critical issues'.[10] This was an optimistic view. Frank's plan for parallel labs was a non-starter, and while General Barnes sought to reassure him that 'most' work would be unclassified, that was soon forgotten when the reality of missile research became clear (by 1950, 90 per cent of JPL reports were classified).[11] On 22 March, the JPL board recommended that Caltech accept Millikan's plan, which a week later they did. JPL had essentially become a weapons lab for the Cold War – against the wishes of its founder but as a tacit consequence of his accepting military funding in 1938.

Other disappointments then compounded the Army takeover. A four-day JPL symposium on upper atmosphere and guided missile research, which brought together scientists from across the US, was the perfect opportunity to showcase the WAC Corporal as a vehicle for the emerging space sciences. Numerous well-known astronomers and atmospheric physicists came to Pasadena – people like James Van Allen from the Johns Hopkins Applied Physics Laboratory and Ernst H. Krause from the Naval Research Laboratory.[12] Surely they'd see the potential of JPL's new sounding rocket? Not exactly. They had already made plans to use the V-2 for research, with General Electric having been awarded a contract ('Project Hermes', the god of thieves) to refurbish and fly the German rockets from their many salvaged parts. Under the auspices of Operation Overcast, GE staff were to be trained by scores of Wernher von Braun's engineers, who had since arrived in Fort Bliss, half of them Nazi Party members, with a few others also having belonged to either the Brownshirts or the SS.[13]

The V-2 cast such a shadow over everything that it was difficult for the WAC Corporal to be seen, far less celebrated. The disparity in size was a factor (the WAC Corporal being not much taller than the fin of the V-2). It didn't help either that the JPL rocket was still classified, while the V-2 was known to absolutely everyone. Even if it had been an operationally poor weapon, the V-2 somehow held its totemic aura in the post-war period, becoming the definitive icon of the early Space Age. (One illustration of its hegemony is that when JPL later started *Lab-Oratory*, a newsletter for its staff recreation club, the rocket masthead featured not one of the lab's own rockets but the distinctive bulge of the V-2.) And, of course, a rocket built to carry a one-ton explosive had plenty of capacity for when the warhead was replaced by an instruments bay. That made it even more irresistible for the new Upper Atmosphere Research Panel, which convened without any representative from JPL, and soon became known as the

'V-2 panel'. (Under James Van Allen's direction, the panel later commissioned a modified variant of the WAC Corporal, the Aerobee, as a workhorse of upper atmosphere research.)[14]

For the home-grown pioneers of American rocketry who had spent their lives fighting fascism, the welcome given to the V-2 and its architects was sickening. Malina couldn't wait to leave ('my brain [is] ... beaten to pulp,' he told Liljan).[15] Every two weeks he'd get a status update on the continuing WAC Corporal tests in which the nose-cone parachute recovery system and a new telemetry system were being trialled. The new Corporal missile was taking shape, with systems of radio guidance being developed by a New Zealander, William Pickering. German rocketry may have had advantages in size, but JPL was far ahead when it came to missile guidance and accuracy, a fiendishly difficult problem that was central to the new security landscape of the Cold War. Precision targeting was technically interesting and, Frank felt, politically dispiriting – all that effort for the pinpoint delivery of indiscriminate annihilation.[16] He chaired his last JPL conference on 7 May, before handing over responsibility to his deputy, Louis Dunn, an ambitious engineer who for all his apparent modesty seemed to relish the opportunity. Like Clark Millikan, Dunn was a true believer in American might.

ONE ASPECT OF ROCKETRY still excited Frank. He and Martin Summerfield, who had since returned to JPL from Aerojet, had been busy working out the mathematical criterion for space flight through and beyond Earth's gravitational field. In their landmark paper, 'The Problem of Escape from the Earth by Rocket', they analysed the complex interplay between: c, the effective jet velocity; ζ, the ratio of propellant mass to initial mass; t_p, the time of powered flight; μ, the ratio of initial mass to maximum cross-sectional area; and C_d, the drag coefficient based on the same cross-sectional area. Based on calculations from the specific impulses of existing liquid propellants – i.e. how many pounds of thrust could be obtained from each pound of propellant – they concluded that a single-step rocket simply could not reach escape velocity, unless by some undeveloped method of nuclear propulsion.

More promisingly, they provided the first detailed mathematical analysis of a 'multiple-step rocket', in which each successive stage would shake off the last, thereby launching from an altitude gained, but not encumbered, by the jettisoned hardware. Earlier pioneers such as Konstantin Tsiolkovsky, Hermann Oberth and Robbert Goddard had sketched some basic parameters for the problem, but none had the benefit of observing high-altitude rocket performance to help frame the permissible conditions for escape. Malina and Summerfield's paper concluded that the optimum step rocket would be 'one in which the ratio of the mass of payload for each step to the mass of the step propelling the payload is the same'.[17] This was important knowledge if you wanted to, say, carry cosmic ray instruments and a transmitter to send data back to Earth.

The primary impact of the paper was Summerfield's realisation that, with the WAC Corporal and the V-2, they actually already had two elements of their step rocket. Notwithstanding their divergent political backgrounds, these rockets could be 'mated' – in other words, a first-stage V-2 could 'bump' a WAC Corporal to a high-altitude launching platform, creating a two-stage rocket. It would be called the BUMPER WAC Corporal. Malina mentioned the idea in a paper written in February 1946 (later published in the journal *Army Ordnance*), but he took care to credit his friend.[18] William Pickering later claimed that it originated in 'early 1946' from a conversation between Clark Millikan and Wernher von Braun.[19] The Army's official project history attributes the initiative to General Holger Toftoy during a June V-2 firing at White Sands, with feasibility studies taken forward by Clark Millikan, Homer Joe Stewart and Wernher von Braun.[20] Already, the names of Malina and Summerfield were being separated from their contributions. It's true that Hoftoy was keen on the idea, adding the BUMPER WAC project to General Electric's Hermes contract in October 1946 as a useful project on which American and German teams could collaborate. Happily for Malina, though not for Summerfield, his leave of absence spared him from having to work alongside von Braun.

A second consequence of the theoretical work behind the 'escape' paper was less obvious: the fact that a payload could reach escape velocity also meant that it could be placed into orbit around the Earth, a conclusion that Frank outlined to General Stilwell's War Equipment Board in January 1946. The idea of an artificial satellite wasn't exactly new. Tsiolkovsky had been alert to the theoretical possibility. The British sci-fi writer and founder of the British Interplanetary Society, Arthur C. Clarke, famously speculated in *Wireless World* about the V-2 as a research vehicle, and ultimately as a bearer of a communications satellite.[21] But unlike Clarke, Malina could for the first time outline the theory based on experimental data, as well as make a concrete proposal for building such a vehicle.

In this instance, the main problem of escape was that Vinegar Joe Stilwell and the rest of the board were not interested in a satellite. (Stilwell hated chairing this committee, commenting that he'd rather 'sit on a tack'.)[22] Nor did he understand Malina's point that the same delivery system that could place a satellite in orbit might also herald the dawn of the Inter-Continental Ballistic Missile.

Frank was hardly surprised by this indifference. In any case, he had already signed a contract between JPL and the Navy's Bureau of Aeronautics (BuAer) to investigate a satellite vehicle, exploring the relation between orbital altitudes, fuel performance, propellant-mass ratio and payload.[23] Malina didn't have the time to do this work himself, so he assigned it to a colleague, W. Z. Chien – Chien Weichang (now Qian Weichang) – JPL staff mostly called him 'Jimmy'.[24] Chien's reputation was a bit like Malina's: mathematically reliable, interested in the visionary stuff and adamantly communist. He had come from China to do his PhD at Toronto, joining JPL as part of its early expansion in 1944 and working with Homer Joe Stewart. The two of them got on famously, despite the improbable rumour that Jimmy Chien had been part of Mao's Long March back in the 1930s.[25] One time, when they were all at a party at Frank's house, Homer Joe commented on Jimmy's penchant for martinis. 'When you're out on the Manchurian Plain,' he shrugged, 'you've got to keep a fire going inside' (Stewart thought this was hilarious; the politics, not so much).[26] Chien wrote a promising report on a liquid hydrogen-oxygen satellite vehicle, after which BuAer contracted Aerojet to test whether the propellant's specific impulse would be borne out in practice.[27] But the ballooning projected cost – $5–8 million – made it politically unfeasible. Besides, who wanted a satellite in 1946?

By the middle of the year, many of Frank's friends seemed to be leaving Pasadena. Jimmy Chien went back to China, taking up a professorial position at Tsinghua University in Beijing.[28] Von Kármán was mostly at the Pentagon. From the Unit 122 scene, Andrew Fejer had moved to Toledo, H. S. Tsien took up an offer of tenure at MIT and Frank Oppenheimer was on his way to the University of Minnesota. But for all these departures, there was one important return. After working unhappily at the Curtiss Wright Laboratory, the Unit 122 leader Sidney Weinbaum was now keen to come back to Caltech. In February 1946, before JPL's Cold War direction had become properly apparent, Tsien lined him up to work on thermodynamic problems in JPL's Materials Section.[29] Sidney was excited to be working at a lab directed by his old friend – until, that is, he discovered that Frank, too, was getting ready to leave.

It wasn't easy for Frank to say goodbye to his colleagues. He called von Kármán and asked that they do a tour of the lab that they had founded, 'so the two of us can have a last look at the place before going away'.[30] With 385 employees and new facilities at the Arroyo, JPL now had an air of stability and respectability, a far cry from the improvised experimentation of a decade before.[31] Here was an extraordinary group of workers advancing a precarious technology, people who so often broke the mould – like Macie Roberts, the supervisor of the rocket 'computers' who went on to hire scores of women for trajectory analysis, thereby cracking open the exclusively male culture of engineering. There were others, too, who shared Malina's desire for a fairer world. At JPL, the conversation was mostly about engineering rather than politics, but there were still plenty of people exercised by questions of equality – enough, in fact, to worry Louis Dunn, the new Acting Director. One colleague described Dunn as 'a suave J. Edgar Hoover', a resemblance that some felt went beyond his heavy-set neck and close-cropped hair.[32]

LONDON AGAIN. Back to summer drizzle, powdered eggs and smoked herring. Back to rationing, with bread and beer in short supply.[33] It was actually good to be there – not least because Frank was now single and looking to the future ('being a bachelor has some advantages').[34] His secondment as a military attaché had none of the urgency and stress of his 1944 posting. The war was over, and the V-2s were no longer falling on England. This allowed him to range beyond a narrow rocketry brief to report back on all matters relating to science and technology in post-war Britain.[35]

To that end, Frank turned up at a meeting of the Preparatory Commission of a UN body that looked set to be called UNESCO, the United Nations Educational, Scientific and Cultural Organization.[36] Something about this UNESCO meeting at Belgrave Square inspired Frank. There was a sense of promise, an orientation to a new world of peace and scientific cooperation – the opposite, in other words, of everything that JPL had since become. At the close of the meeting, he introduced himself to one of the principal speakers, a tall, owlish and pleasantly rumpled biochemist whose political writings Frank had digested back in the Unit 122 days.

Joseph Needham was a man for whom paradox was a kind of signature. Now in his mid-forties, he somehow managed to be a Christian embryologist, a humanities-orientated biochemist and an earnest, wealthy Marxist, all at the same time. In the 1930s, he had joined other colleagues such as J. D. Bernal in playing an important role in the 'Social Relations of Science' (SRS) movement that sought to bring political struggle into the laboratory.[37] His home life was comparably avant-garde. At Cambridge, he eschewed donnish domesticity for a peripatetic and consensual *ménage à trois* with his wife Dorothy, a distinguished scientist in her own right, and her PhD student, Lu Gwei-djen. The relationship with Lu led, in turn, to another – a lifelong passion for the history of Chinese science and civilisation that eventually replaced biochemistry as his main academic calling. Indeed, it was when he was on a trip to China with Lu Gwei-djen in March 1946 that he received a telegram from Julian Huxley, an old SRS friend and now the Executive Secretary of the Preparatory Commission, asking if he would be willing to head up UNESCO's Division of Natural Sciences.[38] Needham accepted immediately. He had for some time tried to get science on the agenda of a body that was initially thought of only as 'UNECO'. The lobbying paid off, and Malina watched from the back of the London meeting as Needham's vision for the science programme was adopted.[39]

Needham's enthusiasm for both transnational scientific cooperation and the need to disseminate new technologies to the non-Western world spoke to all Frank's inclinations. 'You know, I find this fascinating,' said the visitor; 'is there any way I might get into something like this?' The pair discovered an easy rapport and a shared outlook. 'Why don't you come with us?' replied Needham.[40] A formal letter soon followed, with another despatched to Caltech President Lee DuBridge, giving notice that a two-year leave of absence might be required from JPL.[41] After chewing over the idea with his parents and von Kármán (they all approved), Frank accepted the appointment in October.[42] Then things got complicated.

One difficulty was the reputation in Washington of Needham's friend Julian Huxley, who headed the Preparatory Commission and in whose intellectual mould UNESCO was taking shape. Huxley was an evolutionary biologist and public intellectual, well known in Britain for sitting on the BBC's Brains Trust television programme. Nudging 60, he was elfin but energetic, with flat Brylcreemed hair, circular glasses and unruly eyebrows. Back in the 1930s, Frank had read Huxley's dubious book *What Dare I Think?*, which, true to the intellectual fashion of the time, made a case for eugenics. Science, for Huxley, was almost a religion; it was the animating forces for his entire family, from the famous grandfather, Thomas H. Huxley – 'Darwin's bulldog' – to the work of Julian's younger brother Aldous, whose recent *The Perennial Philosophy* was a favourite of Frank's.[43] Julian Huxley may have been more mainstream than Needham, but he was still very much his own man: 'I like that chap,' the old Thomas Huxley once said of his then four-year-old grandson. 'I like the way he looks you straight in the face and disobeys you.'[44] Half a century later, that was what worried the US State Department. They emphatically did not want an atheist fellow traveller setting the course for UNESCO.[45]

The US didn't like Huxley, but they also underestimated him. Having agreed to site UNESCO in Paris, and with the Preparatory

Commission in London, the State Department assumed that they would have a free hand to appoint an American Director General. That assumption was wrong. And even when they realised that Huxley had the nightmare combination of being both popular and qualified, they still wheeled out an American candidate, the former US Attorney General Francis Biddle, with threats to cut all US financial subventions if he wasn't appointed.[46] Frank recognised that if Biddle rather than Huxley got the top job, a post in UNESCO's Natural Sciences Division would not be forthcoming.[47] But for many of the reasons that Biddle was Truman's choice – as a stiff anti-communist Democrat, he'd set up the Attorney General's List of Subversive Organizations – he wasn't especially well liked overseas. The British Foreign Secretary Nye Bevan did his best to persuade Prime Minister Clement Attlee to acquiesce to the Americans, but Attlee stood his ground and supported Huxley.[48] Out-manoeuvred, the State Department punished Huxley's eventual appointment by shaving $1 million from their subvention and insisting that his six-year term be truncated to two.[49] By the time of the formal inauguration in Paris on 4 November, it was clear to the United States that UNESCO was the site of a political struggle that they needed to win.

FRANK RETURNED TO the United States in early December 1946 to pack up his things for the move and to await final confirmation of the UNESCO job. 'The country doesn't appear to have changed,' he wrote to his parents, 'the economic system is just as unjust and the newspapers just as hysterical.'[50] In fact, quite a lot had changed. The campaign for November's congressional elections had been dominated by the spectre of domestic communism – the idea that inside the corridors of the US government a network of party members was working to subvert American democracy. Although such concerns had been around since the New Deal, the gradual revelations from defecting Soviet agents made it seem much more plausible. On the same day as the WAC Corporal flight, 11 October 1945, a low-level Soviet spy called Louis Budenz, one-time managing editor of the *Daily Worker*, quit the party to embrace a new life as a professional witness for the FBI and HUAC.

Even more damaging for the party was the loss, a month later, of Elizabeth Bentley – the so-called 'Red Spy Queen'.[51] Bentley had become disillusioned with her life as a mole. She drank a lot and was distressed to discover that she had been supplying secrets to a Soviet handler rather than, as she thought, the American Communist Party. Worried that the Budenz defection might put her in danger, and that her lover might be a Soviet agent, Bentley handed herself over to the FBI, revealing details of her extensive espionage network. Lina Weinbaum wrote to Frank that Bentley was a 'dirty bitch'.[52] The tide of public opinion was now running strongly towards Hoover's anti-communism, and with some justification: the party which Malina and his friends had used for advancing progressive causes was also, unquestionably, a mechanism for Soviet operations – atomic espionage in particular.[53]

The first major spy crisis of the Cold War actually preceded both Budenz and Bentley, centring on a minor Soviet official called Igor Gouzenko, who slipped out of the embassy in Ottawa with some papers stuffed under his shirt – so many papers, in fact, that he had to hold in his stomach ('otherwise,' his wife said, 'he would have looked pregnant').[54] Gouzenko's defection was initially kept quiet in a bid to defuse international tension, but the documents he bore revealed a party spy ring in the Canadian government, with an extensive reach into the international scientific community. Arrests followed. One British atomic physicist, Alan Nunn May, was shown to have leaked a report on the structure of the Manhattan Project, as well as having passed on samples of Uranium-233 and -235.[55] He was sentenced to ten years of hard labour. Then Gouzenko made an even more startling claim, one which dogged the reputation of British intelligence for decades: that an exceptionally valuable spy for the GRU (Soviet military intelligence) with the code name ELLI was embedded high – *very* high – in MI5.[56] That kind of allegation was almost self-cancelling given that the potential truth of the claim was precisely what made it impossible to uncover.

All of these developments placed the question of loyalty and security at the heart of US politics. By the middle of 1946, Truman's response to the spying revelations and to Soviet expansion in Eastern

Europe was an aggressive strategy aimed at confrontation rather than negotiation.[57] And under pressure from congressional Republicans, the president moved to eliminate any government employees deemed politically suspect. Just two weeks before Malina arrived back in Pasadena, Truman set the machinery in motion, establishing an interagency commission to revise the government's loyalty-security procedures.

Roosevelt had made earlier legislative moves against the party, from the 1939 Hatch Act that had once worried Frank to the 1940 Smith Act, the first peacetime sedition act in American history.[58] However, it was Truman's executive action that really inaugurated the McCarthy era. It took a few years for the Wisconsin senator, Joseph McCarthy, to personify his eponymous '-ism' but the defining events of this time actually happened in the three years from 1946 – before his rise to prominence.[59] The climate of suspicion and interrogation, the binary logic of us and them, was well underway in the year after the end of World War II. As J. Edgar Hoover put it, 'every American communist should be considered a potential agent for Soviet Russia'.[60] That was, of course, unfair. Lots of party members, if they were ever even approached, would have been highly critical of Soviet intelligence. Most were committed to the peaceful political transformation of the United States, and their desire for change was itself evidence of loyalty to the democratic ideal. But there were others for whom that wasn't exactly the case, and at least one of these was now living in Frank Malina's house at South Hudson Avenue.

The fact that accommodation in 1940s Pasadena was in short supply meant that Frank had readily found friends to look after the house when he was in London. That was important, not just in terms of covering the rent, but to satisfy the high maintenance demands of his landlady, the unapologetically prim Miss Bessie Jacob. Miss Jacob, then in her late sixties, didn't entirely approve of Frank, but over the years she had warmed to him (though she wished he'd think 'a little less of science and peace and a little more of your home').[61] She also thawed to Frank's charming friends Mae and Bob Churchill and their young children when they moved into South Hudson Avenue – despite Miss Jacob's initial shock at toys being left overnight on the front lawn.[62]

The FBI's concerns about the Churchills were rather different. Robert, a film-maker, seems not to have worried them. Mae Huettig Churchill was a different story. She was remarkable in many respects, not least that she had written a PhD thesis on the industrial organisation of the Hollywood film industry, later published as a formative book in the field of media economics.[63] Before marrying Robert Churchill in 1944, however, she had spent the preceding decade married to Lester Marx Huettig, an engineer at the Remington Arms Company who worked for Soviet intelligence. FBI informants revealed that, in 1937, Huettig secretly passed on the design blueprints of automatic shell loaders to a party functionary, William Crane, who in turn handed them to Colonel Boris Bykov of the GRU.[64] Informant 'T-3' told the FBI that 'Mrs HUETTIG was present when LESTER HUETTIG made his contacts with Colonel BYKOFF [sic] and undoubtedly knew of her husband's involvement in Soviet espionage'.[65]

In other words, the charismatic young mother living in Frank Malina's house, who wrote him warm, funny letters trawling for details of his love life and lamenting the state of US politics, had direct experience of Soviet intelligence. 'If you can think of a country which is more worthy of allegiance than ours is currently, I'd be interested,' she wrote to him, adding, 'You can see why I'd so much rather talk about sex'.[66] There's no suggestion that Frank knew about Mae's background, but it's an instructive window into the contradictory world of CPUSA, an organisation that was at once alive to the struggle for liberty in modern America and deadly in its intent to acquire technologies of destruction. This was the paradox of communism, and it helped inspire the gathering forces that now rallied towards its eradication.

TRYING TO PACK UP his affairs in California at the end of 1946 was, for Frank, a return to heartache and the prospect of imminent divorce. Even though Liljan was now involved with Lester Wunderman, she and Frank maintained such a close correspondence that a letter to Liljan remained the default medium for his passions, political and otherwise. The inevitability of their parting somehow made them both less guarded. 'Something during the last few weeks has made it possible for me to talk to you in a way I never could before,' wrote Frank, a candour 'made easier by your recent tendency to confide in me some of your innermost musings'.[67] It wasn't all easy. Liljan felt judged for continuing her analysis with Joseph Furst, and there were squabbles, too, about furnishings and household effects.

But at the end of 1946, their letters had a kind of raw warmth. Frank felt that his experiences of the last year had forced him to reflect anew on his organising assumptions, particularly when it came to relationships. Politically speaking, there was no real change ('I believe that my evaluation of the major outline of our society has been essentially correct'). He was still evangelical about the Soviet ideal, even to the extent of trying to convince sceptical friends by sending them Joseph Needham's unapologetically communist *History Is On Our Side*.[68] 'I hope that you will keep me informed ... how you are progressing in your work and political thinking,' Frank told Liljan, 'I will always retain an interest in your development along these lines and I will be glad to help you in these matters.'[69]

Liljan needed no help whatsoever. She was immersed in the New York Communist Party, working on the congressional campaign of democratic socialist Vito Marcantonio.[70] Rather, it was Frank who needed something from her, as he sought to make sense of what had happened between them: 'my real struggles,' he admitted, 'have been in the domain of personal psychology and the search for the meaning of an adult relationship between a man and a woman.'[71] He could sense that Liljan was tiring of their correspondence (hardly surprising after at least six letters in December 1946 alone). 'Even if you do not open them,' he pleaded, 'save them for me and someday I will ask for them to be part of my diary.'[72] He eventually agreed to put an asterisk on the envelope if it contained anything of practical urgency ('information on income tax etc'), which left him free to send her long, rambling essays – and her free to ignore them.

Alone in Pasadena, Frank had a little time to catch up with friends, but he had mostly said his goodbyes back in June. He fleetingly saw the Weinbaums and the Churchills, each couple storing some of his furniture until such time as either he or Liljan might need it back. And that was it. That was the low-key end to twelve years at Caltech. He set off driving the 1,600 miles to Texas to spend time with his family over Christmas and await word from UNESCO, his letters to Liljan putting on a brave front about making the journey alone. The drive, he told her, 'was pure pleasure'. The weather was fine and he greeted the appearance of the sun over the horizon by belting out 'Oh, What a Beautiful Mornin'' from *Oklahoma!*[73] A telegram on Christmas Eve did indeed give him a wonderful feeling:

INVITE YOU TO JOIN US AS A PERSONAL ASSISTANT TO HEAD OF DIVISION RANK AND PAY EQUIVALENT COUNSELLOR STOP HUXLEY PUTTING YOUR NAME FORWARD EXECUTIVE BOARD JANUARY TENTH STOP HOPE GREATLY YOU WILL ACCEPT STOP ALL APPOINTMENTS INITIAL TWO YEAR TENURE STOP=
JOSEPH NEEDHAM[74]

But not everything in Washington was going his way. By 17 January, he had heard nothing about the executive board meeting, and was beginning to worry about whether there were other machinations at work inside the US government.[75] There was nothing for it but to sit tight.

After the Christmas holidays, Frank did some shopping, an unfamiliar pastime, and selected a pair of horn-rimmed spectacles and a car radio – small luxuries that he had denied himself when he was with Liljan.[76] He played and listened to music – often a calming influence. It was a few bars into a Chopin piano concerto when Frank had an epiphany. He felt there needed to be a synthesis of the basic principles of socialist organisation with the 'one world' scientific humanism that Julian Huxley had articulated in his long essay *UNESCO: Its Purpose and Its Philosophy*.[77] What was needed, Frank wrote to Liljan, was 'a definite framework for a world distribution of basic materials for the people of the earth to be able to live now'. It wasn't a vision of ordinary people seizing the means of production but a 'small group of ... well-educated experts' that would 'be gotten together for a year to hammer out the framework'. He thought he could chair and direct a committee comprising an economist, a construction engineer, a psychoanalyst ('Furst?' he wondered), a philosopher and a 'so-called practical politician – a tough world leader type and with a grasp of law'. 'If the UNESCO appointment comes through I will let the idea mellow awhile. If it doesn't, I will get a proposal together and start pushing until I am shown that the idea is not valid.'[78]

Frank's satellite plans may not have taken off, but he was now approaching orbit in his political thinking: 'I seem to be able to withdraw from individual crises, happenings of the moment here and there in the world and sit perched above the Earth as an observer of the whole.'[79] He did wonder 'how are a few to impose a simple plan on which the world can function?', but the question of a democratic mandate was not one that troubled him. 'It will have to be simple – at first oversimplified,' he concluded.[80] Frank was at ease with exercising political leadership as if the world was an engineering laboratory.

Had the US State Department, FBI or the Central Intelligence Group (CIG, precursor to CIA) ever glimpsed Malina's proposal, it would have confirmed all their misgivings about UNESCO and its employees, concerns that now reached the desk of President Truman himself. CIG Director General Hoyt Vandenberg laid out the fact that UNESCO appointments under Huxley and Needham were being given to known communist sympathisers, which, he wrote to Truman, had '*considerable* potential for trouble'.[81] The American ambassador in Paris, Jefferson Caffery, warned that 'should Moscow succeed in gaining control or even influencing the vast machinery proposed in UNESCO, the Comintern would enjoy a perfect cover for all sorts of operations, including highly important economic and scientific espionage'.[82] But it was less Soviet personnel that worried them than home-grown communists from Britain and America – Needham in particular. The US State Department was unhappy that Julian Huxley refused to take its concerns seriously:

Huxley dismisses the matter with the observation that Needham is a 'good' Communist. Pursuant to authorisation from the UNESCO general conference Needham proposes to negotiate an agreement between UNESCO and the World Federation of Scientific Workers ... The announced plans for UNESCO, together with the recent conviction of another British scientist Dr Allen [sic] Nunn May, of giving uranium samples to the USSR, point to the grave dangers implicit should Communists occupy strategic posts in the scientific projects of the UN.[83]

It's true that the World Federation of Scientific Workers (WFSW) was a communist front, one that Needham initially hoped would work alongside UNESCO. But he was also critical of the WFSW – not that such subtleties mattered in Washington, which asked the British security services to get a grip on what they saw as the Needham problem.[84]

In London, the approach of both MI5 and the Secret Intelligence Service, MI6, was more measured. They knew that Joseph and Dorothy Needham were 'strong intellectual sympathisers', but felt that 'there is no direct proof that he and his wife are members of the Communist Party'.[85] MI6 did notice that Needham was appointing staff who 'appear to be either Communists or to have Left-wing sympathies' – Malina being the most obvious.[86] John Cimperman, the FBI's legal attaché in London, ultimately demanded that MI5 share any information that they had on the Needhams.[87] The agency complied with this request, without enthusiasm. Roger Hollis, head of MI5's C Division, despatched a particularly icy note to the FBI, asking that 'the Bureau takes no action on this information and gives it no further distribution without referring the matter to this office'.[88]

One obvious interpretation of the difference between MI5 and the FBI on the question of Needham is that it reflects the more muted sentiments about Communism in Britain compared to the United States. On the other hand, in 1987, the former Assistant Director of MI5, Peter Wright, alleged in his infamous memoir *Spycatcher* that the Soviet super-mole named ELLI – whose existence was first announced by Igor Gouzenko – was the same Roger Hollis who later became MI5's Director General.[89] The idea that the highest-ranking officer in MI5 might have been working for the USSR is so existentially scandalous for the British state that most academic historians have been reluctant to countenance the possibility. The case against Hollis was laid out by the intelligence journalist Chapman Pincher, only to be unmade by MI5's official historian, Christopher Andrew, but neither account is entirely persuasive.[90]

It's possible, though far from certain, that British intransigence allowed Needham to continue at UNESCO and ultimately helped push through Malina's appointment. Perhaps the State Department felt that intervention in the UNESCO Board process would ultimately be counterproductive. At any rate, Frank received a telegram in February saying he should report for work in Paris in April.[91] He was delighted. 'There will be periods of elation and depression in the good dialectic way,' he told Liljan, 'if one can just remember that the trick is to get a synthesis out of them.' The methods of socialist reasoning he once learned in Unit 122 had not been forgotten.

BEFORE EMBARKING ON the *Queen Elizabeth* on 9 April, Frank planned a month of activities on the East Coast, attending a UN conference and arranging appointments with scores of politicians, scientists and academics to talk about UNESCO.[92] Many of his meetings were with UN people at its temporary home at Lake Success, where Frank met up with the French geographer Jean Gottman, the Swiss anthropologist Alfred Métraux, and the doyen of anthropology, Claude Lévi-Strauss (all well disposed to UNESCO). Then he spent a stimulating day with the nuclear physicist Edward Condon, one-time colleague of J. Robert Oppenheimer at Los Alamos, who, together with his Czech wife, was now campaigning for the civilian control of nuclear technology. The same day that Frank was visiting the 'wonderful' Condons – 21 March 1947 – Truman gave a speech to Congress that changed everything.[93]

The interagency commission that Truman tasked with revising loyalty-security procedures drew up recommendations that the president now implemented via Executive Order 9835 – the most devastating anti-communist tool of its era. EO9835 mandated a name-check of all government employees and applicants against files of the FBI, HUAC and other agencies, and in the event of any 'derogatory information' turning up, the FBI could then initiate a full field investigation.[94] An employee didn't need to be disloyal to lose their job under EO9835 – any kind of 'sympathetic association' either with the Communist Party or an affiliate organisation on the attorney general's list was sufficient. And with the sources of allegations carefully protected, anyone could now look askance at a colleague's political views and have them hauled before a 'loyalty review board'. Inevitably the same criteria were set for security clearances, so in institutions like JPL it would take only the most trivial concern over Malina, Weinbaum, Summerfield or Tsien for their existing files to be dusted down and admitted as evidence.

EO9835 made its presence felt almost immediately. Two days after Frank's visit to Edward Condon, the physicist found his associations attacked under the headline 'Condon Duped Into Sponsoring Commie-Front's Outfit's Dinner'.[95] This story, carefully planted by corrupt Republican congressman and HUAC chairman J. Parnell Thomas, was designed to trigger a loyalty review board at Condon's employer, the National Bureau of Standards (he was eventually cleared).[96] Frank witnessed these events with growing apprehension, admitting to his parents that he was 'distracted' by the 'Red' hunt.[97]

Just as loyalty review boards targeted communists at home, Truman's speech to Congress that same month marked a new policy shift concerning communism abroad. Under the Truman Doctrine, the US would give aid to both Greece and Turkey 'to support free peoples who are resisting attempted subjugation' – lest they fall to Soviet influence. Unquestioning support for anti-communist governments – what would later be called 'containment' – was a step away from faith in the United Nations towards a more starkly bipolar world. It was the main object of animated conversation and flat disagreement in the next of Frank's many meetings, this time with a Democratic congressman from Texas called Lyndon B. Johnson. 'He became quite emotional & didn't talk much sense,' wrote Frank in his diary, 'said UN is still a baby etc'. Malina didn't seem surprised to find the future president 'not much acquainted with UNESCO', but acknowledged that his own handling of the meeting had been 'lousy'.

The final appointment was much more encouraging, but this was less an ordinary meeting than a kind of pilgrimage: it wasn't every day that Frank got to talk science, politics and peace with Albert Einstein. The theorist of relativity was relaxed at home in Princeton, in an 'excellent mood and laughed frequently' (as they talked he was having his head modelled by a sculptor and family friend, Gina Plunguian). But Frank noted how Einstein was 'very disturbed by present hysteria and Russophobia', and that he himself had just been visited by the FBI to ascertain the patriotism of two prospective employees.

The physicist heartily approved of UNESCO. He felt that the 'US is acting very immaturely, and does not as yet grasp [the] real issues in the world'.[98] It seems probable that they talked about a rally that Frank had attended the previous evening, organised by the Progressive Citizens of America (PCA) in opposition to both E09835 and the Truman Doctrine.[99] Both Einstein and Malina were strong supporters of the PCA leader Henry A. Wallace, Roosevelt's liberal vice president who had also served as Secretary of Commerce until being fired by Truman for being soft on communism.[100] And it was Wallace who had closed the PCA rally of 19,000 people on 31 March at Madison Square Garden with a rousing address entitled 'Back to the United Nations', a speech that could almost have been written for the particular predicament of Malina and the friends he'd now leave behind. Frank listened as Wallace told how E09835 would 'turn Americans against each other'. Who would stand condemned by this new drive for intolerance? It was 'every American who reads the wrong books; every American who thinks the wrong thoughts; every American who means liberty when he says liberty; every American who stands up for civil rights; every American who speaks out for one world'.[101]

Frank sailed for Europe the following week, apparently without realising that his own liberty depended on leaving the United States. That isn't how it appeared to most observers of his situation. For many of them it looked like a well-timed escape.

THESE TERRIBLE TIMES

WHEN SIDNEY AND LINA WEINBAUM returned to Pasadena in 1946, it did little to loosen a knot of heartache and worry that had tightened during the war years. Family life had many conventional difficulties: relations between them were sometimes strained, they fretted about their academically brilliant daughter, Selina, now 17, who Lina thought worked too hard, and they struggled to find enough money for their rent (the FBI felt that for people who lived in 'an extremely slovenly condition' Lina 'spent a disproportionate amount of her income on food and entertainment').[1] But much of what burdened their household was not at all ordinary. There was the fact that both Lina's parents and all her grandparents had been killed by Nazis in the invasion of Kiev; that she had been unable to contact her brother, Abranya, who was gravely ill with tuberculosis in a Soviet sanatorium; that Sidney's mother, Maria Weinbaum, was unwell, near destitute, and had been taken in by strangers in Kharkov; and that his sister Dounia had recently died.[2] For Lina, the killing of her family was a horror that made it impossible to reconcile her past and present lives. 'It's silly and futile,' she wrote to Frank, 'but I must tell you, I'll always have that pain in my heart that my mother had never seen Selina – (Selina is so much like her, in many ways).'[3] Their efforts to support Sidney's mother were also defeated. Not only was there no prospect of obtaining an exit visa so that she might join them in California, but they eventually received a discreet warning from relatives that if they kept wiring American dollars they risked bringing Maria Weinbaum to the attention of the Soviet authorities.[4]

In the absence of family, the Weinbaums relied on Frank and other friends to take their place. Friendship, Lina told him, is 'one of the few rare things that make this world bearable in These Terrible Times'.[5] She sent him photos of Selina because there was no one else to share them with. Lina wrote to Frank regularly, with Sidney including a short note in a rare moment, like when Frank first left to take up his UNESCO appointment: 'have a good time, keep your eyes and ears open, and do good work' (Sidney commonly used this eyes-and-ears line to define what he considered to be members' duty to the party).[6] There was no getting away from the fact that Frank's departure to Paris had been a blow to Sidney, who accepted the job at JPL before it was clear that the lab was essentially an incubator for Cold War weaponry.

Sidney did make a few new friends at the lab. He gravitated to those on the left, many of whom were optimistic about Henry Wallace's new Progressive Party, which had emerged from the Progressive Citizens of America and enjoyed communist endorsement. But by no means were all of Sidney's relations with colleagues cordial. His immediate boss, the Belgian materials scientist Pol Duwez, grew frustrated with the theoretical direction of his work. Tsien had felt sure that they'd get on well, given that Duwez was a fine cellist and Weinbaum a concert-level pianist.[7] That connection had already been made when they first met in the late 1930s; Duwez had even attended the occasional Unit 122 meeting, perhaps drawn more by the music than the politics, though he had himself fled the Nazis leaving everything behind apart from his cello.[8] These days, however, music was not enough.

Lina wrote to Frank to say that 'Sidney thinks that Duwez is mute and difficult – passive and unwilling to discern current research problems which makes the work drudgery'.[9] Duwez had his own complaints about Sidney. The challenge they both faced was to develop materials for the new Corporal missile that would be light enough to fly and still withstand the heat of propellant combustion

– a heat comparable to an industrial furnace. The nozzle was particularly vulnerable. Weinbaum's colleague Howard Seifert said that controlling the expulsion of exhaust gases through the nozzle was like 'asking a snowman to swallow a cup of hot coffee'.[10] Sidney's approach to this problem was to investigate whether porous metals could be 'sweat-cooled' by means of a liquid or gas oxidant, work he eventually published in the *Journal of Applied Physics*.[11] For reasons unknown, this didn't improve working relations between Weinbaum and Duwez. The latter even asked Martin Summerfield to 'find something else for Sidney'.[12] Summerfield ended up sympathising with Duwez's frustration and hoped that Louis Dunn might eventually fire him.[13]

All of this was just the kind of workaday irritation that can occur in any laboratory, but loyalty-security procedures after E09835 had made everyone jittery, particularly those who had ever attended a Unit 122 meeting or had been close to Frank Malina. Frank's devoted JPL secretary Dorothy Lewis wrote to him saying, 'I don't know how much longer I can stick it out here without you at the helm. There have been many changes and things haven't been going too well for me at the Laboratory since you left … I would rather not go into it here … but I would appreciate being able to discuss the matter with you.'[14] She resigned soon afterwards.

Given that the Loyalty Security Program kept past histories of activism dormant until such time as the FBI or other intelligence agencies ran a check for a clearance or a personnel security questionnaire, perhaps Weinbaum and Summerfield suspected that their names would eventually come to light. At any rate, by the summer of 1947 both of them wanted out not only of JPL, but also of the United States. Lina's letters to Frank began to sound desperate. 'Since UNESCO is understaffed – can't you find for the Weinbaums an occupation there? I am serious. All of us could probably work.'[15] They considered taking Selina out of her chemistry degree at UCLA, suggesting to Frank that she could study at the Sorbonne instead. Martin Summerfield also wrote to Frank, complaining about the climate of 'anti-Russian hysteria' and intimating that he was 'looking for some kind of professional or teaching appointment in France', ideally under the auspices of UNESCO.[16]

Why did they both want to leave now? The picture is complicated, but the likely catalyst for Weinbaum and Summerfield's parallel
glance at the exit was the experience of a mutual friend, the chemical engineer David Altman, whom Malina had personally hired, and
who had since been promoted to JPL's Chemistry Section Head. At
some point in the first half of 1947, David Altman found himself
suspended, his security clearance revoked. This was significant for
two reasons: one was that Altman had come to JPL from Berkeley's Radiation Laboratory, which had for some time been in the grip
of concern about communist infiltration (later brought to light by
HUAC and in the very public hearings of J. Robert Oppenheimer);
the second was that everyone was now worried about contagion.[17]
It was precisely because Altman was known to be close to Summerfield (his academic referee) and Weinbaum (a 'close associate') that
they, too, came under renewed scrutiny. The idea of political infec-
tion spreading from Berkeley's Rad Lab put Army Intelligence on a
state of high alert, never mind the fact that a culture of left radicalism
had been endemic to Caltech rocketry from the outset. The alarm-
ing thing for Summerfield and Weinbaum was that, unlike them,
Altman had never even been a member of the Communist Party, and
had never been much of an activist. Not that these distinctions even
mattered very much now.

When Altman travelled to Washington to appeal his clearance at
the loyalty review board, he found no serious case against him. Army
Intelligence was unwilling to reveal that their investigation of Altman's
past was founded on his friendship with two Rad Lab scientists, Martin
Kamen and Joseph Weinberg, both prime espionage suspects and both
acolytes of J. Robert Oppenheimer.[18] Kamen was left-ish but mostly
shied away from political affiliations and, like Altman, had never
been a party member. He had, however, sometimes played the viola
at wartime benefit concerts for groups like the Anti-Fascist Refugee
Committee and the National Council of American-Soviet Friendship, and at one of these he accompanied his friend, the noted violinist
Isaac Stern.[19] Kamen was an exceptional physicist who co-discovered
Carbon-14, the isotope later used in radiocarbon dating. But sheer
brilliance wasn't – unlike with 'Oppie' – enough to protect him.

In July 1944, Kamen lost his job. Stern had innocently introduced him to some Russian consulate officials in San Francisco while
under surveillance from Army Intelligence. That left him blacklisted
from professional science and unable to find work other than as an
inspector in the shipyards. Kamen couldn't even continue scientific
collaboration as a hobby, being told that he might disappear 'by accident' if connections with the Rad Lab were not severed.[20] (If that
sounds paranoid, note that some consideration is now given to the
possibility that Oppenheimer's communist lover, Jean Tatlock, was
murdered by US Army Intelligence the same year.)[21] David Altman
got mixed up in Kamen's troubles because they both attended meetings
organised by Oppie himself, to discuss support for the CIO-affiliated
union, the Federation of Architects, Engineers, Chemists and Techni-
cians (FAECT), which was then organising at the nearby Emoryville
factory of the Shell Development Company.[22] Oppie seemed not to
grasp that supporting a communist-dominated union might prove
controversial for atomic weapons scientists.

Union meetings were one thing, but the security services were
even more concerned about Altman's association with his fellow Rad
Lab physicist Joseph Weinberg and wife Muriel, both party members.
The FBI recorded Muriel as having mailed party literature to Altman
on 6 July 1943.[23] A few months earlier, Weinberg was alleged to have
spoken freely about his highly confidential research to a local party
chairman, who passed on the details to the Soviet GRU.[24] There's no
evidence that Altman had anything to do with Weinberg's loose talk,
but after E09835, any kind of contact was deemed suspicious.

For Altman, it wasn't fair, but the past intrigue at the Rad Lab
was real, and it was close at hand. Nor was he the only link between
Berkeley and Pasadena. Weinbaum's friends Eugene Brunner and
Frank Oppenheimer had also moved from Caltech to Berkeley, and
were involved in FAECT. Brunner, who was working as a hydro-
dynamicist at Shell, was never part of FAECT's Stalinist faction, unlike
his colleague George Eltenton, a party member who attempted – via
an intermediary, Haakon Chevalier – to recruit Oppie to transmit
technical information to the Soviets.[25] But these were the contrasting
realities of the Communist Party: ruthless efforts at atomic espionage

were mixed up with well-meaning support for organised labour. This wasn't coincidence, it was strategy. Eltenton and Weinberg perfected this ambiguity as a kind of camouflage, leaving Kamen, Altman and Brunner fully exposed to the systemic forces of anti-communism.[26]

David Altman's Rad Lab history was never given an airing at his loyalty review board hearing. Luckily for him, all this information was still deemed too sensitive. He won his appeal and was reinstated at JPL with lost salary.[27] But this suggests that Weinbaum and Summerfield probably didn't know much of what was going on behind the scenes. Still, they must have realised that if Altman could be suspended from JPL for his political associations, they might easily be next.

FRANK MALINA SETTLED INTO his new life in Paris, insulated from the stresses of loyalty review boards. 'The hysteria against Russia and the Communists in the US seems remote here,' he wrote to his parents.[28] He once again went to hear Henry Wallace, who was now on a European tour ('typical French crowd – full of excitement').[29] But even Frank realised that he should probably exercise some caution in what he detailed in correspondence. 'You forgot to mark your last letter "PERSONAL",' he lectured his parents in his weekly missive home, 'it doesn't matter too much since we don't send atomic secrets to each other'.[30] He did, however, write to them quite frankly about his political interests, and J. Edgar Hoover already had an informant embedded in UNESCO, keeping an eye on Malina. The FBI briefed the State Department about Frank's movements, emphasising that he was a known party member, and activating a passport 'refusal notice' on his file, to be enforced when he next applied for a renewal.[31]

At UNESCO, work soon resumed the hectic pace of JPL, with Frank often rushing down the grand corridors of UNESCO's Hotel Majestic at 2 a.m to confer with Joseph Needham. The building, just off the Champs-Élysées, had long since lost any vestige of hospitality. It had previously been occupied by the German military high command; now it was full of anti-fascists (a development that would have been more pleasing were American rocketry not full of Nazis). Joseph Needham, for instance, worked at a monumental desk that had belonged to the German admiral in charge of Axis naval operations.[32] From elsewhere in the gloomy interior, plans for the annihilation of Jews had been conceived and enacted.[33] No one in such a building could forget what was at stake in the eradication of fascism and the dividend of a peaceful settlement between East and West. 'There is much to be done,' Frank realised, 'if we are to be effective in stopping another war.'[34] (Not everyone shared his optimism: one of his Unit 122 friends, the cynical chemist Gus Albrecht, wrote to offer him a $10 bet that another war in ten years was unavoidable.)[35]

Frank's work for peace was often more satisfying than his engineering career. In the summer of 1947, he represented UNESCO at the World Youth Festival in Prague (motto: 'Youth Unite, Forward for Lasting Peace!'). He was delighted to return to home territory. 'I was really surrounded by Communists,' he told his parents. 'The US government boycotted the whole festival because the majority of the youth in the World Federation for Democratic Youth [is] so Left.'[36] That was one reason, certainly, along with the fact that the WFDY was an instrument of Stalinist foreign policy.[37] In any case, the US still had a clandestine presence at the festival – their informants were here too, sending updates on Frank's activities.

It was while he was in Prague that someone congratulated him on getting his medal. What medal? He thought he was having his leg pulled until it was explained to him that he had been awarded the French REP-Hirsch *Prix d'Astronautique*, a prize named after astronautical pioneers Robert Esnault-Pelterie and André-Louis Hirsch, in recognition of Malina's early theoretical work.[38] The award had actually been made in 1939, but notification had been delayed by the war. Frank recalled submitting a paper for their consideration – diligently translated and typed by Liljan even before they were engaged. Now their divorce had just come through.[39]

That wasn't the only instance of rocketry making a comeback into Frank's professional life. The nature of work at UNESCO was that it consisted of interminable meetings on scientific diplomacy and cooperation, meetings that would strain the attention of even the most devoted peacemaker. Pencil sketches of rocket motors would sometimes appear in the margins of his notebook. Other UNESCO scientists had similar compulsions. When Julian Huxley's secretary went to transcribe his notes from one meeting, she discovered that the zoologist had been furtively tabulating the sex ratio of ducks sitting on the lake outside.[40]

One highlight of UNESCO was being introduced to a young English woman from the personnel department. Marjorie Duckworth was a sociology graduate and a former captain in the British Women's Auxiliary Corps who had been inspired to join UNESCO after hearing a radio broadcast by Julian Huxley. All of this suggested someone with her own ideas and direction, an influence on Frank that was perhaps first manifest when he cautiously started to digest the work of Max Weber.[41] It wasn't to his liking, even in short spurts: 'I am studying sociology,' he told his parents, 'usually sitting on the toilet. That's where much of so-called social science deserves to be studied today'.[42] Frank wasn't going to let go of his positivist empiricism for anyone, not even Marjorie. He described her as 'English, very fine and full of common sense', noting that 'we see eye to eye on most things in this world'.[43]

Frank was in the midst of 'getting acquainted' with Marjorie when he decided to grow a beard. 'I do not know if I look like an Italian revolutionary or an Indian mystic,' he wrote home, 'I guess I am something in between.'[44] But the beard turned out to be quite useful. 'It makes me look older and in this kind of racket it has certain psychological advantages.'[45] The beard had its debut at a meeting of the British Association for the Advancement of Science, followed, more controversially, by the General Assembly of the World Federation of Scientific Workers in Prague – the front organisation that so upset the State Department (returning to Czechoslovakia, Malina was generally positive about the Stalinisation of his homeland since the communist takeover in February 1948).[46]

Word of the new look spread quickly. 'There is a rumour that you appeared in London in a Barbarian's beard,' Lina Weinbaum wrote to him. Lina was invariably direct with Frank, and that could be tricky, especially when she dealt with more pressing matters than facial hair. She told him in detail of the worsening situation at JPL, in which Sidney was 'not with Pol [Duwez] and not with anyone else

– just at his desk. He is going to pieces.' 'I hope he will be able to get some sort of fellowship to work at research in the Sorbonne or what have you.' That was a little pointed. Then she complained about all of their old comrades. Tsien was 'rather peevish ... inclined to intellectual snobbishness'; 'associations with Martin are nil'; while Pol [Duwez] 'really deserves spanking, only there is no one to administer it'. Lina resented the fact that Frank was slow to respond to their current problems, particularly their need for an escape route. All of this spilled out in a postscript to what was an already desperate letter. 'Frank, why are you so noncommittal, it is not like you ... do write, won't you? I don't give a damn if I am not diplomatic with you, life's too short, it is just too bad if they're not, at least, very few people with whom one does not have to be diplomatic.'[47] Then Selina Weinbaum followed up to ask if he knew 'of any way, through UNESCO perhaps, that I could get a scholarship and fellowship of some sort?'[48] It was less direct than her mother's letter, but it had the same anxious tone: 'please Frank, if you can think of anything, it will make a dream come true, write soon.'

Throughout 1948, the Weinbaums saw darkening skies at every turn. On 29 June the leaders of the CPUSA were handed indictments for violating the Smith Act, which aimed to penalise anyone advocating the violent overthrow of the government. The act was of course founded on the assumption that American liberal democracy had reached its endgame, and that any communist who believed in revolution must intend revolutionary violence. The trials that followed, known as the Foley Square trails, would be among the longest in American history. In California, a systematic blacklist was instituted after ten Hollywood writers and directors were cited for contempt of Congress after refusing to cooperate with HUAC. The Hollywood Ten were eventually jailed. Then in August, a former party member, Whitaker Chambers, appeared before HUAC to accuse a high-ranking State Department official, Alger Hiss, not just of party membership but also of being involved in espionage. The allegation turned out to be true. Chambers also told the FBI about Mae Churchill's former husband Lester Huettig being a spy. The machinery of investigation and prosecution was duly cranked up. All of this cast a shadow over the Weinbaums, but the first real sign of the storm gathering over-head appeared in the guise of someone else's misfortune.

There had been an afternoon back in May when the Weinbaums had had a visit from their friend Pauline Gibling Schindler, a noted musicologist and party member. (An important figure in the arts world, she had been married to architect Rudolph Schindler, and was subsequently the lover of John Cage.)[49] She came with another friend, Jenny Marling, and her two young daughters, who were visiting the US from the Soviet Union. Marling, an American ballerina and the widow of the Russian playwright Alexander Afinogenov, was being followed everywhere by the FBI, who didn't believe her purported mission of 'cultural exchange' from the USSR. Marling was both a true believer in the communist ideal and smart enough to understand that in Moscow anything less could be dangerous. Her closest friend, Lina Prokofiev, wife of Russian composer Sergei Prokofiev, claimed that Jenny's 'life began after she read Lenin's speeches'.[50] There were rumours in Russia that she worked as a low-level spy for the People's Commissariat of Internal Affairs (NKVD). The FBI certainly thought that she was 'probably engaged in espionage activity' in the US, which is why their agents were now sitting in a car outside Sidney and Lina's house.

Perhaps Marling knew she was being followed. At any rate, she no longer felt comfortable in the United States, and had told her daughters that they were moving back to where their father was buried.[51] The Weinbaums just loved her. She talked to them of contemporary art and writing in the Soviet Union, and left behind some translations of Shakespeare and the poetry of Boris Pasternak.[52] Lina in particular had wanted to see Marling in the hope that she could find out anything about Lina's ailing brother, Abranya, her last surviving family member. Neither Sidney nor Lina had any idea of Marling's alleged ties to the NKVD. She stayed until 9.20 p.m. and then went home – followed, of course, by the FBI.[53]

No more was heard of Jenny until the Weinbaums received notice that in September she had died onboard the ship back to Russia. As the *Pobeda* (*Victory*) was within sight of Odessa, the cinema projectionist was winding a film at speed when the friction ignited the celluloid. The flames spread first to his own clothing then to other film canisters and phonographs, down to the carpeted hallway and up to the plywood bulkheads, tearing into the living quarters, and then into the steering and radio rooms.[54] Jenny's daughter Alexandra, who was playing in a corridor, last saw her mother behind a wall of flames. Over forty people died. There are suggestions that Jenny Marling's name had already been added to a Union of Soviet Writers list of people to be arrested – a fate for which the odds of survival were lower than a shipboard inferno.[55] Lina was distressed to hear the news – 'another senseless tragedy', as she put it to Frank.[56]

Marling's visit, however, had left the Weinbaums with more than the poetry of Boris Pasternak. Why, the FBI wondered, had this suspected spy come to visit *them*? Even now, there was too little known about Sidney to attribute any particular significance to the visit – the information just sat on file – until the third week of January 1949. That's when the investigation took off. An informant walked into the Bureau's Los Angeles office and closed the circuit, igniting the entire security and judicial apparatus against the engineers who had made American rocketry possible.

1. Theodore von Kármán at Caltech, around 1938. The first director of the
Guggenheim Aeronautical Laboratory, he ranked himself third, behind Isaac
Newton and Albert Einstein, in the pantheon of scientific greats.

2. The nativity scene of American rocketry: methyl alcohol and gaseous oxygen test, Arroyo Seco, Halloween, 1936. From left to right: Rudolph Schott, a pith-helmeted Apollo M.O. Smith, Frank J. Malina, Edward S. Forman, Jack Parsons.

3. Frank J. Malina and Jack Parsons setting up the rocket motor, Arroyo Seco, Halloween, 1936. 'Very many things happened that will teach us what to do next time.'

4. The happy couple: Frank and Liljan Malina on their wedding day, 24 June 1939.

5. From right to left, Helen Parsons, Jack Parsons, Liljan Malina and unknown friends, *c.* 1939–40.

6. Cartoon of Theodore von Kármán and his Caltech circle as drawn by Liljan and Frank Malina for von Kármán's birthday, 1940. From left to right: Homer J. Stewart, Duncan Rannie, H. S. Tsien, M. A. Biot, Theodore von Kármán, Frank Malina, Pipö von Kármán, William R. Sears, William Bollay and Albert Lombard. On the walls are protraits of Tera Biot, Margaret Rannie, Liljan Malina and Mabel Sears ('faculty wives I didn't like', said Liljan).

7. JATO discussions around the wing of the Ercoupe, March Field, August 1941. From left to right: Clark B. Millikan, Martin Summerfield, Theodore von Kármán, Frank J. Malina, and pilot Capt. Homer A. Boushey.

8. Martin Summerfield catches some sun (left), along with Frank J. Malina, Walter B. Powell, Paul H. Dane and Theodore von Kármán, after testing liquid JATOs on the Douglas A-20A at Muroc Field, Mojave Desert, April 1942.

9. Aleister Crowley making the sign of Pan, 1910. The title of the book, *Perdurabo Magister*, reflects a name he took on entry to the Order of the Golden Dawn. He used over 100 pseudonyms including 'BAPHOMET', '666', 'To Mega Therion', and famously, 'The Beast'.

10. The Gnostic Mass at Agape Lodge No 1 of the Ordo Templi Orientis, undated but around 1939. The high altar in this Hollywood attic occupies the wall nearest Boleskine House – 5,000 miles away at Loch Ness, Scotland. From left to right: high priest, Wilfred T. Smith, high priestess, Regina Kahl, and deacon, Luther Carroll.

11. Liljan Malina finishes a poster in support of Russian War Relief, 1942. 'I break my neck getting them out, big ones in colour too, and then do you think they would even call up to thank me? Not on your life!'

12. Frank and Liljan Malina, undated, but likely to be early 1945, around the time Liljan bought a gun.

13. Frank J. Malina and his 'little lady' WAC Corporal sounding rocket at White Sands Proving Ground, 11 October 1945.

14. Jack Parsons and Marjorie 'Candy' Cameron, 1946. 'I have my elemental!' Jack wrote to Aleister Crowley.

15. L. Ron Hubbard and Betty Northrup aboard the Allied Enterprises schooner *Blue Water II*, in Miami, Florida, June 1946. 'Suspect Ron playing confidence trick. Jack evidently weak fool,' claimed Aleister Crowley.

16. Frank Malina tries out a new look with beard and glasses, 1948. 'I do not know if I look like an Italian revolutionary or an Indian mystic.'

MY LOVE
1948

17. Marjorie Malina portrait, captioned by her husband, 1948.

18. Frank Oppenheimer, his lawyer Clifford Durr, and Jackie Oppenheimer outside the HUAC hearing, 14 June 1949.

19. The US Medal of Merit, in recognition of JPL's war effort, being awarded to Louis G. Dunn, Clark B. Millikan, and Robert A. Millikan, 21 March 1949.

20. The launch of the Bumper WAC Corporal, round 8, 24 July 1950, the first such launch from Cape Canaveral. On 24 February 1949, an earlier Bumper 5 reached the then record of 244 miles.

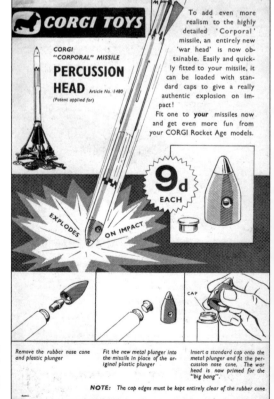

CORGI TOYS

CORGI "CORPORAL" MISSILE

PERCUSSION HEAD Article No. 1480
(Patent applied for)

To add even more realism to the highly detailed 'Corporal' missile, an entirely new 'war head' is now obtainable. Easily and quickly fitted to your missile, it can be loaded with standard caps to give a really authentic explosion on impact!

Fit one to **your** missiles now and get even more fun from your CORGI Rocket Age models.

9d EACH

EXPLODES ON IMPACT

| Remove the rubber nose cone and plastic plunger | Fit the new metal plunger into the missile in place of the original plastic plunger | Insert a standard cap onto the metal plunger and fit the percussion nose cone. The war head is now primed for the "big bang". |

NOTE: The cap edges must be kept entirely clear of the rubber cone

21. Fun for the nuclear family, 1959: an advert for the British Corgi Toys' percussion war head – 'easily and quickly fitted to your missile'.

22. Jailed for four years: Sidney, Selina and Lina Weinbaum outside court, Los Angeles, 12 September 1950. 'I'm sure my father never did anything to be ashamed of and I'm proud of him,' Selina told reporters.

23. William Perl arrested by FBI agents, 14 March 1951.

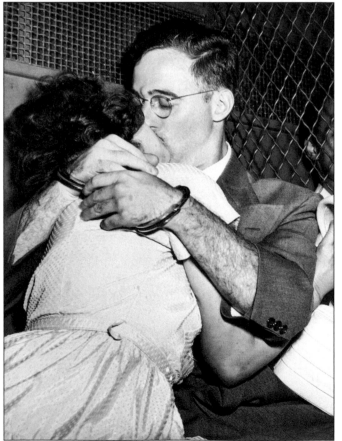

24. Ethel and Julius Rosenberg embrace in a prison van outside Federal Court after arraignment on atomic spy charges, 23 August, 1950. They were executed by the US Federal Government at sundown on 19 June 1953.

25. Hsue-Shen Tsien (Qian Xuesen) pictured in 1955, before his return to China as part of a diplomatic exchange of prisoners. 'I do not plan to come back,' Tsien told reporters. By the following year, he was leading China's ballistic missile programme.

26. Frank Malina takes a break from working in his studio, Boulogne-Billancourt, 1956.

27. Frank J. Malina (l) being shown photographs of early JPL by the then current director, William H. Pickering (r), 31 October 1968. Louis G. Dunn glares down from the display, top right.

28. Frank Malina at home working on *Leonardo*, Boulogne-Billancourt, Paris, 1970s. One of his Lumidyne art works is visible in the background.

DEFINITE LEFTISTS

THE DETAILS OF EXACTLY what happened next are sparse, but a number of different FBI reports all refer to the same climactic moment. The snow had just melted from a freak cold snap that had blanketed Los Angeles when Louis Dunn – JPL's new director, a racist white South African – made an appointment with senior FBI agents and declared that his lab had been infiltrated by communists. Dunn hadn't accidentally stumbled into a secret lair, catching the suspects red-handed. He just had a hunch. He explained that JPL had 'the largest library of secret military information on the West Coast'. Until the previous month security regulations were 'appallingly lax', and he had been concerned for the safety of the secret material on file.

Library access was one concern, but what really bothered Dunn was the behaviour of 'certain individuals ... whom he termed "definite leftists"', whose Army clearance may have been obtained before the tighter post-E09835 regulations came into force. Dunn felt he was now in possession of information, admittedly of 'a somewhat non-specific nature', that made him doubt the *loyalty* – not just the politics – of four Jewish scientists (Sidney Weinbaum, Martin Summerfield, Paul Chambré and Mitchell Gilbert), two Chinese scientists (H. S. Tsien and W. Z. 'Jimmy' Chien) and a secretary called Patricia ('Patty') Line, a close friend of Malina's.[1]

There's no evidence that any of these people had current ties to CPUSA, though Weinbaum, Summerfield and Tsien had been members a decade before. So what was it now that had so raised Dunn's suspicions? Apparently, some of those in the 'leftist group' had been 'very active in the support of Henry Wallace during the past election'. If Dunn considered supporting Wallace's presidential bid to be scandalous, it was quite a widespread scandal given that he polled over 100,000 votes in Los Angeles and 1.2 million nationwide. (Admittedly, that still left him on the fringe of American politics.) The Wallace manifesto promised a less adversarial approach to US–Soviet relations, the end to all racial segregation and a system of universal national healthcare – demands that were generally dismissed as extreme and unworkable. There was a good deal of public anger at the audacity of even asking for such things, an anger stirred by the fact that Wallace got electoral support from the Communist Party.

Time magazine described how Wallace 'ostentatiously rode through cities and towns with his Negro secretary in the seat beside him' … 'more like an agitator than a Presidential candidate'. It wasn't just that he 'chose the homes of Negro supporters for meals and overnight stops', but that he also refused to speak in any segregated setting.[2] Politely petitioning for the removal of segregation was one thing, but Wallace, and his running mate Glen Taylor, refused to observe its restrictions. For Louis Dunn, this kind of politics was beyond the pale. His own racial views were well known in the lab. Asked by a journalist about his transition from South Africa to Caltech, Dunn replied that 'the thing I missed the most … was servants … they were always as thick as fleas on the family ranch'.[3] That was the kind of comment he was happy to make on the record.

Dunn was to find worse goings-on at JPL than the mere presence of Wallace voters, however. He had overheard Mitchell Gilbert and Patricia Line expressing 'a very critical attitude on the House Un-American Activities Committee hearings', particularly the recent HUAC treatment of Dr Edward Condon. (Even President Truman was upset when criminal fraudster and HUAC chair J. Parnell Thomas traduced Condon as 'one of the weakest links in our atomic security'.) Dunn thought Chambré was 'secretive' and, though he had since left JPL, there had been rumours of clearance difficulties. Dunn had absolutely nothing on Weinbaum, Summerfield, Tsien or Chien – just 'non-specific observations' and 'unrecalled remarks'. That didn't matter. Under E09835, a search of the security indices soon turned up 'considerable information'. Weinbaum's name immediately raised the link to the investigation of Jenny Marling, a suspected Soviet spy. That's the moment when the FBI felt they might have a case.

Richard B. Hood, the LA Bureau's Special Agent-in-Charge who had earned his stripes trailing writers Bertolt Brecht and Thomas Mann, titled his memo to Hoover: 'COMMUNIST INFILTRATION OF JET PROPULSION LABORATORY, CALTECH, ESPIONAGE – R' where 'R' naturally stood for 'Russia'.[4] For Hood, it was a tantalising prospect. Since E09835 had come into force, two and a half million federal employees had been screened, with 8,000 field investigations, producing only 200 suspects – and yet no evidence of espionage.[5] This time, however, the lead felt promising.

Events now moved rapidly. The FBI heard from the Army Counter Intelligence Corps (CIC) that they had, once again, suspended David Altman's security clearance. Dunn fired him. Then the FBI broke into his apartment, or, in the idiom of the bureau, 'source T-16 made available ALTMAN's personal effects'.[6] Agents rummaged through his scientific books and papers, finding 'in a secreted place ... several books of Communist reading material'. Then they turned to his correspondence with friends ('the *tone* of the letters indicated that they had ... been ... Communists').[7] They found a few scientific reports, including 'Temperature Distribution for Various Heated Fluids Flowing Turbulently in a Heated Tube', which had been checked out of the JPL library. This they deemed suspicious. It wasn't classified, but the agents figured it could have been. Weinbaum was listed in Altman's address book. They also took this to be suspicious.

On 23 February, the FBI broke into the Weinbaums' house on South Arroyo Boulevard. Sidney and Lina weren't at home, and the agents took care that their presence would not be noticed on their return. One invasion of privacy begat another: a wiretap on their phone and a covert listening device – a bug, at the time, was cutting-edge technology – extended the FBI's reach into Sidney and Lina's every conversation. This wasn't cheap to install, and the evidence was probably inadmissible in court, but they hoped it would reveal 'traffic' to determine 'whether he [Weinbaum] is engaged in Soviet espionage'.[8]

Despite all this activity, Louis Dunn still worried that his concerns might seem 'rather nebulous and non-specific'. So in a subsequent interview he upped the ante by insisting that when 'all available information' was pieced together, 'he felt there was a "ring" of individuals who might present a serious hazard'.[9] Sidney Weinbaum, of course, was at the centre – at least within JPL. In Dunn's mind there was something wrong about Sidney's reputation for being 'a very heavy reader'; it 'was hard to reconcile with the fact that he was generally considered as lazy and not an outstanding research man'. So Dunn took it upon himself to snoop on Weinbaum's JPL library borrowing. The director could find no explanation for why Sidney was reading widely 'outside of his current and past research field', from work on jet-propelled underwater missiles and guidance systems to oceanographic survey. 'This type of reading,' he told the agents, 'was generally confined to men who were young, ambitious and seeking to broaden their overall knowledge.'[10]

Perhaps Dunn had forgotten Weinbaum's reputation for scientific curiosity, or the recent fallout in the Materials Section that had left him underemployed. Or perhaps he hadn't. Just as Henry Wallace had anticipated, Sidney now found himself among those reading 'the wrong books'. Agents were alarmed to find a paper in Sidney's house called 'The Solution of Secular Equations' by Jack Sherman, which, they felt, 'may contain matters involving Atomic energy'.[11] Jack Sherman had, like Sidney, been a PhD student of Linus Pauling, and shared his interest in mathematical approaches to the nature of the chemical bond. The only connection to atomic energy was the recurrence of the word 'atom'.[12] But that's not how it seemed to the agents, who asked Louis Dunn to widen his enquiry into Sidney's library record to determine what reading counted as legitimate. Dunn duly came in to the bureau office on 15 March, answered a few questions, then dropped his next bombshell.

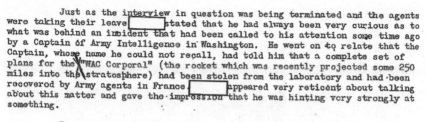

Just as the <u>interview</u> in question was being terminated and the agents were taking their leave ▮▮▮ stated that he had always been very curious as to what was behind an incident that had been called to his attention some time ago by a Captain of Army Intelligence in Washington. He went on to relate that the Captain, whose name he could not recall, had told him that a complete set of plans for the "WAC Corporal" (the rocket which was recently projected some 250 miles into the stratosphere) had been stolen from the laboratory and had been recovered by Army agents in France. ▮▮▮ appeared very reticent about talking about this matter and gave the impression that he was hinting very strongly at something.

By reason of the timing, past discussion, etc., the agents formed the distinct impression that ▮▮▮ was hinting very strongly that FRANK MALINA had knowledge of this incident. Particularly when ▮▮▮ was being questioned as to what bulk such a document would entail, etc., he very casually called to the attention of the agents the fact that this had happened prior to the time he was ▮▮▮ The distinct

APART FROM BEING the nucleus of a rapidly expanding international spy investigation, everything about Frank's life was going well. Frank and Marjorie got engaged in November and married on 12 March 1949, the happy news trickling back to the FBI via an intercepted cablegram from the Weinbaums: 'How perfectly nice. We are happy for you and are sending our best wishes for your very good life together.' As it turned out, their marriage coincided with another – the German V-2 to JPL's WAC Corporal. In the absence of the engineers who had inspired it, the BUMPER programme, which had ticked over since late 1946, continued to investigate the feasibility of separation in a multi-stage rocket. Would separation even work at higher velocities and altitudes? Could the materials of the missile skin survive the temperatures of such speeds? Flight testing in the spring and summer of 1948 got off to a good start with BUMPER 1, which, though only partially charged with propellant, successfully demonstrated separation. But BUMPER 2 malfunctioned, BUMPER 3 exploded and BUMPER 4 veered out of control and crashed in the desert. With this kind of record, it's a bit surprising that BUMPER 5 – with a fully tanked WAC Corporal as second stage – got as far as the launch pad.

On 24 February 1949, BUMPER 5 sat in its gantry in White Sands Proving Ground just a few hundred metres from where Malina had tested the WAC Corporal in October 1945. The BUMPER looked like a very different beast. As the crew awaited flight authorisation, the haze from the V-2's liquid oxygen wreathed the base, giving the scene an infernal atmosphere. The modified WAC second stage appeared as little more than an artist's pencil balanced on top.

At 3.14 p.m., lift-off.

The first stage went smoothly, and when the V-2 shut down after 64.5 seconds and 18.3 miles, the combined rocket was travelling at 3,600 miles an hour – already faster than Frank's WAC Corporal flight.[13] The second stage WAC separated cleanly and surged into the mesosphere. The sheer power of the V-2's 'bump' gave the WAC an unprecedented momentum as it pushed beyond the extreme cold and into the thermosphere, where oxygen and nitrogen molecules become excited by solar radiation. Two minutes from take-off, and the WAC was reaching 5,150 miles an hour, the first human vehicle to reach Mach 5: hypersonic flight.[14] The rocket soared on, passing the point at which an earlier White Sands V-2 had, on 24 October 1946, taken the first photograph of Earth from space. It showed a grainy corner of the planet, wrapped in cloud against the jet black void – not an iconic image, though perhaps it should be, given its radically unfamiliar perspective. It would take time for our species to adjust to this view. But the WAC Corporal from BUMPER 5 was going much, much higher.

Telemetry data – itself untested at this altitude – indicated that the WAC Corporal topped out at 244 miles, by some margin the furthest reach of any human object beyond Earth. It's an odd, liminal vantage point on the brink of blue and black, a place that's not quite earthly yet still subject to its gravitational pull.

This first journey away from *Terra Mater* was a sublime escape, one freighted with the ambivalent hope and fear that comes with leaving home for the first time. Engineers are perhaps not prone to expressing such separation anxiety, but there's a yearning, an anticipation that future journeys will go higher, further and faster. It's a profound point of departure. It can't be undone.

In this early race to get into space, there was no exact finish line, no clear threshold over which a rocket could definitively pass to universal acclaim. It wasn't quite like terrestrial exploration, where the Northwest Passage could be navigated or not; or even like the subsequent TV success of the Apollo moon landings. The BUMPER

WAC Corporal might seem like an abstract triumph, an exploration of 'extraterrestrial space' even as the journey itself expanded our conception of the Earth and its limits.

Back then, some commentators cautiously acknowledged it as the first time that humans had reached into space (by modern definitions, the BUMPER WAC was only the first time that the United States had managed that feat).[15] The altitude of the rocket may now look quite modest – it's still within Lower Earth Orbit, at roughly the height occupied by the International Space Station, but it was more than double that reached by any V-2. The BUMPER WAC never entered into orbit, but it got comfortably within the range of Sputnik's orbit eight years later, which is when the Space Age is usually thought to begin. Of course, this wasn't just Frank Malina's success; in practical terms, he wasn't even involved in BUMPER. But his development of the WAC Corporal and 'The Problem of Escape' paper should have amply qualified him as a lead author in its achievement, not least because the flight successfully demonstrated staging – the solution to the problem of escape. The FBI seemed to recognise this even if no one else did. BUMPER 5 gave their investigation a renewed urgency.

LOUIS DUNN HAD NOW spelled out to the FBI that he 'had definite reservations about the loyalty of Malina and feels that in Malina's position with UNESCO, he is in an extremely good position to engage in espionage activity'.[16] Dunn 'very strongly hint[ed]' that Malina 'either himself took the set of plans or at least made it possible for someone else to do so'.[17] This was a remarkable accusation – without evidence – from someone who Frank regarded as a friend. Just before he had left for Paris in April 1947, Frank had met up socially with Louis Dunn and David Altman in New York. Political differences had never prevented the relationship from being cordial. Frank had even recommended his deputy as his replacement. Dunn showed apparent appreciation of this support, and expressed his hope 'some day we can get together in some remote corner and really raise <u>hell</u>'.[18]

One week after fingering Frank as a spy, Louis Dunn, along with Clark B. Millikan and Robert A. Millikan, was awarded the Medal of Merit, the highest civilian decoration for 'exceptionally meritorius conduct' in recognition of JPL's work during the war. It's perhaps understandable that the photograph of the occasion shows the new director itching for a smoke, looking ill at ease accepting an honour for what were the achievements of his predecessor.

The FBI now had two priorities. They had to work out whether espionage was really happening, and they had to acquire evidence of Communist Party membership in order to pursue prosecutions or to leverage potential witnesses. The first task was awaiting intelligence from the wiretaps and microphone surveillance on the Weinbaums, but neither delivered very much. On the first day, the bug captured Sidney telling Lina about some 'heart pain', but he reassured her that it was 'nothing serious' … just 'nervousness'.[19] The second task was also difficult. Despite the thick file on Weinbaum, 'close scrutiny' revealed that any evidence of his party membership came from only two sources. There were the party records for 1938/9 leaked by William Ward Kimple to his LAPD Red Squad boss William Hynes, but these were a decade out of date. And there was the testimony of a single confidential informant, whom the FBI knew to be 'a character' – which, all things considered, was a generous assessment of Jack Parsons. Other informants stated that 'information received from him should be treated with extreme reserve'; the fact that Parsons would not come across as a credible witness in court was definitely a problem.[20] The FBI kept him under general observation, but at the moment his reputation was an obstacle.

Agents now prepared to move on the more marginal players in Unit 122, hoping that the key suspects – Weinbaum, Summerfield and Tsien – would not be tipped off. The FBI brought in Louis and Shirley Berkus, who were never at Caltech, nor even in science, but had somehow ended up in Unit 122 and now ran a furniture store in Alhambra. The agents found them 'decidedly uncooperative'; Louis Berkus 'would not even admit his own membership in the Communist Party'.[21] But they fared better with the folk-dancing physicist Sylvan Rubin, now working on smog control at Stanford. At first he denied his involvement, but he soon folded under pressure. When asked who would 'be the type even to this day to place first allegiance to the USSR rather than the US ... hesitated and said "I think – SIDNEY WEINBAUM"".[22] Before they could get much further, however, the FBI got wind that Army Intelligence was planning to downgrade Weinbaum's security clearance, making his employment at JPL untenable. Given that this move would likely alert Sidney to his being under investigation, the FBI persuaded Army Intelligence to hold back while the bureau, in collaboration with Dunn and JPL management, plotted their next move.

After months of surveillance of the Weinbaums, the FBI had uncovered no evidence of espionage. To blacklist Sidney would be easy enough, but an espionage prosecution might be too difficult. In that event, there was always perjury, the strategy then being used against State Department employee Alger Hiss. The FBI were more than willing to undertake the meticulous legal groundwork to cultivate, then successfully prosecute, the non-disclosure of party membership. But with Sidney there was a problem: his personnel security questionnaire (PSQ) at JPL, which was issued when he joined in 1946, was now over three years old – beyond the reach of the statute of limitations on perjury.[23] That's when JPL's personnel director, Walter Murphy, stepped in to help his friends in the FBI. He asked Sidney to fill out a new PSQ, making the same request of his office mates just to make sure that it all seemed routine.[24]

Of course it was a trap. A new PSQ was not needed for any purpose other than to get Sidney to *not* list his party membership, as to do otherwise wasn't much of an option. To admit membership would mean losing his job, obviously, and there was a good chance he'd never work in science again. But as the Foley Square trials made clear, there was little to stop a prosecution under the Smith Act. Might they try to reverse his citizenship? To admit party membership was effectively to waive Fifth Amendment constitutional rights that guarded against self-incrimination. Yes, one could confess joining the party as a youthful political indiscretion, but the price of atonement was to name names or face contempt of court. Cooperating with the FBI effectively meant heaping ruin on one's friends and their families. Sidney the chess champion must have seen that his opponents had him squarely in check. At any rate, he filled out the form on 8 April, specifying only the American Physical Society and Sigma Xi under his 'organizational memberships'. That gave the FBI something to hold over him in interview.

Throughout the spring of 1949, the FBI pursued Dunn's claim of communist infiltration on all fronts. Technical surveillance on Martin Summerfield began on 7 April but didn't yield anything useful.[25] Jack Parsons was being watched in his job at North American Aviation, where he had struggled with his own clearance.[26] H. S. Tsien was in the midst of moving from MIT back to Caltech, where he was to take up a new appointment – the 'Robert Goddard Chair of Jet Propulsion' – and then he, too, joined the surveillance circuit, after falling under official investigation in May 1949.[27] Agents reasoned that if they could just prove Tsien's party membership, it could be the 'basis for a possible deportation proceeding against him' by the Immigration and Naturalization Service (INS).[28]

With news of Sidney's clearance denial about to break, and his likely dismissal from JPL imminent, the FBI finally decided to bring him in for questioning. Louis Dunn called him at the lab to say that agents wanted to interview him the following day.[29] Another JPL informant described how Sidney immediately phoned Lina in 'a very distressed state'. He also sought legal advice from a friend, Spencer Austrian, a chess player and party member who owned a factory making children's clothes.

At 9 a.m. on 27 April, Sidney arrived at the FBI's LA field office to meet with Special Agents Arthur C. Wittenberg and H. Rawlins Overton. They started by soliciting Sidney's life story, from his upbringing in Russia, the murder of his father by Bolsheviks, his flight to Poland, to America, to Caltech, to his work at Bendix, Curtiss Wright and JPL. The conversation eventually turned to his work on heat diffusion in metals.

'Do you have any pet ideas that you are exploring and any particular field that you are interested in?'

'No. Right now I am very much interested in the field of heat diffusion.'

'That takes all of your time?'

'Yes.'

'You don't have – shall we say …'

'I don't have an inventive mind.'

'… you don't have anything that is disconnected with the field that you are now on?'

'No.'

The interview moved on to different topics – his listed membership of scholastic societies, his previous interest in working in Russia, the status of his mother – each one charged in ways that everyone knew and no one acknowledged. And 'have you talked to anybody since the war who knew anything about conditions over there?' Sidney paused, perhaps wondering what the FBI had on Jenny Marling.

'No, not that I can think of.'

The agents noted that 'soon after that line of questioning the subject made a rather impassioned statement to the effect that he had always regarded himself as being a loyal citizen of the United States'. Then the questioning resumed.

'What did you think of the HITLER-STALIN Pact?'

'I did not think the Russians were ever going to help the Germans. Probably they were stalling for time. I felt badly about it at the time.'

'As near as you can recall during the thirties ... you belonged to no organisations on the campus in Pasadena or elsewhere with the exception of the Chess Club?'

'In those days I was not even active in the Chess Club.'

'Do you vote regularly? And I am not interested in whom you vote for.'

'Yes. I vote regularly.'

'Do you take any interest in politics?'

'No.'

'Did you ever go to any meeting that was advertised as "White" and turned out to be "Black"?'

'No. I think possibly I did go to one in Pasadena but I don't remember anything about it.'

And so it went on, hour after hour. What about when Einstein was at Caltech, all these movements that arose after his speeches? 'You didn't interest yourself in any of those little movements that sprang up either on or off the campus?'

'No. I have had no contact of that sort.' Then the agents pulled out Sidney's recently signed PSQ and pushed it across the table. 'Did you think of anything that was left off the form that would have a bearing on the matter just discussed?'

'No.'

'Are you sure of that doctor? ... Do you recognise the name of SYDNEY EMPSON?'

'No.'

'Are you sure you don't recognise it?'

'I don't. My first name is SIDNEY'.

'You did not ever use that name in connection with any group you belonged to?'

'No.'

'As a matter of fact we know that you did, doctor. You know what we're driving at, don't you?'[30]

At this point Sidney decided to say no more without legal counsel. Special Agents Wittenberg and Overton reminded him that he was there 'voluntarily', but whenever Sidney got up to leave, the atmosphere grew menacing: 'you're putting your head in a noose,' warned Wittenberg.[31] The agents chose not to transcribe those portions of the interview where they leant on him to name names, writing only that 'it was ... pointed out to the subject that ... he would be in a position to give information of value as to the membership'. They even promised to help find him another job if he cooperated.[32]

Before Sidney had left the field office the previous afternoon, the agents impressed on him that he should return the next day, after he 'had time to think over the advisability of telling the complete, full story'.[33] So on the morning of 28 April, Sidney called in sick at JPL, then phoned the FBI to arrange another meeting at 2 p.m. Now that he'd had time to think, Sidney told them that he felt insulted that they did not believe him and that he wanted no further discussion with them. He was out of the building within twenty minutes. Later that day, he sat down to write to Frank Malina and J. Robert Oppenheimer, the two people Sidney suspected might be the FBI's ultimate targets.[34] After some abrupt pleasantries, Sidney told Frank that he was

> recovering from the oddest (and shocking!) experience of being grilled for two days in succession and by the FBI. I expect I will be leaving the Lab very shortly but now I am mostly feeling incensed by the fantastic accusations they have hurled at me. I am not good at writing long letters but I felt a letter from me was long overdue even if it is a short one. Best wishes to you and your wife, Sidney.[35]

Sidney must have figured that even if the FBI read the letter, it was still a kindness to forewarn his friend of potential trouble ahead – especially given that Frank and Marjorie were planning to travel to the United States. It's not clear, however, that Frank received the letter in time to act on its message.[36] What Sidney wrote to Oppie had a similar intent, though with the additional hope that his old friend might find him another job.[37] That wasn't likely – Oppie wanted to put as much distance as possible between them – but he also needed to make the situation safe. If Sidney cooperated with the FBI by detailing Oppie's attendance at a party meeting in 1937, or his involvement with FAECT, for example, it could have life-changing consequences.

That weekend, Sidney heard the bell ring. Half expecting that it might be FBI agents, he cautiously opened the door to find the tall, lean figure of J. Robert Oppenheimer standing on the porch. It was his first visit since the 1930s, when he used to drop by while walking around campus.[38] This time, however, Oppie came with a friend, Ruth Tolman, a clinical psychologist and the widow of the Caltech physicist Richard C. Tolman, with whom he maintained an odd, intense not-quite-affair. The company was deliberate, felt Sidney. 'He came with a protector for the interview … and we talked together … I couldn't talk completely openly. But it was clear to Robert Oppenheimer that there was no danger to him from me.'[39]

One week after Sidney's FBI grilling, HUAC convened 'Hearings Regarding Communist Infiltration of Radiation Laboratory and Atomic Bomb Project at the University of California, Berkeley', a series of special sessions held at the Old House office building in New York.[40] The rationale for these hearings predated the FBI's interest in JPL and Caltech, but the nature of the suspicion fell on similar targets: physicists, communists and Jews. The whole process was chaired and driven by John S. Wood, the same congressman that Liljan Malina and her New York friends in Citizens United had targeted back in 1946. Wood was the kind of politician who fondly described 'the threats and intimidation of the [Ku Klux] Klan' as 'an old American custom, like illegal whiskey-making'.[41] Beside him on the committee sat the dome-headed Red-baiter J. Parnell Thomas, a young congressman from California called Richard M. Nixon, and HUAC's investigator, Bill Wheeler (that's the same Bill Wheeler who went to ball games with Malina's analyst, Phil Cohen). The committee was ostensibly investigating the party's infiltration of Berkeley's Rad Lab, but, as with many such hearings, there was a sense that HUAC picked up where FBI enquiries had run out of steam. Clear evidence of espionage would naturally result in criminal charges; in its absence, HUAC provided a spectacle in which witnesses could be publicly degraded without the usual protections of a courtroom.

Consider the HUAC grilling of David Bohm, a Berkeley physicist caught up in the controversy around FAECT.[42] After being asked about his career, HUAC's senior investigator, Louis J. Russell, cut to the chase: 'Mr Bohm, have you ever been a member of the Young Communist League?' Bohm replied with a line he had clearly practiced at home. 'I can't answer that question on the ground that it might tend to incriminate and degrade me, and, also, I think it infringes on my rights as guaranteed by the first amendment.'[43] It was the kind of response that often riled the committee; on the other hand, getting satisfactory answers was hardly the point of this circus. The humiliation lay in presenting the question, in drip-feeding suspicion to the newspapers for maximum public impact and personal loss. HUAC reserved its most aggressive tactics for defensive witnesses like Bohm who appealed to the constitution, but the committee were also after many more people than could ever be called to appear in person. Even to have one's name mentioned at HUAC, everyone knew, could be the end of a scientific reputation. A line of questioning might have less to do with a particular witness than to signal a new horizon of enquiry.

This, at any rate, was what happened with David Bohm, who, after being questioned about the basics of his professional resumé, was asked, 'were you acquainted with or did you become acquainted with an individual named Frank Malina?'[44] Jacob Dubnoff? Sidney Weinbaum? These questions weren't difficult for Bohm – he had never met any of these Caltech scientists. (It was harder when they asked him about the Berkeley communists he did know, like Joseph Weinberg and Steve Nelson.)

The real significance of these questions was not apparent to Bohm. But for the first time, the names of both Frank Malina and Sidney Weinbaum were made public in an enquiry about the communist infiltration of military science. Ever since Dunn's inside information, the FBI had considered Frank Malina their primary target. So far, they had been thwarted by the fact of his safely living outside the United States and outside the reach of HUAC. But on the first day that Frank's name was aired in a congressional committee, he wasn't in Paris. He had just arrived in Texas to introduce Marjorie to his parents.

DEGRADATION CEREMONIES

THE MALINAS ARRIVED IN BRENHAM, Texas on 21 May 1949 to spend a couple of weeks with Frank's family before travelling north to Pasadena.[1] At Caltech, there were many old friends who were keen to meet Frank's new bride, and a few who wanted to talk privately about the dangerous political climate. The Weinbaums had hoped that the Malinas might stay with them, but Mae Churchill had already written to Frank to say that she and Bob were taking a holiday at the beach, so his old house at South Hudson Street would be available.[2] That suited Frank perfectly, though returning to his former marital home might have been a bit odd for Marjorie. Frank cabled Martin Summerfield to ask that he meet them from their flight at 11.25 p.m. on Monday 6 June. Martin wanted a chance to talk about everything. He had even planned a party for Frank and Marjorie, until Aerojet's Andy Haley insisted that they could both be hosts if the party was held at his own mansion on South Orange Grove Avenue on Thursday evening.

The next day, Tuesday, Frank called the Weinbaums. Lina happened to be entertaining a friend when the phone rang, the sort of friend who reported everything she heard to FBI agents. When Frank asked if 'everything was alright', Lina put on a brave face, saying that Linus Pauling had offered Sidney professional shelter at his

Crellin Laboratory if events 'came to the worst'. Things were clearly more serious than Frank had realised. He asked if he and Marjorie could still come round so that the four of them could chat privately, warning Lina she would need to 'keep things quiet'. 'We are not,' said Lina curtly, 'going to start a hippodrome at this moment.' She went on to detail the 'mysterious things going on' – strangers around the house – all of which made Frank even more apprehensive. 'Oh, well, you have lived out of here too long, dearie,' Lina reassured him, 'you are practically a hick.'[3] Frank and Marjorie drove round to the Weinbaums for some lunch, after which Frank and Sidney went on a long drive.[4] There was a lot to catch up on, and they may have already suspected that the house was bugged.

On the Wednesday, Frank met up with Martin Summerfield. Presumably they talked about his falling out with Sidney, their worry about clearance difficulties and a new job that Martin was considering at Princeton. Of course, none of them had any idea that most of their troubles originated with Louis Dunn, who had offered Frank the use of his car while giving strict instructions to JPL's security guards to not let Malina past the front gate. Frank and Marjorie borrowed the Churchills' car instead.

The party at Haley's on Thursday 9 June was a memorable affair for Frank and Marjorie.[5] So many old JPL and Aerojet friends were present to wish them well. In particular, Frank hadn't seen Jack Parsons for years, and was genuinely pleased to catch up with him. Jack was outwardly friendly, and buttonholed Frank to see if he could find him a new job in Paris.[6]

Things were not going well for Parsons. He'd been having clearance difficulties, and had recently denied attending Unit 122 meetings at an industrial employment review board. Meanwhile his marriage to Candy was disintegrating: he initiated divorce proceedings now that she was pursuing art and ardour with other people in Mexico ('extreme cruelty', according to Jack).[7] Candy's absence was, he felt, one more example of his persecution by women, a sense of victimhood that was given a signature twist by his belief that women had also ruined his previous lives, including those as the New Testament heretic Simon Magus, the child-killing French nobleman Gilles de

Rais, and the Scottish aristocrat Francis Stewart.[8] All of this troubled history was revealed to him in the course of a seventeen-day ceremony that began on Halloween 1948 – the beginning of a series of rituals that he called his 'Black Pilgrimage'. Some twelve years after the first rocket motor tests, the big revelation was not about nozzle shape or ignition method; rather, he saw a 'heavily robed and veiled' figure who showed him how his past life as de Rais had been blighted by the 'stupidity' of Joan of Arc; and that his life as Stewart had likewise been compromised by the 'unworthy instrument' of Geillis Duncan (a maid accused of witchcraft in Scotland's North Berwick witch trials). His past lives had ended in failure – all because of women.

'Will you fail again?' the figure asked him.

'I will not fail,' replied Parsons.[9]

Jack tended to absolve himself of blame for how his life had turned out. 'I have been stripped of wealth, of honour, and of love … as it was foretold.'[10] Yet he now believed that he was the primary power in the Thelemic transformation of the world. 'The links are certain,' he wrote; 'The Beast 666, the Pole Star 132 [Wilfred T. Smith], Grim Saturn [Crowley's appointed OTO head, Karl Germer], the dark passionate star Regina [Kahl], the bright deceitful star cassup [Betty Northrup], the disastrous star of the White Scribe [Hubbard] and the wandering star [Candy], now nameless in whom you [BABALON] were incarnated.'[11] These relationships were in fact astronomical alignments, pointing to his own magickal destiny, a messiah complex that reached its apex when he reinvented himself as BELARION ARMILUSS AL DAJJAL, ANTICHRIST – 'Belarion, Antichrist', for short – the man who would lead entire nations to 'accept the Law of the Beast 666'.[12] In this name, he took the oath of a Magister Templi, solemnly witnessed by Wilfred T. Smith, pledging to fulfil the work of The Beast.[13] Jack's inflated sense of rocketry achievement was now projected onto his astral ambitions. 'It has seemed to me,' he wrote to Karl Germer, 'that if I had the genius to found the jet propulsion field in the US, and found a multimillion dollar corporation, and a world renowned research laboratory, then I should be able to apply this genius in the magick field.'[14] Crowley was no longer around to tell him otherwise.

All of this was going on in the background when Frank met Jack at Andy Haley's party. The struggle to navigate the earthly realm of jobs and relationships had left Jack in a dark place, but he still managed to put on a front for his old friends. Andy Haley always had a soft spot for Jack. Yes, he'd been a liability at Aerojet, but once he had been safely eased out of the company, his maverick ways seemed colourful rather than sinister. It helped that the church-going lawyer was as ignorant of Aleister Crowley and the Ordo Templi Orientis as he was of Malina's membership of the Communist Party. It helped, too, that Haley was every bit the convivial host. What was that weird thing that Jack used to do? Hymn to Pan! Jack! Jack! Hymn to Pan! Haley cajoled Jack into a performance of Crowley's ode as nothing more than a fun party piece. For Jack, however, it was the anthem for his broken life and the intimation of his magickal destiny. Climbing on to a balcony above the assembled crowd, Jack channelled The Beast, stamping out the final lines:

> And I rave, and I rape and I rip and I rend
> Everlasting, world without end,
> Mannikin, maiden, Maenad, man,
> In the might of Pan.
> Io Pan! Io Pan Pan! Pan! Io Pan!

'I'll never forget Jack doing this,' Frank said later.

It was the last time they ever saw each other.[15]

ON 14 JUNE 1949, Frank and Jackie Oppenheimer were called before HUAC. The Malinas had by this stage moved on to New York, and they read the news story with growing alarm. Frank wrote to his parents to ask them not to talk to journalists.[16] If anything, the full hearing was even more sensational than the papers could convey. Frank Oppenheimer had been called up, accompanied by his lawyer Clifford J. Durr, one of the few attorneys willing to represent 'red' clients at HUAC, and who later acted for Rosa Parks (his services had been arranged by Oppie, who still managed to keep the growing crisis at arm's length). The session started with the usual preliminaries

– name, place of birth, educational background, brief resumé – before HUAC's investigator, Louis Russell, cut to the heart of the matter.

'Have you ever used the name Frank Folsom?' asked Russell.[17] As ever with these hearings, so much intrigue was invested in the question that no one on the committee was prepared for Oppenheimer's answer.

'I have never used that name,' replied Oppenheimer, 'except to write it down on a card.'

'Except what?' Congressman Burr P. Harrison stirred.

'I have never used the name except that it was written on a card, on an application card.'

'On an application card?'

'Yes.'

'What was the application filed for? What was the purpose of it?' asked Russell, resuming the direction of the enquiry.

'It was an application for membership in the Communist Party.' A wave of rapt attention spread across the room. Frank Oppenheimer had only just started, and already his testimony was dynamite.

'Are you acquainted with Frank J. Malina?'

'Yes, I remember that name.'

'Do you recall in what connection you were acquainted with him?'

'No, except that I knew him in Pasadena.'

'Was he known to you as a member of the Communist Party?'

'I do not wish to talk about the political ideas or affiliations of my friends, but I knew him in Pasadena.' As the line of questioning had grown more awkward, Oppenheimer's voice dropped.

'I can't hear you,' said Congressman John S. Wood.

'I knew Mr Malina in Pasadena.'

'He said he did not want to talk about the political affiliations of any of his friends. Is that what you said, doctor?' clarified Harrison.

'That is right.'

Russell now turned to Wood in a sombre tone. 'Mr Chairman, Frank Malina is a very important subject to the committee, and perhaps I can refresh Dr Oppenheimer's memory in some respects on his associations with Frank Malina, after which I would again like to ask if Frank Malina was known to him as a member of the Communist Party.'

'It is a simple question,' agreed Wood, 'whether you knew him as a member of the Communist Party.'

'It is a simple question but I feel I know nothing evil of Mr Malina. I know of no evil act of Mr Malina and do not want to discuss his political opinions or affiliations.'

'That is not an answer to the question,' said Wood. The committee tried various other tacks, but the witness maintained the same stance, that he did not wish to answer this question. 'Your preference is not an answer,' said Wood, growing exasperated. They cornered Oppenheimer into a blunt refusal to answer, all of which contributed to the general excitement about who this Frank Malina was – 'a very important subject in our investigations,' said Russell, one more time for the benefit of the press. The questions continued. Did he know Sidney Weinbaum? Martin Summerfield? Oppenheimer admitted to being acquainted with these people but again refused to answer whether he knew them through the Communist Party. But when Russell got to other names, such as David Bohm and Joseph Weinberg, Oppenheimer's testimony changed from refusing to answer whether he knew them through the party, to a simple 'no'.

'When you say "no",' asked Russell, 'would you like to base your answer on the fact you have no knowledge that Joseph Weinberg or David Bohm were members of the Communist Party?'

'That is right.'

'In those two cases, you are not refusing to answer,' noted Richard Nixon. He had been a seasoned attorney before entering politics, and also had experience skewering Alger Hiss at HUAC the previous year. He immediately pounced on the significance of Oppenheimer's shifting testimony.

'I am not refusing to answer.'

'You are stating,' said Nixon, 'you have no knowledge of their membership in the Communist Party.'

'That is right.'

'In the other cases you have declined to answer?'

'Yes.'

It was quite true that Oppenheimer did not know Bohm and Weinberg through the party (he had dropped out by the time he moved to Berkeley). But the clear implication of Oppenheimer's different responses suggested that he likely *did* know Malina, Weinbaum and Summerfield through the party in Pasadena, even if he'd rather not talk about it.

In the afternoon, Jackie Oppenheimer was sworn in for her own testimony. And once again, the HUAC investigator Louis Russell steered the questions towards Malina: 'did you ever know Frank Malina under the name of Frank Palma?'

'No.'

'Did you ever know Frank Malina under the name of Frank Parma?' At this point some of the congressmen on the committee were left trailing in Russell's wake.

'What is the purpose of these questions, Mr Russell?' The interruption came from Morgan Moulder, the representative from Missouri. 'I am somewhat in the dark as to what we are driving at.'

Russell helpfully clarified:

> In the case of Frank Malina, we have information, in connection with another investigation, which indicates that he is a member of the Communist Party, and in certain other capacities was acquainted with Mr and Mrs Oppenheimer … it might be important to the committee to establish the fact that he [Malina] was a member of the Communist Party. We have the number of one Communist Party card issued to him.

Short of waving Malina's party card in the air, the committee had achieved their aim of administering its greatest punishment. As the writer Victor Navasky has argued, these hearings were essentially 'degradation ceremonies' (the term comes from the sociologist Harold Garfinkel, describing how certain rituals mark and punish deviance, reaffirming the importance of norms and rules).[18] It wasn't enough to expose the Oppenheimers, and Malina too, as communists – they wanted to enroll Jackie in her own subjection.

'Do you believe that the Un-American Activities Committee is a greater menace to our free institutions than the Communists are?'

'As far as this committee or any congressional committee, I think they are a good idea,' replied Jackie. The whole thing was humiliating.

The session had started with Frank being sworn in at 11.30 a.m. and ended with Jackie stepping down at 4.45 p.m. It had been a difficult day, but they did what they had planned to do – cooperate to a limited degree in the hope of being left in peace. The Oppenheimers expected everyone to behave well. And if the worst came to the worst and Frank lost his job, they could always sell one of their Picassos and move to their Colorado ranch. Sidney Weinbaum had always bristled at how, back in the 1930s, the Oppenheimers had argued that Unit 122 should be completely open – easy for them, they were rich enough not to have to worry about being careful.

All of this would have been less galling had Sidney not just taken the personal risk of holding firm with the FBI, going out of his way to protect and reassure J. Robert Oppenheimer. And now here were Frank and Jackie abandoning their implicit compact that membership would always be denied, leaving Sidney more exposed than ever. Denial of membership wasn't a tactic for self-preservation – if anything, it was more risky – but it was the only way of not proliferating the fallout for other party members. This is something that they had all understood and agreed upon. It was a case of putting the needs of the many over those of the few, a discipline that Frank Oppenheimer had followed – until now. In 1947, for instance, when the Washington *Times Herald* had accused him of being a communist, Frank Oppenheimer told the *Minneapolis Times* 'I am at a loss to account for such a trumped-up story.'[19] He had given a similarly robust written denial to his higher-ups at the University of Minnesota, saying that the *Times Herald* claim was 'a complete and unequivocal falsehood'.[20]

Frank Oppenheimer realised that there would be a fallout from his changed tactics at HUAC. His own university would be unhappy at the 1947 denials, but in the longer term he hoped to get credit for telling the truth. Before leaving for HUAC, he submitted a letter of resignation to his departmental chair at the University of Minnesota, not quite expecting that it would be accepted. That was a misjudgement.[21] When he then tried to apply for other jobs, they were often offered only to be withdrawn at the last moment. The FBI finally spelled it out to him that until he cooperated more fully, naming names, he wouldn't ever find a job. That, of course, was the whole point of the party's discipline about denying membership in the first place. Once you cooperated, there was no end to it – unless, like Frank and Jackie Oppenheimer, you didn't need to work and could fall back on a 830-acre ranch in Colorado.[22] Frank wrote to a friend to say that 'the sight of the meadow and the mountains seems to tingle and refresh the deepest parts of the nervous system'. Being able to escape to the high country was a balm for their souls.[23]

SIDNEY WEINBAUM'S REFUSAL to cooperate with the FBI had left him badly marked. 'We'll be after you,' they told him, just in case he held out any hope that his ordeal might be over.[24] His clearance denial and sacking from JPL did not come as a surprise, but it was an encouragement that their old friend Linus Pauling had promised to immediately re-hire him at his Crellin Lab. There was a problem, however. Pauling was undoubtedly one of Caltech's brightest stars, but the institute's new president, Lee DuBridge, was not sympathetic. Aside from their scientific differences (the physicist DuBridge was privately unenthusiastic about Pauling's molecular interests), their main tension was about the role of politics in academia. There was a lot they agreed on, including that big business should have an assured place in universities. Pauling had even gone so far as to name his son Crellin after the steel magnate who funded his laboratory – that's hardly the act of a revolutionary.[25] But the chemist did become a bit more radical throughout the late 1940s, and his high profile meant that Caltech's name was sometimes caught up in campaigns against atomic weapons or his suggestion that US military spending would be better directed to UNESCO. DuBridge didn't like this at all. He had to face the institute's conservative trustees and donors, who couldn't understand why such views should be tolerated, especially since the Board of Regents at the University of California had recently instituted a loyalty oath for staff. Caltech may be a private university outside such public scrutiny but, they felt, that was no reason why it should harbour communist sympathisers. How could Pauling offer Weinbaum a job after he had been outed at HUAC? Wasn't this evidence that maybe Pauling was himself a communist?[26]

Having heard about the HUAC's interest in JPL, Lee DuBridge moved to shore up Caltech's reputation and clear the air once and for all. If Pauling wanted to employ Weinbaum on unclassified research at Crellin, that was his business, but he could only do so if Sidney could prove he was as blameless as Pauling claimed. And the true test of political virtue, thought the president, was for Sidney to appeal the loss of his clearance at the Army's Industrial Employment Review Board – a solution that would appease restive trustees if he won, and appear vigilantly anti-communist if he lost. Sidney could tell that DuBridge was channelling external pressures, but if he wanted to remain at Caltech, and indeed in science, he didn't have much of a choice.[27] Once again, he found himself in check.

The day was warm and humid when the board convened on 23 September.[28] 'The procedure was long, the attitude civilised ... the questions – abominable,' wrote Lina. Her abiding impression was of oppressive heat (she claimed that the mercury hit 102°F, though weather records suggest only 81°F).[29] And then it was her turn on the stand. She wasn't even employed at Caltech, couldn't possibly be subject to a loyalty review, but if the board were to clear Sidney, they felt it necessary to be satisfied of *her* good character. 'It turned out I was the black sheep,' she explained to Frank Malina. The problem was her hospitality to Jenny Marling the year before. By the end of the hearing, Sidney and Lina were devastated by 'all this filth' – the morass of intimate evidence that left them wondering who among their friends was passing information to the FBI.[30] At 3 p.m., they finally emerged exhausted from the hothouse to learn that the USSR had just tested their own atomic bomb.

The failure of Sidney's appeal a couple of months later probably wasn't much of a surprise – he was out of a job and, after twenty years, out of science.[31] It's possible to imagine circumstances in which this would have been the end of the trail for the Weinbaums. Yes, Sidney would be blacklisted and broke but perhaps they'd eventually find menial work. Yet with every passing week, the political climate was changing in ways that made ordinary unemployment seem like the least of their worries. Ever since Louis Dunn's allegations and the HUAC hearing, the tide of geopolitical events was running against anyone suspected of communist association. It wasn't just the Soviet atomic test. A week after Truman's announcement, Mao Zedong declared the founding of the People's Republic of China after the victory of the communists over the Kuomintang forces. Then at the pivot of the twentieth century, January 1950, the perjury trial of State Department employee Alger Hiss heightened public anxieties about the extent of Soviet penetration into the US government. His conviction on 21 January saw Hiss sentenced to five years in jail, Senator McCarthy seizing the opportunity to make an infamous speech about 'enemies from within', in which, with the immortal line 'I have here in my hand ...', he brandished a list of alleged communists embedded in the State Department. Hiss, it seemed, was only the tip of the iceberg.

That same month, the German physicist and Manhattan Project scientist Klaus Fuchs confessed to being a Soviet spy, telling MI5 interrogators that the Soviets had an agent embedded in Berkeley. That ultimately led to the most important arrests of twentieth-century counter-intelligence: Harry Gold, David Greenglass, and Julius and Ethel Rosenberg. Sidney could see that pressure was building for the FBI to score a pushback against these forces. He could see, too, that even without evidence of espionage he was the perfect fit for a high-profile West Coast anti-communist trial. His denials of party membership at the clearance board could constitute fresh grounds for a perjury prosecution that would justify and focus the specific concerns about communism infecting universities. The scrutiny wasn't really *about* Sidney. Public anxieties about communists needed to be focused on individuals. By virtue of Frank Malina's good fortune in moving to UNESCO when he did, it was Sidney who took his place.

For Frank, there was a sense in which his friend's life was the road not taken – Sidney had stayed in the United States and that made all the difference. Both of them had shared the same political history, worked at JPL, been named at HUAC, and had turned up in William Ward Kimple's records. The main distinction was that

Sidney remained in the US and had now sworn under oath that he had not been a member of the Communist Party, while the FBI had 'friendly witnesses' who suggested otherwise. Without the dead weight of a denial under oath, Frank was able to slip freely in and out of the United States in 1949. It was a little awkward at UNESCO now that Frank's name had appeared in the papers, but he was kept busy launching an Arid Zone Research Program, an echo of 1930s soil erosion work, for which he again recruited von Kármán.[32]

Life at home with Marjorie was rich and full. Aerojet landed regular dividends and Frank found time to start painting: pastels mostly, with colours emerging like a desert spring across their Parisian apartment. By November 1949 he was finishing a new painting just about every day.[33] And then into this happy state came the most welcome news of all – Marjorie was pregnant. 'We are very excited about the "baby",' Marjorie wrote to her in-laws, 'particularly Frank who already has the look of a proud father.'[34]

Yet below the happy hum of their Parisian life, the Red Scare rumbled on. Among those most taken aback by Oppenheimer's HUAC allegations was Andy Haley, who had recently left Aerojet to become counsel for another congressional committee. 'I have been extensively ribbed because the [HUAC] stories broke after Martin and I were joint hosts for the party in your honour … I have never heard either of you indicate that you were or are Communists, so you can imagine my bewilderment.' Haley's surprise was real, but underneath his wry tone was a genuine concern. 'It is none of my business,' he added, 'but you might write the Committee a simple factual and straightforward statement that you never have been and are not now a member of the Communist Party … the Committee is composed of a decent bunch of fellows … I believe they would correct the record and exonerate you.'[35]

Frank tried and failed to draft a long reply to Haley. He scribbled out long arguments in blue biro about having 'done nothing illegal and have always been loyal to the American people', but in the end it was easier to sidestep these comments and scrap the letter. The political struggles of the late 1930s were just too difficult to translate into the McCarthy era.[36]

THE FBI ARRESTED SIDNEY at home in Pasadena on Friday 16 June 1950. A simultaneous New York swoop caught David Greenglass, a 28-year-old member of the Rosenberg spy ring. The cases weren't linked, but it didn't hurt to give the impression that they might be. The *New York Tribune* reported the arrests as following the return of agents 'from London last week after questioning Dr Klaus Fuchs, German-born British atomic scientist who is serving a fourteen-year sentence for atomic espionage'.[37] Sidney was not arrested on espionage charges – he was arraigned on perjury and 'fraud against the government' for not having disclosed party membership – but blurring these details was all part of the degradation.

Reporters soon turned up at the Weinbaum home on South San Marino Avenue and pestered Lina with questions. 'It isn't true,' was all Lina could say, but the full story was accompanied by publishing close-up images of Sidney after his arrest, their exact address, along with names and distinguishing details of Lina and Selina. *Time* magazine sent the news worldwide, prompting a distant acquaintance of the Weinbaums from England to write to J. Edgar Hoover saying that it was all Lina's fault. Mrs Janet Knight Cheney of Dorset had been a friend of the Weinbaums' former nanny, and wanted the FBI to know that she thought Mrs Weinbaum was 'a spoiled personality … wanting what then seemed out of reach'; 'passionately homesick for Russia, disappointed in America'; 'rather posey, an exhibitionist'; 'liked to have photographs of her taken in extravagant poses, in extravagant clothes'.[38] The Weinbaums soon learned that political trouble very quickly became personal. No sooner had Sidney been granted bail than their devout, church-going landlord evicted them from their family home.[39]

On campus, everyone was unsettled. Linus Pauling told a newspaper that he had 'the greatest confidence in Dr Weinbaum. He is a high-minded man and a fine American', adding – absurdly – that 'he has never seemed at all interested in politics'.[40] The low-level anxiety that had dogged all who had ever been involved with Unit 122 was particularly taking its toll on H. S. Tsien. Did he and his wife Jian Ying really want to raise their two young children in the United States in this climate? The question of whether to return to China was further prompted by FBI grillings on 25 May and again on 6 June, interviews which, as with Weinbaum, involved the presentation of relevant 1938 party records. It was clear that Sidney was in trouble, and that because Tsien had lined up his friend for the JPL job in 1946, the suspicion was spreading. Tsien denied his party membership but conceded that while he had previously trusted his friend, 'he now had reservations due to the recent clearance difficulties on Weinbaum's part'.[41] But then Tsien lost his own clearance. He took that as a personal affront. On the day of Sidney's arrest, the US Air Force Office of Special Investigations (OSI) phoned the FBI to pass on intelligence that Tsien 'had resigned from Caltech and was leaving for China'.[42] Tsien hadn't officially resigned but he had gone to see Dean of the Faculty Earnest Watson to speak his mind.

'I'm going home,' Tsien told him, to Watson's evident surprise.

'Aren't you happy here?'

'I was brought up to believe that if you are a guest you do nothing to offend your host. I'm a guest in this country,' continued Tsien, 'I'm an unwelcome guest. I'm going home.'[43]

One week after Weinbaum's arrest, the threat of domestic communism was displaced by news that North Korea, supported by China and the Soviet Union, had invaded the South. With the Cold War warming up, it wasn't an easy time to be a Chinese national inside the US security establishment, yet Tsien still seemed in two minds about leaving.[44] He agreed to appeal the loss of his clearance – probably under some pressure from Caltech – and planned to make his case before the Industrial Employment Review Board in Washington. At the same time, he kept his options open by booking international flights and packing the contents of his home and office into crates to be shipped to China. The hearing was due on 23 August; his flight to Hong Kong was on the 28th. Sidney Weinbaum's trial was scheduled for the 30th. It was crunch time.

After flying to Washington, Tsien decided to get some last-minute advice from a friend, Dan Kimball, a former executive vice president of Aerojet who had switched to the other side of the military-industrial complex and was now undersecretary of the Navy. Tsien felt that Kimball would be both sympathetic and influential, which he was – just not in the right way. He successfully got the hearing postponed to give time to prepare the legal defence, but Tsien didn't conceal from him the fact that he intended to return to China, adding 'I don't want to build weapons to kill my countrymen'.[45] Seeing what this move would mean for the future missile capability of communist China, Kimball was blunt: 'Tsien, I'd rather see you dead than back in China'.[46] They parted with Tsien agreeing that he wouldn't do anything without further discussion, but as soon as the engineer had left the office, Kimball called the Justice Department to advise that Tsien must not be allowed to leave the United States.

Tsien stepped off the plane in LA to be met by an INS agent with written instructions forbidding his departure. Kimball's call, as it turned out, was secondary to the fact that the customs officials had already impounded Tsien's belongings when one of the packers had informed his boss that the shipment contained papers marked 'secret' and 'confidential'. Also discovered were nine notebooks of newspaper clippings of esponiage trials and details of the US atomic energy programme running to 400 pages.[47] The first that Tsien knew about his papers was when the news story broke, from the *Los Angeles Times* to the *New York Times*: 'Secret Data Seized in China Shipment'.[48] The implication was that Tsien, too, might be a spy. Tsien was detained for questioning for fifteen days. The problem was less about the expertise in Tsien's papers, which were in any case declassified or out of date, but in having such an expert, a foreign national, with the liberty to move. After all, as Caltech's president Lee DuBridge understood it, 'he wasn't going back to China to grow apples'.[49]

SIDNEY'S TRIAL OPENED on 30 August 1950. He pleaded not guilty to four counts of perjury and two of making false statements. Prosecuting attorney Ernest A. Tolin told the jury that he would not only prove that Sidney was a member of the party but also that he acted as its recruiter. There would be a range of witnesses, Tolin said, from former comrades, to JPL management and Army Intelligence. First up was Gustav Albrecht, a one-time Unit 122 member who now taught physics at Chapman College. Albrecht told the jury how he and Weinbaum had shared an office at Pauling's Crellin Laboratory back in the 1930s, that they had been friends and colleagues and that he eventually 'succumbed to Dr Weinbaum's urging and consented to join the Party'. Gus Albrecht's testimony was damaging, yet he also admitted that his former friend had once claimed *not* to have been a party member in a strict sense, because he did not want his CPUSA membership to jeopardise getting a university position back in Russia. (This chimes with Linus Pauling's belief that 'Sidney seems to have felt that he was innocent, but it seems to me that he may have been relying on some technicality'.)[50]

It was awkward for Pauling to have his old research team at loggerheads in court. He hovered above the fray, and was out of town during the trial itself, yet took the personal risk of coordinating a small group of Caltech faculty to raise funds for Sidney's defence. Many staff contributed to this fund, though not all wanted their names used and none were as generous as Pauling himself, who chipped in $500. One scientist, the crystallographer Henri A. Levy, gave a modest donation but then asked for it back when he learned that Weinbaum had hired the attorney Ben Margolis, a party stalwart famous for defending the Hollywood Ten.[51]

Not every witness was as willing as Albrecht. That same afternoon, Jacob and Belle Dubnoff told Tolin and the newspapers that despite being subpoenaed they would not take the stand against Sidney. Yet the very next morning Jacob dutifully took his place in the witness box, looking and sounding burdened, mumbling answers to Tolin's questions and often having to be told to speak up. He fudged as much as he could, but Dubnoff still admitted to being a member and to being the unit's treasurer. He identified his own membership dues book and talked about his party name. He 'assumed' Sidney was a member from his presence at meetings. The imprecision in his testimony didn't, as Sidney saw it, lessen the betrayal, but it would have been hard not to feel sympathy for the indignity of Dubnoff's position.

'Have you … been told directly or indirectly that unless you came in here and testified as desired that you might lose your job at the university?'

'That has been implied, yes.'

'And you want to save your job at the university, don't you?'

'Well, naturally.'

'As a matter of fact, if you lost your job at the university, you don't know what you would do, do you?'

'That is true, yes.'

'So you are in a pretty desperate position to hang on to your job at the university, aren't you?'

'...'

'No further questions.'

Margolis took a similar line of attack against Frank Oppenheimer, who admitted that the FBI had visited him on his Colorado ranch and had made a 'general appeal' to him to do his duty as a citizen and testify against Weinbaum (in fact, Oppenheimer had told the FBI that he was 'terrified' of 'being prosecuted for perjury').[52] In the course of that appeal, Oppenheimer told the court it was 'suggested that if I wanted a job again in the universities, that this was the way that university people would see it'.

'In other words, this is the quid pro quo for getting a job back, is that right?'

'It was not stated in such black and white terms.'

'But that is how you understood it, wasn't it?'

Oppenheimer sat mute.

'That is all.'

Like Dubnoff, Frank Oppenheimer tried his best to limit his cooperation, testifying only that Sidney had been present at Unit 122 meetings but insisting that because meetings were open to both members and non-members, he 'couldn't say' into which category the defendant fell. All the old comrades had different strategies for dealing with Tolin's subpoenas. Albrecht had moved right and had little compunction about becoming a witness. Dubnoff and Oppenheimer testified to protect themselves but still tried not to give away anything that was too damaging for Sidney. That meant lying under oath, though not in ways that were readily detectable.[53] Louis and Shirley Berkus, who had previously been uncooperative with the FBI, indicated at first that they, like the Dubnoffs, would refuse to appear.[54] But then Judge Ben Harrison made it clear that he would hold any witness who took the Fifth Amendment to be in contempt of court. The Berkuses knew he was serious. By July 1949, sixteen communists had been convicted in California alone – many jailed – for invoking the Fifth Amendment.[55] So the Berkuses revised their strategy and claimed not to be able to identify Sidney Weinbaum. That proved effective, and they were soon dismissed.[56] Only Eugene Brunner refused to play ball in any way. He was willing to appear, but not willing to lie under oath or give full answers to Tolin's 'have you ever been ...?' questions.

Eugene Brunner had been part of Unit 122 in the early days before leaving Caltech to work for the Shell Development Company in Berkeley. He and his wife Adeline, a nursery school teacher, were committed party members and maintained a strict discipline regarding public admission of membership. Eugene had already endured more than his share of FBI surveillance, most likely because of his attendance at FAECT trade union meetings. Their daughter Alice remembers that in the run-up to the trial her parents treated the phone as if it was tapped, that they routinely burned any ephemeral papers – fliers, pamphlets and the like – that might get them into trouble, and that their house was constantly felt to be vulnerable. (One day they found that a nail had been driven into the frame of their front door, preventing its tight closure.) Such was their worry that both parents might be subpoenaed and that they might lose custody of their children, that in the summer of 1950 Adeline Brunner and kids escaped to live with a family friend in rural northern California.[57]

All of this meant that when Eugene Brunner entered the witness box, he had a good idea of what was at stake both for his own family and for Sidney. When Tolin asked about his membership of the Pasadena party, there was a lot riding on his reply: 'I must decline to answer that on the ground that it might incriminate me.' Just as he had promised, Judge Harrison was in no mood for the Fifth Amendment. He found Brunner in contempt, asking the deputy marshal to lead him away, handcuffed, to a six-month sentence in the county jail. Brunner appealed and was released after a few days, but it was two years before the legitimacy of using the Fifth Amendment as a defence in political cases finally was tested in the Supreme Court. In the end the court ruled in Brunner's favour. Nevertheless, the price of not testifying against Sidney Weinbaum was that this Caltech-educated physics PhD never worked in professional science again. He spent the next fifteen years repairing TVs.[58] (He eventually found work in New York as a translator of Russian physics journals.)

The final day of Sidney's trial saw a small moment of drama. During his summing-up to the jury, US Attorney Ernest Tolin cast doubt on the source of a claim heard earlier in the day, that Sidney's own father had been against the Russian Revolution and had, as a consequence, been killed by the Bolsheviks. Nerves were frayed after a week of listening to hostile witnesses pore over the question of whether Sidney had been a member of the Communist Party, and whether he had sought to conceal this fact from JPL. Parsing the fine truths of political affiliation was one thing, but minimising what had happened to his father was more than Sidney could bear. He jumped to his feet at the counsel table and shouted: 'I DON'T LIE ABOUT MY FATHER' before slumping back in the chair. Harrison pounded his gavel only for the distress to propel Sidney up a second time. 'I don't think it's fair to cast such things upon me.'[59]

There was a lot that wasn't fair about a perjury case that was itself so full of white lies and obfuscations, mistruths and partial recollections, all of them squeezed out of witnesses under threat of the blacklist and the county jail. Even if the jury was sympathetic – as some were – the raw, involuntary truth of Sidney's outburst did not help the rest of his testimony. Sidney sat still, hands clasped, as the jury found him guilty and Harrison sentenced him to four years in prison. He showed no emotion as the verdict was read out, but as the court cleared, and Lina and Selina embraced him, the strain was evident. Of course, the press were everywhere. 'I'm sure my father never did anything to be ashamed of and I'm proud of him,' Selina told reporters. 'I am sure it will all come out alright in the end.'[60]

THE COFFEE CAN

QUITE EARLY ON in the investigation of Sidney Weinbaum, the FBI decided that Jack Parsons should never appear as a government witness. There was no way that they'd put at risk such an important anti-communist trial by letting Parsons be cross-examined by a defence attorney. It worked out quite well for Jack that his own supreme unreliability acted as a kind of protection.

But almost everyone else who had attended a Unit 122 meeting now worried about subpoenas; if they were called as a witness, they strategised about how to avoid contempt of court on the one hand while not destroying Sidney Weinbaum on the other. Not Jack; FBI agents found him eager to cooperate. Right at the beginning of the Weinbaum enquiry, Jack furnished agents with a list of Unit 122 members, which, combined with William Ward Kimple's records, became the FBI's inventory of potential interrogees. Perhaps this cooperation isn't so remarkable: Jack Parsons and Ed Forman had always felt upstaged and displaced by Sidney, annoyed that their original rocketry group of Malina, Tsien and Summerfield had been taken over by someone who exceeded their own learning in every field.

What is more surprising, however, is that Parsons had been a regular informant for the FBI since 1942.[1] It's not clear that anyone knew this. Exactly why agents first approached him is also hazy, but it seems that the OTO had been on their radar back in March 1942, when they intercepted cryptic cables between Aleister Crowley and the official head of the OTO in the US, Karl Germer.[2] A few months later an anonymous letter signed 'A Real Soldier' (some suspected a disgruntled Regina Kahl) was sent from Texas to the Pasadena Police Department, alleging that

> meetings were being held at Mr Parsons' home by a Mr [Wilfred] Smith ... in which 'Crowleyism' and Sex Perversion were being advocated and taught, together with 'Survival of the Fittest'. The letter went on to state that the writer hoped they (the Police) would do something to clean out this nest of enemy aliens and their revolting practices.[3]

The Pasadena police interviewed Parsons and, given he was working on a classified project, shared their information with the FBI and the US Army Counter Intelligence Corps (CIC). Aside from Paul Seckler's grand theft auto, none of the subsequent LAPD or FBI enquiries into the OTO amounted to very much. What's notable, though, is that the impetus behind the original FBI interest in the OTO was out of a concern about fascism (that, after all, was the implication of the anonymous letter). And in the early days, it's true that Crowley had harboured a soft spot for Hitler as a possible means of spreading the law of Thelema.[4]

The FBI realised that, while the OTO did not constitute a fascist threat, Parsons could still be a useful informant, especially when it came to his socialist colleagues. Jack may have been worried about losing his security clearance, or perhaps about the fate of the fledgling Aerojet. But even that doesn't explain the fulsome extent of his cooperation when FBI agents interviewed him on 4 November 1942. Back then they were curious about the OTO, but they particularly wanted to know about Frank Malina. Parsons had told the FBI that, yes, he was 'well acquainted with the subject and knew considerable [sic] about the personal life of the subject as well as his business'.[5] But then Jack went the extra mile. When asked if he thought that his friend was loyal to the United States, Parsons declined to give an affirmative answer, but volunteered instead that Frank 'was associated with groups of "pinks" at Cal Tech about three years ago'. What do you mean by pinks? the agents asked.

Parsons told them that 'the group was radical but I believe them to be merely philosophical,' adding that 'they would never be active in doing anything to change the form of Government which we now have'. Other FBI files suggest he went further: Parsons' 'original allegations left no question as to the fact that these were Communist Party meetings,' wrote one agent.[6] At the end of this interview Jack back-pedalled, insisting that Malina 'had not been active at all in any radical or Communist activities in the past three years', a claim that he likely knew to be false.[7] But the damage was done. Jack Parsons had not initiated the surveillance of Frank Malina but he certainly chose to give it added momentum.

By 1948, however, Jack's cooperation with the FBI was no longer enough to protect him. They had always taken what he said with a degree of scepticism (other informants insisted that his testimony should be treated with 'extreme reserve')[8] – Jack was useful to the bureau but not *that* useful. And as he was now contributing to the SM-64 Navaho, an intercontinental cruise missile under development at North American Aviation, the FBI and CIC resumed their enquiries into Jack's background and soon rescinded his security clearance. 'I have no idea of the reason for this action,' Jack complained to Theodore von Kármán; 'possibly it is simply because I am not enough of a rubber stamp personality.'[9] Jack didn't actually believe this. He freely admitted to Karl Germer that he knew the trouble stemmed from his OTO activities and from the circulation of Crowley's *Liber Oz.*[10] But either way, he wanted out of the US, 'to begin a career elsewhere in a more liberal atmosphere, as soon as possible'.[11]

In the meantime, Jack once more tried his best to please the FBI. He met with Special Agent Arthur Wittenburg in January and again in March, telling him about Weinbaum's role as Unit 122 leader, but also about the involvement of Frank Malina and the Dubnoffs (though he fudged a little about Tsien, saying that he had no reason to doubt his loyalty).[12] The agents, however, found it hard to interpret

Jack. They took seriously his laughable claim that he now worked as a 'physicist', but they had no idea just how tormented and isolated he had become. His letters from this time are full of morbid thoughts, wishing, for instance, that Candy could understand his 'ache to dissolve, to pass away, to go, to be one – to drink utterly of the cup men call death or madness – to be away, at rest, at peace'.[13] He told Karl Germer that he had now 'entered magickal regions where there is no possibility of outside help or assistance. The oaths are taken, the bridges burned, and there is no ... turning back.'[14]

WITHOUT HAVING FIXED a plan of escape from the United States, Jack now followed up other possibilities – including a suggestion of von Kármán's. The Boss put Jack in touch with Herbert T. Rosenfeld, president of the southern Californian chapter of the American Technion Society, an organisation that assisted the flow of scientific information to the fledgling state of Israel. Von Kármán had already been the conduit of aeronautical expertise to Israel's air force, but Rosenfeld also wanted the new state to develop a rocket programme, and asked for Parsons' help to write a proposal for the construction of a new explosives plant. This opportunity, Jack figured, might be just the ticket he needed out of the United States. (Candy, with whom he was partly reconciled, may have been less keen given her strongly anti-Semitic views; she wanted to live in fascist Spain.)[15]

But there was a hold-up: Rosenfeld had become unwell, and the project stalled. So Jack pursued sidelines working at a gas station and setting up a mail-order religion called 'The Witchcraft', a 'religious, benevolent and fraternal organization pledged to the ideals of love, of liberty and universal brotherhood'. He offered a two-page manifesto and a Hubbard-style instructional course in 'the secrets of alchemy, magic, metaphysics, yoga, and all occult science' – at ten dollars, it was cheaper than Dianetics.[16] Mail-order magick became less important, however, when Jack somehow persuaded the Industrial Employment Review Board to restore his security clearance. This in turn opened up a respectable job in chemical plant design at Hughes Aviation. But he still desperately wanted out of the United States. This is why, at Andy Haley's party, Jack had shamelessly badgered Frank Malina about potential jobs in France. Jack kept an eye on opportunities, all the while hoping that the Israel project might yet come through. 'I should rather live in Paris,' he wrote to Candy, 'but am on to a big thing here which I had better stay with until it breaks.'[17]

By September 1950, against the background publicity of the Weinbaum trial, Parsons took action in his bid to escape. Once again he approached Rosenfeld, who was receptive but requested a detailed set of costings for the creation of a rocket programme and explosives plant in Israel. Jack may have recently been trying to sell the secrets of alchemy for ten bucks, but with Rosenfeld he wanted to emphasise his scientific credibility. He could usefully lift information on costs from similar work he was doing at Hughes. But to really make sure that he was taken seriously as a scientist, he planned to send through a series of reports that he had co-authored while at GALCIT and Aerojet. First, he'd need to get them copied, which wasn't straightforward given that much of the work was still classified as 'confidential'. He'd taken them to Hughes to be stored for safekeeping (classified material obviously couldn't be kept at home). Still, as far as Jack was concerned these were personal copies of papers that he wrote himself. He intended to recruit a secretary at Hughes, Blanche Boyer, who would have the necessary clearance to handle confidential documents and could type up his proposal and supporting material.[18] To that end, Parsons withdrew the documents from Hughes on 15 September, flashing his property pass to the security guard before delivering the whole package to Boyer at her house the following day.[19]

How much Blanche Boyer knew about the typing request before Jack handed it to her isn't evident. But as soon as she saw the contents of the package, she realised that Jack was in violation of security procedures. She alerted the authorities at Hughes, who in turn called the FBI, who descended with the energy of a full espionage investigation.[20] Jack naturally claimed that it was all a misunderstanding, and that he had intended to run the proposal past the State Department *after* the document was typed. But it was too late for that.

As the FBI pursued their enquiries, Jack found bits and pieces of work with powder companies. He even started his own concern, Parsons Chemical Manufacturing Company, in North Hollywood.[21]

There was an echo of Aerojet days: Jack played the respectable businessman, resumed his relationship with Candy, from whom he was now divorced, and returned to live on that most familiar street, South Orange Grove Avenue. They took a lease on the old coach house of a now demolished mansion at 1071. It was a step down in size and prestige but they liked it, and, for a short spell, it permitted a return to visitors and parties, sharing their notoriety with a younger generation.

Yet the FBI investigation still wound on, and it wasn't until October 1951 that Jack learned he would not, after all, face charges. The Assistant US Attorney in LA, Angus D. McEachen, declined to prosecute on the grounds that there was 'a lack of sufficient evidence of intent or reason to believe that information ... was to be used to injure [the] United States'.[22] There was no question that he had breached security regulations at Hughes – the evidence was overwhelming – it's just that Israel was not the Soviet Union and Jack had been motivated by money rather than ideology. It was still a remarkable outcome given the energy expended on trying to prove other espionage investigations at GALCIT and JPL. Jack's relief, however, was short-lived. By January 1952, the Industrial Review Employment Board wrote to him to say that they were again revoking his clearance. 'You do not possess the integrity, character and responsibility essential to security of classified military information,' they told him. The Hughes episode had left the board with the impression that he 'might voluntarily or involuntarily act against the security interest of the United States and constitute a danger to national security'.[23] Without a clearance, Parsons' rocketry career was over.

A FEW WEEKS LATER Jack went to visit Jane Wolfe, who, at nearly 75, would have been the grande dame of the OTO had it not been for the fact that the OTO at that point scarcely existed. Agape Lodge had blown apart after the excommunication of Smith. Parsons and Smith no longer spoke, though Jack still counted Jane as a friend. He told her that there was no longer anything left for him in Pasadena, and that he planned to go to Mexico with Candy.[24] When Jack returned to see Jane the following month she noticed that he avoided eye contact. In fact, it seemed as if his gaze 'more or less constantly moved now here, now there, now back again'. More than once, he told her that he was frightened.[25] 'Everything is dead,' he said.

Jack approached the preparations for the Mexico trip with an unaccustomed thoroughness, dividing their possessions into those they'd take with them and those they would arrange to be sold. He clearly still had some thought of future work given that he asked von Kármán for a letter of reference. Rocketry was not an option, but the Boss could still testify to Parsons' uncommon knowledge of explosives, albeit with the understatement that he was 'better for independent work than for office work with many employees about'.[26] In the months preceding their planned departure, Parsons found employment at a variety of powder companies developing pyrotechnics and special effects, such as the tiny charges used in films to simulate a bullet hitting a body. Downstairs in the coach house, the laundry was configured as an improvised laboratory and storage facility for the many chemicals Jack used to fulfil these bespoke orders. But these, too, were now mostly packed up in anticipation of the move, and of the fact that two of their friends were taking over the lease to 1071 in early June.

Friends at this time noticed different things about Jack. Jane was encouraged to see that he seemed more 'relaxed and happy'. George Frey, a friend from North American Aviation, observed how he 'thoroughly cleaned up all his personal obligations', carefully 'bidding each and everyone good-bye'.[27] The fact that it was all so meticulous was a bit out of character. Frey sensed that Jack didn't particularly want to go to Mexico, and supposed he must be somehow afraid of Candy. Jack never said much explicitly. By far the biggest glimpse of his interior life was given to New York Thelemite Karl Germer, who wasn't exactly a friend. Jack corresponded with him as the closest living embodiment of The Beast. He wrote candidly of his attempt to converse with his own Holy Guardian Angel, which, in Crowley's esoteric philosophy, is the central work of an adept. The letter worried Germer for obvious reasons:

The operation began auspiciously with a chromatic display of psychosomatic symptoms, and progressed rapidly to acute psychosis. The operator has alternated satisfactorily between manic hysteria and depressing melancholy stupor on approximately 40 cycles, and satisfactory progress has been maintained in social ostracism, economic collapses [sic] and mental disassociation.

These statements are mentioned not in any vain glorious [sic] spirit of conceit, but rather that they may serve as comfort and inspiration to other aspirants on the Path.

Now I'm off to the wilds of Mexico for a period, also in pursuit of the elusive H.G.A. before winding up in the guard (room) finally via the booby hotels, the graveyard or —? If the *finem*, you can tell all the little practicuses that I wouldn't have missed it for anything.[28]

On 17 June 1952, the day before Jack and Candy were due to leave for Mexico, an urgent request came through from the Special Effects Corporation. The exact details are not known, but Jack apparently agreed to fulfil the order as Candy did some last-minute shopping for the trip. Outside the coach house, their battered Packard with trailer was loaded up with everything they needed for their fresh start. At about 5 p.m., their friend Greg Ganci, one of their new tenants, came down to see Parsons working in the laundry, pouring chemicals from one test tube to another. Even at the best of times, these were not ideal laboratory conditions, but with flasks and materials packed up in boxes, Jack was having to make do with what was to hand. It was entirely in keeping with his long history of tinkering and his disdain for scientific rules that he may have been using a coffee can as the receptacle in which to mix fulminate of mercury. Improvisation, try-it-and-see, was at the heart of Jack's scientific method. On the other hand, no one had a greater feeling for the unforgiving properties of what he was preparing. No one knew as well as he did that as a primary explosive, mercury fulminate has an absolute intolerance for friction, heat or shock of any kind.

Those who heard the sound at 5.08 p.m. said that it was almost like two separate explosions within a fraction of second. A moment later, part of the wall was gone, there was a gaping aperture in the ceiling, the windows were blown out and the heavy garage doors had come off their hinges. A nearby greenhouse was floored by the blast. And as the smoke cleared, a new layer of debris blanketed the laundry floor with fragments of plaster, splintered and distorted wood, shards of glass from windows, test tubes and the brown reagent bottles. A water heater had shrunk in the corner, its pipes buckled. And beside it, a heavy laundry sink lay on the floor concealing the shattered body of Jack Parsons.

The new tenants were first on the scene. They tried to prop him up and heard him make an audible groan. It was indescribably shocking. As they waited for the police and an ambulance, they spotted some hypodermic needles spilling out of the upturned trash can. They didn't want the police thinking that they were drug users, so they quietly disposed of them.[29]

Greg Ganci left to find Candy, who he assumed would be in a house two miles away. She and Jack were staying there with Jack's mother after moving out of the coach house. Candy wasn't there when Ganci called. He had no option but to break the news to Jack's mother that her son was seriously, perhaps fatally, injured. He promised to return when he heard more, leaving her in the company of her friend Helen Rowan. Candy, meanwhile, had arrived at the coach house, which was now swarming with reporters and photographers.

By the time she made it to Huntingdon Memorial Hospital, Jack was dead.

In the aftermath of Ganci's visit, Ruth Parsons drank steadily. She was in a manic state. Her friend Helen, who was confined to a wheelchair with arthritis, tried to calm her. She called the doctor, who prescribed Nembutal, delivered by another elderly friend. No sooner had Ruth taken the prescribed two pills when she screamed 'I'm going to kill myself! I can't stand it!' and swallowed the bottle's remaining contents.[30] Helen Rowan, unable to get up, watched helplessly. When the doctor arrived, Ruth Parsons was dead too, just hours after her son.

The next day's tabloid *The Mirror* ran the full-page headline 'WEIRD BLAST IN PASADENA: Scientist Killed, Mother Suicides', printing a photograph not only of the decimated coach house but also

'SINISTER MAN' CROWLEY AND HIGH PRIEST OF CULT

John W., Parsons (above) Pasadena rocket scientist killed in mysterious blast Tuesday, today has been uncovered as high priest of a weird love cult. He headed the strange black magic sect founded by British "Sinister Man" Aleister Crowley (left). (See story, Page 3.)

of Ruth Parsons, slumped in a floral armchair, her body guarded by the family dog.[31] Of course, the story was a gift for the papers. Photos revealed a box found at the scene of the explosion inscribed with mysterious cabalistic symbols. In the days that followed, the public were titillated with fresh details: 'SLAIN SCIENTIST WAS PRIEST IN WEIRD SEX CULT'.[32] And then, inevitably, they found the connection to Crowley, a 'British witch doctor'. They printed Parsons' 'I height Don Quixote' poem, and were given access to old police files that included the accusing letter from 'a real soldier'. The indeterminate cause of the explosion – misfortune, murder or suicide – gave the story another edge, though the settled explanation has become that the coffee can just slipped out of Jack's hand. George Frey didn't believe this; neither did Jane Wolfe.[33] Wilfred Smith, his feelings still raw, wrote to Germer that Jack 'took the easy way out'.[34] Helen Parsons Smith's interpretation – that 'The Gods stepped in before he met a worse fate' – was slightly more generous, though not to Candy, whose company she deemed more ghastly than the grave.[35]

TAKING THE FIFTH

MORE THAN A DECADE had passed since Malina and Parsons had last been close, but Jack's death still came as a shock. Frank looked back to their early partnership with a degree of nostalgia, as being the basis of his most important work. Yes, Jack had been difficult – Jack had *always* been difficult – but his passing made Frank more aware of the singularity of what they had achieved together. This conviction found expression in bits of writing he had started in an attempt to chronicle the history of their jet propulsion success.[1] With the field moving so quickly in his absence, he thought it was important to secure their contribution and that of von Kármán. Even now, the Boss and his prodigy remained exceptionally close.

Frank's life in Paris was quiet considering the chaos that was unfolding among his friends. Admittedly, there were a few tensions at UNESCO, particularly with his new boss, Pierre Auger, a French physicist who replaced Joseph Needham as Head of Natural Sciences. But Frank also had a sense that his horizons were widening, a situation no doubt helped by Aerojet delivering a massive $3,900 dividend. 'I keep telling Margie that I want to retire,' he wrote to his parents, 'but she thinks I should still earn a salary.'[2] That was hardly surprising given that they were soon expecting their first child and had hopes of buying a house in Paris. The baby, a boy they named

Roger, was a delight. Frank recorded the progress of his newborn son with laboratory precision, carefully placing the infant on the scales before and after feeding ('there always seems a diaper on or off so that one doesn't know what one is weighing').[3]

In the midst of this domestic contentment, Frank received a letter from the FBI requesting him to report for an interview at the American consulate. He had little idea about the extent to which bureau agents had been following him for years. Jack, it turned out, was the least of his troubles. Frank didn't know, for instance, that FBI attention seemed to start back in 1942, when Aerojet secured a top-secret contract from the US Army Air Forces for an ultra-high-speed rocket-propelled interceptor.[4] Nor did Frank know that there were FBI informants embedded in his department at UNESCO, or keeping track of his visit to the World Youth Festival in Prague. He was clueless that, when he and Marjorie visited Brenham in May 1949, the local sheriffs – family friends – were observing them and reporting to the FBI. Deputy Sheriff Hugh Burk, who lived just along South Baylor Street from the Malinas, told agents that Frank had 'grown a long "House of David" beard' that had 'created quite a bit of comment everywhere he went in the town'.[5] Sheriff Tieman Dippel, who went to school with Frank, accused him of 'look[ing] like a Bolshevik'.[6] Dippel had 'no specific information concerning the ideals or beliefs of FRANK MALINA, JR. but stated that he was considered as a rather radical type person during his early childhood' (the fact that 'he is wearing such a long beard at the present time is in itself, according to DIPPEL, some evidence that he has "screwball ideas"'). So when Frank was invited to interview with the FBI, they had been watching him closely for eight years. The FBI ostensibly wanted to see whether he might testify against Tsien. The files, however, give the impression that the agents were just curious to see what he might do or say.

Most of the details of the interview on 1 November 1950 are still redacted; those that remain make it clear that it did not go well for the FBI. Perhaps Frank had some inkling of trouble. In any case, he took the Fifth Amendment and refused to make any statement about Tsien on the grounds of potential self-incrimination.[7] And that was that. Frank returned to his Arid Zone work at UNESCO, apparently out of harm's way.

The punishment came just a few months later. Frank had by now realised that travelling back to the United States might not be sensible. It was awkward when the birth of Roger prompted Frank's father, Frank Malina Sr, to suggest that they might all return to be closer to their kinfolk. Frank put the issue in political terms. 'If we returned to the U.S. now I do not see what good I could do,' he explained. 'I have no intention of starting rocket research again, especially for war purposes ... I would end up in the army or in jail, and I don't see too much purpose in those two possibilities.'[8] But the question of travel was soon settled for him when his application for a passport renewal was refused. Frank's 'further travel', the letter said, 'would be contrary to the best interests of the United States'.[9]

Marjorie, in particular, was shocked. When the news came through she and Frank went for a drive to talk it over. Frank worried that Marjorie might leave him, though that was the last thing on her mind.[10] He could still travel on his UN laissez-passer, but that would only last for as long as he stayed at UNESCO.

THE FBI WERE NOW going after everyone from Unit 122. The spadework in the Weinbaum investigation had given them a head start with Martin Summerfield, who, having recently moved to Princeton, was being pursued by the Newark FBI office. Their plan was to confront Summerfield with the combined testimonies of Jacob Dubnoff, Gus Albrecht, Richard Rosanoff and Sylvan Rubin – all former members who were notionally cooperating with the FBI.[11] Dubnoff, however, was proving tricky. Having been compelled to testify in the Weinbaum trial, he tried his best to fudge any useful detail regarding Summerfield. Yes, he'd seen him at meetings of Unit 122, and thought he dropped out sometime in 1939 or 1940, but he couldn't be certain that Summerfield had been a member. How, agents wondered, could Dubnoff know that Summerfield had 'dropped out' yet be unsure if he had even been a member? Dubnoff insisted that the only time he could make an affirmative statement about someone else's membership was if he could see the card for himself.[12] He refused to make any kind of signed statement. His wife Belle took the same line. Their tactic frustrated the FBI and certainly shielded Summerfield: 'even though they are cooperative,' agents explained to Hoover, 'it is often impossible for them to be specific as to the membership of other individuals ... apparently ... [they had] completely forgotten'.[13]

The FBI could at least depend on the ex-LAPD Red Squad Captain William Hynes and his undercover sidekick, William Ward Kimple. According to their party records from 1938, Summerfield's party name was probably 'Fred Kane'. But this didn't get them very far.[14] In fact, building the case against Summerfield was an uphill struggle. The FBI still maintained a mail intercept, but their technical surveillance back in 1949 revealed nothing of any use.[15] By April 1951, the FBI was dismayed to realise that the central plank of the evidence against Weinbaum, the personnel security questionnaire (PSQ), would apparently not work with Summerfield. The Atomic Energy Commission PSQ that Summerfield had signed in April 1947 was of an old kind, the wording of which would not permit prosecution for fraud against the government for not listing party membership. In many ways, Summerfield got off lightly, but revoking his security clearance still made it almost impossible for him to continue his existing astronautics research at Princeton.[16]

Tsien wasn't so lucky. The good news was that US Attorney Ernest Tolin (Weinbaum's prosecutor) told the press that he believed Tsien had acted in 'good faith' in packing his belongings, and that they would not be seeking a charge under the Export Control Act.[17] The bad news, however, was that the INS wanted to deport him for contravening the new Internal Security Act, a freshly minted anti-communist dragnet associated with its principal sponsor, Democratic senator Pat McCarran. Among other things, it made legislative provision to place communists in concentration camps.[18] Truman had vetoed the McCarran Bill but Congress enacted it anyway just weeks earlier. In a strange way, the government's contradictory impulse to both deport and detain Tsien paralleled his own confusion about whether he wanted to stay or go. Now, however, the decision had been taken from him.

The deportation hearings opened on 15 November 1950 in an unremarkable government building in downtown Los Angeles. The room on the first floor was tiny, perhaps ten feet by twenty, with cold green walls, drab brown lino and drawn venetian blinds.[19] It didn't look like a court, and it didn't have many of its legal protections, but the cast of witnesses and the slant of the enquiry gave proceedings more than a passing resemblance to the Weinbaum trial. There was the obligatory appearance from the LAPD Red Squad's professional witnesses, William Hynes and William Ward Kimple. There were the questions about Unit 122 meetings, particularly those in the homes of Malina and Weinbaum. Tsien talked about the animated discussions they used to have, and conceded that, yes, they *could* have been communist meetings but that he was in no way a communist himself. In other words, he didn't *knowingly* attend a meeting of the Communist Party. Weinbaum's hostile witnesses, Richard Rosanoff and Jacob Dubnoff, were once again wheeled out for Tsien. In a later session, they showed Tsien a photograph of another witness they hoped would testify against him – former Unit 122 chemist Richard N. Lewis. Did he recognise this person? Tsien said he didn't, though perhaps the face was vaguely familiar.

By this time, Richard Lewis was at the University of Delaware teaching chemistry on a limited contract, but with every reason to think that it might be renewed. He was understandably wary when INS agents first approached him about Tsien at the beginning of October 1950. Lewis's views about the Communist Party had changed over the years. It wasn't that he felt ashamed of his former association, but with a diminished loyalty to the party, he wasn't going to lie to protect it. As with many in Unit 122, the big change had come with Pearl Harbor back in 1941. 'In my own mind,' recalled Lewis, 'that was the time I left the Party.'[20] By the end of the war, he felt 'out of sympathy with ... [the] objectives [of the USSR] ... which then seemed to be unlimited extension of Soviet domination by undemocratic means'.[21] Yes, Lewis said privately to the INS agents, he had been a member of the Communist Party, and yes, he had known Tsien at Caltech. But that was all he was willing to say.

On 29 January 1951, when called to testify at Tsien's ongoing hearings, Lewis refused. He had no intention of playing the friendly witness, so he declined to answer certain questions on the grounds of self-incrimination. Of course, there was no way that taking the Fifth would go unpunished. The retribution came three weeks later in the form of an indictment on a classic FBI fallback charge: 'concealing a material fact and making a false statement' by not listing his party membership on a personnel security questionnaire, in this case for a job at General Electric in 1948. Lewis was now facing unemployment with a family of two small children to support and, if convicted, a possible jail sentence of ten years. He urgently needed legal representation, but in the hysterical climate of 1951 it was difficult to find a lawyer who would risk the reputational contamination of a red client. Lewis didn't want Tsien to be deported, but he also didn't want to end up in prison like Sidney Weinbaum. On 9 March, he decided to cooperate. He signed individual statements concerning the membership of no fewer than twenty-two of his former Pasadena comrades.[22]

One of them was Frank Malina. Another was Liljan, who was now living a very different life as a young mother in Glen Cove, New York, married to Lester Wunderman, an advertising executive at Maxwell Sackheim and Company. 'I've never been able to understand,' she later wrote in her diary, 'why I was never interrogated ... why it was not until 1951 ... that the FBI caught up with me in Glen Cove'.[23] There had been times over the years when she felt she was being followed, as when an appointment book had gone missing from her car, but mostly the FBI left her alone. And that's because until the confession of Richard Lewis, none of Unit 122's new friendly witnesses had ever told the FBI that Liljan had been a member. She had, in fact, stayed in the party longer than any of them, dropping out only in 1948 (the New York members were 'quite another breed of political animals from the genteel, intelligent people I had known in Pasadena').[24]

Soon after Lewis's statement, Liljan watched a black car full of FBI agents arrive unannounced at her house on Buckeye Road. It's not clear that four agents were strictly necessary to take details from a mother at home with her baby son, but they doubtless wanted to make an impression. They asked her about Malina and Tsien but Liljan told them nothing. She didn't deny that she had herself been a member but she didn't admit it either, and said nothing about anyone else. That's not good enough, said the agents, you either cooperate with us now or we're going to visit your husband at his office. Oh, and we'll tell him the names of all your lovers. 'They *stormed* in there,' Liljan recalled. But Lester played it cool: he told the agents that he already knew everything about his wife. They were just quietly living out their lives in Glen Cove.[25]

The cooperation of Richard Lewis had a much greater impact on Tsien. At a further session of the INS deportation hearing on 11 April 1951, Lewis explained that during the three and a half years when he had been a member of Unit 122, he had attended meetings about once a week, and had seen Tsien 'at the majority of these meetings'.[26] Two weeks later, the INS determined that Tsien was an alien who had been a member of the Communist Party. He should, as a consequence, be deported. The State Department, meanwhile, maintained their prohibition on his departure. No arm of government seemed able to break the stalemate. Tsien just had to stew in his own private confinement, unable to cross the county line out of Los Angeles until such time as his technical knowledge became a little less cutting-edge and he could be permitted to return to China.

TSIEN AND SUMMERFIELD were high-value targets, but Frank Malina was the real prize. The pressure came from the top. In February 1952, J. Edgar Hoover himself asked the Department of Justice to furnish an opinion about whether Malina could be prosecuted for violating statutes about 'making a false claim' – concealing membership on an old PSQ.[27] It took six months for the Assistant US Attorney in LA, Angus D. McEachen, to conclude that it wouldn't result in a conviction. First, he didn't believe that the wording of the PSQ strictly required Malina to list his party membership; second, he was doubtful that William Ward Kimple's copied party records would convince a jury; and third, he saw no witnesses who could authoritatively testify to Malina's membership (odd, given Richard Lewis's categoric statement that Malina *was* a member).[28] One other thing helped Frank: Jack was dead. The FBI were deeply frustrated that a potential witness, even one as unreliable as Parsons, had been taken out of the equation.[29] Any statements that Parsons had given to the FBI over the years would now be deemed inadmissible.

In the absence of Parsons, the FBI sought out his old friend Ed Forman, who was now living in Phoenix, Arizona and was more than willing to take up where Jack had left off. Forman relished the opportunity to undermine his old colleague; he and Malina were 'natural born enemies' he told the agents. He talked of the 'bull sessions' with Malina's left-wing friends at GALCIT, and about 'the many occasions [in which] they attempted to indoctrinate him'.[30] He described Frank and Liljan's house as being littered with 'subversive' pamphlets and texts, and of an 'indoctrination meeting' at which Forman had finally '"sounded off" … indicating his distaste … for any Communist ideology'. Forman's fulsome cooperation extended beyond naming the names of those who attended this meeting; he also accused Frank of 'hold[ing] up the progress of jet rocket propulsion units by at least two years … by virtue of inserting numerous technical objections and requesting considerable unnecessary analyses'.[31] Forman left agents with the impression that, while communism was bad, mathematics was worse.

J. Edgar Hoover was furious that Assistant Attorney General McEachen didn't share his own zeal in fighting the communists. He wrote to the Los Angeles FBI office to say that McEachen's delinquency would not do, and they should demand that he reconsider his opinion – which he did, this time agreeing to secure an indictment against Malina.[32] Then there was another problem – the statute of limitations on 'making a false claim' was due to expire on 31 December 1952. A federal grand jury returned a true bill against Malina on 17 December and Judge William C. Mathes issued a bench warrant for his arrest, with a bond set at $10,000.[33] Frank Malina, the person who had arguably done more than anyone to initiate the US space programme, was now formally considered a fugitive.

In order to execute the warrant for his arrest, Washington wanted to force Malina back to the US (extradition would be ideal, but the State Department did not want to ask if there was a possibility that France might decline). To that end, the State Department pressurised UNESCO to abolish Malina's position, even going so far as to say that the continued employment of Malina, as well as a few other American nationals suspected of subversive ties, 'jeopardizes US participation in UNESCO'.[34] Director General Jaime Torres Bodet, the replacement for Julian Huxley, resisted the onslaught, but two weeks later he suddenly resigned in protest at a proposed cut to the UNESCO budget. No one at the top could protect Malina now.

Back in the US, the Senate's McCarran Committee was investigating any American UN employee who might be suspected of subversive associations – dozens were hauled before the committee – but even this process, which cut to the heart of questions of national versus supranational sovereignty, was overtaken by another executive order from Truman. Malina may have dodged the worst of EO9835 that besieged his friends, but EO10422, issued on 9 January 1953, caught him broadside. It authorised the creation of the International Organisation Employee Loyalty Board, requiring a full field investigation of all existing employees, who would now be expected to fill out a new and extensive PSQ, complete with an elaborate fingerprint chart. Finally, Frank Malina had to face The Question: 'ARE YOU NOW OR HAVE YOU EVER BEEN A MEMBER OF THE COMMUNIST PARTY...?' On 6 February, Frank received a memo from Charles Thomson, US representative to UNESCO, asking that four completed copies of the PSQ be submitted on Friday 13 February.

Frank took a day to think about it and then tendered his resignation.

IT WAS NOW THREE YEARS since Louis Dunn's allegations to the FBI about a potential spy ring at JPL, and the FBI had chalked up a list of wins. Malina lost his passport and his UNESCO job, Summerfield his security clearance, Tsien was under house arrest and Weinbaum was in prison. And that was just the JPL crowd. Other scientists of Unit 122 hadn't come out particularly well: Eugene Brunner had been jailed, released and blacklisted; Andrew Fejer had clearance difficulties. Cooperating with the FBI hadn't helped Frank Oppenheimer, and although Richard Lewis was eventually acquitted by a jury for PSQ fraud, he never worked in academia again. Jacob Dubnoff and

Gus Albrecht found that their willingness to testify didn't help their careers, but it certainly lost them their friends.

The punishment of Sidney Weinbaum, however, was of an altogether different kind. He was taken a thousand miles away from Lina and Selina, to McNeil Island Penitentiary in Puget Sound, a notorious high-security facility for America's most prized undesirables and J. Edgar Hoover's public enemies. Conditions inside McNeil were grim. Sidney initially shared a cell with nine other inmates, with 'neither privacy nor possibility for concentration,' he told Linus Pauling.[35] His old friend wrote when he could but never quite knew what to say apart from to complain about his workload and give updates on problems of molecular structure. Sidney tried to be gracious in reply, but it wasn't easy now that prison wardens refused him access to biochemistry books.

Sidney must have wondered why, among all his friends, he was the one in prison. 'It is not the injustice … that bothers me, but the inanity, the stupidity … of putting me … in prison, while all the crooks and racketeers are … amusing themselves in the free world!'[36] It seemed arbitrary. There was a sense in which he was there as a kind of substitutionary atonement for the others who were out of reach. Yet the price paid by Sidney was more than shared by Lina, even if her punishment made no newspaper headlines. For one thing, it was almost impossible for her to visit him. Even if she could scrape together the money to travel, visitors were only permitted via a special government boat on alternate Sundays. Lina found that her life had become quiet without being restful, its pace stilled by a new social isolation. The world of 1950s McCarthyism was, as the historian Ellen Schrecker has put it, 'less a world of fear than of silence'.[37] It was lonely.

With few friends and no money, Lina subsisted on horse meat and brown rice ('a healthy diet,' she said to Frank, trying to sound upbeat).[38] Sidney's impending parole board hearing in December 1951 was a particular strain, as if the hope of an early release revealed anew the unfairness of his original imprisonment. He tried to reassure Lina that everything was okay. 'Did [the judge] … shake my belief in humanity?' he asked her. 'No. Did he take you away from me? No. Did he break me, did he actually make a felon of me? No.'[39]

In February 1952, they heard the board's decision: parole denied. 'I might have thought I was here for perjury,' Sidney wrote to Pauling, 'but I was frankly told a few days ago that my "philosophy" ... is my crime.'[40] His philosophy was not a matter of historical materialism or dialectical reasoning; it was his refusal to name names. Lina didn't try to conceal her rage at the penal system's 'curdled milk of human kindness'.[41] It felt personal. 'Ah, if Sidney were a "good" boy and would "cooperate" with them – that would be different,' she explained to Frank.[42] That wasn't the end of it. Sidney had applied for a transfer to Tule Lake, which would be slightly more accessible to Lina and Selina: this, too, was denied after an intervention from Washington. He was kept at McNeil, made to strip cable for wire salvage or nail boxes in the box factory, serving his sentence alongside the recently convicted Los Angeles mobster Mickey Cohen.

The strain of the parole refusal took a particular toll on Lina. She visited Sidney at the end of August, and though he did not notice any changes in her, he could tell from an undertone in subsequent letters that something was wrong. For two weeks he received no word at all and then was called into the prison office and handed a letter that came from Selina. He knew, without reading, that the crisis had come.[43] The details are hazy but clearly Lina had had some kind of breakdown that forced Selina to admit her to the Department of Mental Hygiene at Stockton State Hospital.

Even before her parents and grandparents were all killed by Nazis, before her husband was imprisoned as a perjurer, Lina had a history of emotional instability. But the circumstances leading up to Sidney's arrest, particularly the FBI's technical surveillance, might have tested the sanity of anyone. Designed to eavesdrop on communists with double lives and split allegiances, the bug and the informant brought about precisely what they purported to find: disturbed and fractured selves. Lina had often been oppressed by the feeling that one or more of her friends had informed on them. They're 'seeing spooks at every turn,' said one unnamed friend to the FBI, who both boasted that Lina trusted her completely and still insisted that 'Mrs Weinbaum's allegations with respect to tampering with her mail, etc. are based purely on suspicions arising from her very disturbed state of mind'.[44]

It wasn't easy to keep a firm hold on reality with friends who could betray her to the FBI yet in the same breath insist she was paranoid.

As the crisis unfolded in the autumn of 1952, Sidney felt utterly helpless. He had received only one letter from Lina in Stockton. It made no acknowledgement that she was unwell, but claimed instead that she was the victim of circumstances. 'I have … learned to smile at my own misfortunes and to face the future with confidence,' Sidney wrote to Pauling, 'but this latest blow has undermined my morale considerably. As a citizen and as a human being, I am more indignant than ever about being here.'[45]

For the next year, there is a gap in our knowledge of what happened to the Weinbaums. Lina may not have been able to write, or maybe her letters have not survived. But it's apparent from her later correspondence that she was never quite the same. Even twenty years after this date, her letters are rambling and accusatory, distressed and disjointed. She wrote frequently and erratically about nuclear testing and atomic fallout – a common enough anxiety at the time – but in her case, one that was further contaminated by other insidious, unseen threats.

What we do know is that Stockton State Hospital in the 1950s was an institution pioneering new scientific techniques for the care and management of psychiatric patients. Electroconvulsive therapy (ECT), a new instrument developed by Caltech scientists, was extensively used.[46] Hundreds of patients were subjected to lobotomy; clitorodectomy was prescribed for women whose behaviours – indiscreet masturbation, for example – were deemed unacceptable.[47] It's not known what happened to Lina Weinbaum other than that she was subject to both the horrors of emotional breakdown and the vagaries of its treatment. From later letters, it's evident that she felt that Sidney's sentence was being applied to her, that she considered her confinement to be political and that it was somehow all Sidney's fault. She grew suspicious of scientists, chess players, instruments and machines.

Inside McNeil, Sidney counted down the days. He played the piano in the prison chapel, taught basic inferential statistics to other inmates (mobsters involved in gambling were his most attentive students) and advocated for prisoner welfare. The parole board eventually released him, three years into his four-year sentence.

He finally left McNeil Island in September 1953, at the end of that 'queer, sultry summer,' as Sylvia Plath put it in the opening of *The Bell Jar*, 'the summer they electrocuted the Rosenbergs'.

THE ROSENBERG CONNECTION

AT NEARLY 70, Theodore von Kármán was slowing up a little, but he had no plans for retirement. In fact, he was keen for a career change. To that end, he proposed an Allied scientific advisory group that would eventually become NATO's Advisory Group for Aerospace Research and Development (AGARD), based in France.[1] The Boss and his sister Pipö had always been big Francophiles; indeed it was in their company at a Bastille Day ball in Los Angeles that Frank had first met Liljan. The development of AGARD, alongside his continuing Washington role, allowed von Kármán to spend more of his time in Paris, which he came to regard as a kind of home, as well as with the Malinas, whom he treated as family. Whenever von Kármán was in France, he and Frank would meet at least once a week at his Paris hotel suite, where they'd chat and play chess late into the night.[2]

After one return to Paris in October 1950, Frank noticed that the Boss seemed particularly exhausted. His work at the Pentagon was certainly demanding. 'His last trip to Washington really wore him out and raised his blood pressure,' Frank remarked to his parents.[3] Perhaps no one other than Frank had an inkling that this anxiety was about more than the stress of setting up AGARD. That summer the FBI had been making some significant inroads into breaking up Soviet espionage in America. Von Kármán always seemed above the fray when it came to political involvement. He had never even attended a Unit 122 meeting. As a long-trusted advisor to the Army Air Corps, he occupied a stratospheric level of military seniority, well above the usual range of intelligence and counter-intelligence operations.

But not long before the Boss left the US in October, the Army CIC had sent through a memo to the FBI entitled 'Possible Communist, California Institute of Technology'.[4] Its concern was probably triggered by the Weinbaum investigation, but this time it didn't refer to any of the usual Unit 122 suspects. The CIC had recently been speaking to Professor Zoltan Bay, a Hungarian physicist now at George Washington University, who told them that Theodore von Kármán had been a member of the Hungarian Communist Party in 1918 and, on moving to Aachen, had transferred his membership to the German Communist Party.[5] This revelation should hardly have come as a shock, but it did. It was a matter of public record that von Kármán had held a ministerial position as Undersecretary for Universities in the short-lived communist government of Béla Kun.[6] In fact, he had even served under the Leninist philosopher György Lukács, one of the most important figures in Western Marxism and at that time the Commissar of Education. But it seems as though von Kármán had been coy about mentioning his historic membership when first emigrating to the United States. This left the FBI wondering about the politics and loyalty of von Kármán now.

That same month, the FBI discovered Jack Parsons' attempt to pass classified documents to the American Technion Society, a deal set up in part by von Kármán. The coincidence of these discoveries was bad news for the Boss: the FBI initiated a routine security investigation 'to determine whether this individual is a threat to the internal security of our country'.[7] But the more they looked, the more they saw von Kármán's name tied to other open investigations: Parsons, Tsien, Malina and Summerfield. And even these unfolding concerns turned out to be minor compared to a new lead that was emerging.

THE SUMMER OF 1950, the summer of Weinbaum's arrest and trial, had seen a run of counter-intelligence victories for the FBI. It had started earlier that year when Klaus Fuchs, a senior physicist working on the British atomic bomb project, confessed that he had been a Soviet spy when working on the Manhattan Project in 1944 and 1945. His confession led the FBI first to Harry Gold, Fuchs' primary go-between with the KGB, who in turn pointed them towards David Greenglass, a skilled machinist who had worked in a Los Alamos laboratory making components for the implosion-design plutonium bomb. For the FBI, these revelations were like dominoes: line them up right, keep them falling and the whole Soviet intelligence network could be dismantled. Greenglass fell hard at 1.40 a.m. on 16 June, confessing his own role and carelessly implicating his wife Ruth, as well as his sister Ethel Rosenberg and brother-in-law Julius Rosenberg.[8]

Julius and Ethel Rosenberg were soon arrested, but they did not confess. Today, they are known for being the most infamous spies in American history, a notoriety that stems from their conviction for transferring nuclear secrets, for their 1953 execution in Sing Sing, and for being the object of liberal hand-wringing about the apparent injustices of anti-communism. There's no longer any doubt about the guilt of Julius Rosenberg – Ethel's case is more complicated – but their usual reputation as brokers of A-bomb secrets tends to overlook the fact the Julius was much more important to the Soviet KGB for his non-atomic intelligence.[9] He ran a network of agents embedded in US military science, many of whom he had known since his student days at New York's City College, or from the Steinmetz Club, the campus branch of the Young Communist League.

Only in recent years have historians understood that the network's greatest success was less the leaked component designs of the atomic device dropped on Nagasaki than the near-total access that one Rosenberg agent had to the world's pre-eminent aeronautics expert – Theodore von Kármán.

That agent was Bill Perl, born William Mutterperl, who worked as a research assistant to von Kármán at Columbia University between 1947 and 1948. A tall, sharply dressed aeronautical engineer, Perl enjoyed a similar relationship with the Boss in the late 1940s as Frank Malina had a decade earlier. He helped prepare lectures and provided technical support, while von Kármán gave doctoral supervision and mentoring. He'd drive von Kármán around in a dark-blue Plymouth sedan that really belonged to Pipö, but which the Boss would often lend to Perl for long periods.[10] Von Kármán's colleagues came to know Perl as the trusted assistant, and were used to him handling routine correspondence. All of this might have worked out very well were it not for the fact that Perl was the single most valuable asset in Rosenberg's network, an agent so committed and discreet that not even his own wife knew he was a spy.[11]

As is often the case with espionage, the placement of Bill Perl was a combination of luck and design. He worked at the National Advisory Committee for Aeronautics (NACA), the US government's primary research and development lab for aircraft and the precursor organisation to NASA. And it was at NACA, first at Langley Field, Virginia and then later at the Lewis Flight Propulsion Laboratory in Cleveland, that he started to plunder classified material from their extensive library. Perl even tried to get Rosenberg a job at NACA, though that didn't quite work out (it didn't need to; Perl was more than effective on his own).[12] He passed on details of ninety-eight technical studies, including the Westinghouse 19A, the first wholly American jet engine, and a 12,000-page blueprint for the Lockheed P-80 Shooting Star, the US Army Air Forces' first jet fighter, widely used in the Korean War.[13] The supply of documents was so continuous and so valuable that Julius Rosenberg had to make special arrangements for other agents in his network to assist with the Herculean effort of photography.

It wasn't always plain sailing for the Rosenberg ring. The 1945 defection of Elizabeth Bentley set them back, as did Rosenberg's discharge from the Army when they discovered his party membership. This meant that the network had to be put on ice for a few years: an unwelcome disruption. But it also allowed agents to maintain their covers until such time as it was safe to resume their mission. It was during this spy sabbatical that Perl, encouraged by his bosses at NACA, started work on transonic flow with von Kármán at Columbia's Pupin Physics Lab. The Boss no longer needed the burden of PhD students, but Perl was such an exceptional scientist – and came recommended by no less than the Nobel laureate Isidor Rabi – that in 1946 von Kármán took him on and made him his research assistant the following year.[14]

At Columbia, Perl more than lived up to his billing as a hard-working student. He saw von Kármán two or three times a week for an hour or two, spending the rest of his time in the lab or socialising with other engineering comrades from the Communist Party. The similarities with Malina's life in Pasadena are striking. The Rosenberg cell observed the same rules about not acknowledging its existence, about always denying one's membership and professing no knowledge regarding the membership of anyone else. As in Pasadena, meetings rotated around homes.[15] It's not clear whether members had party names per se – after all, this was not a regular unit – but as Soviet agents they obviously had code names. The six-foot three-inch Perl was called 'Gnome'.[16]

THERE WAS ONE DAY just before Frank Malina permanently left the United States in 1947, between his visits to Albert Einstein and Lyndon Johnson, when he met up with Perl and the Boss at von Kármán's suite at New York's Gotham Hotel.'[17] Frank found them working on von Kármán's presentation for the prestigious Wilbur Wright lecture.[18] This encounter is the only certain occasion on which these three met socially, but given that Perl spent a semester at Caltech in 1946 (around the time that Malina was finishing up on the couch), it seems likely that they were acquainted earlier.[19] Beyond that, there's no evidence of any link between Unit 122 and the Rosenberg network. Yet the superficial similarities between these groups is significant. From the outside, both are composed of clever, high-minded and progressive young people. Frank Malina and Sidney Weinbaum doubtless thought of the Communist Party as a force for good in the world, as did Rosenberg, Perl and their comrades. The reality of the party, however, was of a Janus-faced entity that used the cover of benign campus communism to operate intelligence networks that genuinely *were* subversive.

One year after this meeting with Malina, Bill Perl had finished his doctorate and come to the end of his contract as von Kármán's assistant. The Boss was based in Washington and New York but, as ever, was an inveterate traveller (in 1948, he was in Paris nine times, London three times, Los Angeles twice, plus Palo Alto, Wiesbaden, Zurich, Milan, Rome, Madrid and Lisbon).[20] During his absence it was often left to Perl to conduct his business, including corresponding on von Kármán's stationery and receiving classified material to file on his behalf.[21] Perl had his own security clearance at NACA. In fact, he not only had the keys to von Kármán's filing cabinet, he also had the combination code to his safe. So in 1948, when Rosenberg eventually got the nod from his Soviet handler to reactivate the spy network after a hiatus of three years, Perl's carefully cultivated placement set the scene for one of the most important feats of espionage in the twentieth century.

Sometime between 26 June and 9 July 1948, but probably over the Fourth of July holiday weekend, Perl let himself into von Kármán's office and checked out thirty-five test reports. In all likelihood, he also emptied von Kármán's safe.[22] The operation was meticulously planned. Perl might well have been seen by others in the Pupin Lab; no one remembered – it was just so ordinary for him to be there on von Kármán's business. He then lugged the material into von Kármán's Plymouth and drove down Manhattan to an apartment at 65 Morton Street.[23] There Julius Rosenberg, Morton Sobell, Perl and an unidentified fourth man worked through the weekend with two Leicas that were usually kept hidden underneath the floorboards.[24] It took the four men seventeen hours to record 1,885 pages of material on 35mm film. By the time they were finished, the box into which Rosenberg and Sobell packed the film cannisters was so heavy that it took both of them to lift it. Even in a pre-digital age, millions of dollars of research and years of concentrated aeronautical expertise could essentially fit into one box: blueprints, wind-tunnel data, comparative tests of different designs – it was a treasure trove of scientific labour.

Sobell and Rosenberg struggled with the box out of Greenwich Village and deposited it with other Soviet agents on a quiet railroad platform in Glen Cove, Long Island, not far from the Killenworth estate for USSR diplomats.[25] Perl, meanwhile, returned the original material to von Kármán's safe and went back to his job at NACA – a low-key ending to one of the most effective espionage operations ever mounted.[26] Its significance only became apparent years later, when Soviet MiG-15s could run rings around American propeller aircraft in Korea and, later, Vietnam – all thanks to a tail design obtained by Perl.[27]

In January 1950, however, the Rosenberg network ran out of luck. Klaus Fuchs confessed his role in atomic espionage to British MI5 agents, setting off a chain reaction of confession and revelation that led the FBI first to Harry Gold, and then to David Greenglass, who was arrested the same day as Sidney Weinbaum. Greenglass was brash and arrogant and soon spilled the beans about the Rosenbergs, eventually testifying in the trial for what Hoover called the 'crime of the century'. This testimony sent his own sister to the chair. The KGB did its best to protect Perl by sending $2,000 cash to flee the country, but he preferred to stay put and hoped to ride out the storm in the United States.[28] In hindsight, that was a mistake. When Perl was called before the Rosenberg Grand Jury in August 1950, he proved slippery on the subject of his relationship with Rosenberg and Sobell – denying that he knew them, claiming not to even recognise their photograph – but then vaguely conceding that he was acquainted with Sobell.[29]

All of this is why, in October 1950, Frank Malina was so concerned about Theodore von Kármán's raised blood pressure. Amid the blaze of publicity from the Rosenberg trial, there was something particularly stressful about von Kármán's position. And the Boss couldn't hide it: when Tsien was in the middle of his difficulties with the INS he asked von Kármán for a letter of support as a character witness, to the effect that Tsien had never been a communist. 'I'm sorry,' said the Boss, 'but this is not the right time – I'm in trouble myself. Find somebody else!'[30]

Von Kármán's situation was difficult because even the most generous and apparently credible interpretation of events was quite damning: that he had been oblivious to the political inclinations of his closest colleagues, both with Perl at Columbia, and also with Malina,

Summerfield and Tsien at Caltech. And such political naivety was somehow the case even though he had himself once been a member of the Communist Party. Then there was also his role in encouraging Parsons to cooperate with the American Technion Society. Assistant Attorney General James McInerney asked J. Edgar Hoover to clarify whether the cases of Perl and Parsons might be related.[31] The FBI thought not. They were right, of course. But it's at least a little surprising how quickly the bureau satisfied itself with von Kármán's claim that he was unaware of Perl's subversive activity and that 'he had no reason to question Perl's loyalty'.[32] Perl was arrested in March 1951, and in May he was convicted of two counts of perjury – for denying his 'acquaintance and association' with Sobell – a case that was easier to prosecute than espionage. His five-year prison sentence was markedly lighter than that of the Rosenbergs.

THE EXECUTION OF JULIUS AND ETHEL Rosenberg in June 1953 is one of the Cold War's most divisive episodes. Whatever we might say about the many people affected by anti-communism in the McCarthy era, Julius and Ethel Rosenberg were the only ones to actually lose their lives as a direct consequence. For many, they became national martyrs. Frank Malina told his parents that the execution was causing riots in Paris and universal ill will towards the US, adding that 'some think that McCarthy will be the American Hitler who will come to power if there should be another Depression'.[33]

In the late twentieth century, even American leftists who had long since given up on the USSR still believed in the Rosenbergs. Their co-defendant, Morton Sobell, spent eighteen years in prison for espionage, but on his release he joined the Rosenbergs' orphaned sons, Michael and Robert Meeropol (they took the name of their adoptive family, the communist songwriter Abel Meeropol), in campaigns to proclaim the injustice of the entire trial.

Yet the martyr mythology of the Rosenbergs has taken a beating over the years. There was a wave of fresh disclosure based on Freedom of Information requests in the 1980s, and a second wave in 1995 when the US government admitted the existence of intercepted Soviet cables – the Venona decrypts – that confirmed the existence and identities of the Rosenberg spy ring.[34] A weight of forensic historical scholarship has since been devoted to establishing their guilt. And then after nearly six decades of assiduously maintaining his innocence, Morton Sobell suddenly changed tack at the age of 91. In 2008, the *New York Times* journalist Sam Roberts asked him directly if Rosenberg was a spy. 'Yeah, yeah, yeah. Call it that. I never thought of it as that in those terms,' he shrugged.[35] Sobell also acknowledged his own role – 'I did it for the Soviet Union' – and that of Perl in the theft from von Kármán's safe.[36] It should be clear by now that the case is essentially closed.

There's one aspect of the story, however, that has been left relatively unexamined. It's an obscure detail that promises no great revelation for our understanding of the Rosenbergs. It amounts to nothing more than a dangling thread. But it's a thread that if pulled in a particular way, causes the story of Frank Malina and Theodore von Kármán, Sidney Weinbaum and Unit 122, JPL and its rockets to unravel a little. At the very least, the fabric of this story starts to look rather different.

The thread in question is an interview that the FBI agents conducted with Julius Rosenberg's brother-in-law, David Greenglass on 17 July 1950. By this stage, Greenglass was cooperating to save his wife Ruth from indictment, leaving the Rosenbergs to look out for themselves. He told the FBI everything he knew about his recruitment by Julius Rosenberg and the atomic information he passed on via Harry Gold to the Soviets. He testified that his sister Ethel had typed his notes – it wasn't true – but it was part of an immunity agreement that would protect Ruth.

Greenglass went on to divulge all the details he could remember about other potential members of the Rosenberg ring. For the most part, Julius had been careful to protect his sources, but Greenglass could remember him mentioning that one of his Soviet contacts had been a consultant on a dam project in Egypt. When Greenglass had accumulated a stack of unpaid bills in 1948 – insurance, income tax, Ruth had been in hospital – Julius had helped him out, saying that the consultant friend had agreed to tide him over.[37] The implication was that this consultant was someone who was sympathetic to

Rosenberg, aware of his Soviet work, and willing and able to offer clandestine financial support. This revelation immediately sent the FBI scurrying to the State Department to go through lists of American personnel who might have worked on such a project.

There were other details that Greenglass remembered about the consultant. He had the impression that he was a civil engineer, had a lot of money, that he might even have made $200 a day in consulting fees, that he was probably single, that he worked for the government but had special permission to do some consultancy on the side. He recalled that Julius had described this ally as a 'sophisticate'.[38] Greenglass also felt that Julius did not know the consultant through his usual networks of old City College contemporaries.

Nothing happened very quickly, but an ongoing investigation trickled on, gathering information on all potential American contractors who had worked on the Aswan low dam. By October 1952, the FBI had a formal suspect – a consultant who in 1947 was paid a lot of money by the Chemical Construction Company and by Hugh L. Cooper & Company to calculate the practicability of using rectangular outflows as part of a redesign for hydroelectric power. The suspect had worked closely with William Perl: he was Theodore von Kármán.[39]

THE IDENTITY OF THE 'unknown consultant' was – and indeed is – anything but straightforward. For one thing, David Greenglass may not have remembered correctly. We now know, of course, that Greenglass was a serial liar, whose deceit brought a death sentence on his own sister. But in this case, he had no particular reason to mislead – on the contrary, he had every incentive to cooperate – and much of what he told the FBI at this time aligned with other evidence. Another complication is that Greenglass's details about the Aswan consultant are not a perfect fit for any of the FBI's longer list of suspects. When the bureau re-interviewed David and Ruth Greenglass in 1952, their stories changed a little; for instance, Ruth rowed back from the claim that the consultant earned as much as $200 a day.[40] That was significant for the FBI's other prime suspect – the less well-off William Perl.

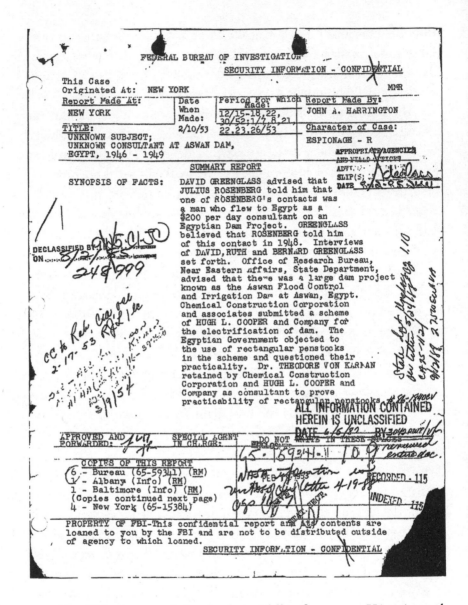

The FBI discovered that in the middle of 1947 von Kármán took on the Aswan consulting job for which he was paid $4,800. He in turn gave much of the calculative work – a theoretical analysis of a water-hammer problem in rectangular penstocks – to his assistant Perl, for which he paid him only $152.40.[41] The FBI eventually concluded from this evidence that the mystery consultant must be Perl. After all, he did do some consultancy work on the Aswan dam and, from what they could then tell, he was likely part of the Rosenberg network. Before they reached this conclusion, however, the pressure on von Kármán during 1952 was intense. In October, when the FBI activity on this case was at its peak, Frank again remarked that von Kármán was showing his age, and had 'an attack of hives on one of his legs'.[42] But by February 1953, it was over. Robert Lamphere, one of the FBI's key special agents, had decided that 'the most likely suspect is William Perl and that the case is now properly closed'.[43]

Yet there is a lot about this decision that doesn't quite make sense. Greenglass said that the consultant was single (von Kármán, not Perl), and that Rosenberg did not know the consultant through City College (which was exactly how he did know Perl). This man 'did not worry about money because he could also pick up money on the side as a consultant'; the implication at the beginning was that the consultant was quite wealthy – which hardly applied to Perl and his $152.40 fee but definitely did for von Kármán, sitting on a substantial consultancy income as well as his Aerojet stock. Greenglass reported that the man was a 'sophisticate'. Given that Perl grew up in the Bronx and came from exactly the same kind of Jewish immigrant community as Rosenberg and Greenglass, that seems like an unlikely descriptor. Von Kármán, on the other hand, came from a relatively wealthy Hungarian family, from a line of distinguished rabbis. His scientific and artistic circles were in every sense rarefied.

There are also aspects of Greenglass's testimony about the mystery consultant that don't fit von Kármán but do fit other potential suspects, such as the fact that the consultant was known to Rosenberg through the union FAECT. Russell McNutt, for instance, was a communist engineer who lent money to Rosenberg and was a member of FAECT.[44] (McNutt, however, was not a government employee and was working for a business in Venezuela the whole time that this was supposed to have taken place.) And by far the biggest problem for any retrospective indictment of von Kármán is that he is nowhere mentioned in the notebooks of Alexander Vassiliev, a former Russian KGB operative and historian who had unique access to the KGB archives that detail Rosenberg's other contacts.[45]

As with so much about the Rosenberg network, the Communist

Party and the murky world of Cold War espionage, this is not something that can be readily resolved. It seems unlikely that Theodore von Kármán gave support to the Rosenberg network. His correspondence to the unknown consultant can be put down to coincidence. It's strange, however, that so many of his doctoral students and postdocs in Pasadena should have been involved in the Communist Party (Malina, Summerfield, Tsien and probably Andrew Fejer). It's also hard to believe that von Kármán did not know about this: he introduced Weinbaum at parties as 'my friend dealing with chemistry and communism', and would attend concerts, organised by party front groups, raising money for the International Brigades going to Spain.[46] As someone who probably concealed his own historic party membership, von Kármán could hardly be unaware that such associations were significant for people involved in classified military research. And surely the biggest coincidence of all is that even after leaving Pasadena's political hothouse, he should acquire a new doctoral student and close assistant at Columbia, William Perl, who would turn out to be a Soviet agent. Unsurprisingly, von Kármán chose not to mention Perl in his 375-page autobiography.[47]

Theodore von Kármán may not be a household name today but it is forever attached to the established boundary between Earth's atmosphere and outer space – the Kármán line. It was von Kármán who realised that at altitudes of around 100 kilometres the forces due to orbital dynamics exceed those of aerodynamics: the atmosphere becomes too thin to support aeronautical flight and only centrifugal force prevails.[48] In practice, however, the Earth's atmosphere doesn't have a neat edge, it just becomes progressively thinner. The Kármán line is a conceptual threshold only: 100 kilometres is slightly arbitrary, as in fact a recent proposal suggests revising it down to 80 kilometres.[49]

Something similar could be said of von Kármán's political past and that of Unit 122. There are limits to what we can know about this history, and for practical purposes it may be necessary to assume that these engineers acted with integrity rather than with subterfuge (unless, as with Perl, there is substantial evidence to the contrary). Yet the closer we look, the less certain these distinctions become. It's possible that Theodore von Kármán was involved in espionage.

Probable? No. But did he turn a blind eye to it? Was he cavalier in disregarding the potential threat from campus communism among his own friends and students? There aren't easy answers here, but even raising these questions presents Unit 122 in a rather different light.

It may have been that Louis Dunn's concerns were well-founded – after all, the idea of communist espionage at JPL is hardly far-fetched. It's more plausible to think that his bigotry fuelled a frenzy of loyalty-testing that saw JPL scientists unfairly jailed, blacklisted and ostracised. The real headache would be if Dunn's accusations were motivated by political prejudice but then also turned out to be accurate. In the absence of evidence, however, that doesn't seem like a workable hypothesis. Even so, it's notable that whatever befell the rocketeers of Unit 122, they fared a good deal better than their Russian counterparts. The USSR's leading rocket engineer, Sergei Korolev, along with numerous colleagues, spent years in the Gulag. Mikhail Tukhachevskii, a prominent sponsor of Soviet rocketry, was shot in a military purge. Ivan Kleimenov, director of a rocket institute, was similarly executed; so too Georgii Langemak, a Russian pioneer of solid propellants. Unlike the Rosenbergs, they didn't work for a rival superpower. All of them were tortured.[50]

The challenge then is to make sense of how America's leftist rocketeers negotiated two contrasting realities: first, that campus fronts and discussion groups of like-minded colleagues really did provide cover for Soviet espionage networks in the service of Stalin's militarised expansionism; second, that the forces of anti-communism, often animated by racism and anti-Semitism, worked to protect structures of privilege and discrimination and ruined countless lives in the hunt for the domestic communist enemy. To invert the familiar metaphor for McCarthyism, the witch-hunt may have been delusional but at least some of the witches were real.

LIGHT ENGINEERING

BY THE 1950S, Frank's old friends and colleagues had come to regard him as a consummately nimble operator, someone with the instincts to anticipate the Red Scare and sidestep its consequences, someone who could outfox the FBI to live a good life. That wasn't the case at all. The forces of anti-communism were well ahead of him. He was just lucky.

Frank had ended up outside the United States for two reasons, neither of which were J. Edgar Hoover. He sought a fresh start after a broken heart, and he wanted the chance to turn swords into plough-shares by working at UNESCO. It was there that he met Marjorie. Together they forged a new life and family with Roger, and then, in May 1952, another son, Alan. A black cocker spaniel puppy called Rocket eventually completed the picture, snoozing at his master's feet as Frank worked on new paintings and sat out the Red Scare, in partial ignorance of the forces arraigned against him. But all of this was upended at the beginning of 1953, when Truman's execu-tive order forced Frank's resignation and put him in a seemingly impossible position. On the one hand, his residency in France was conditional on his UNESCO employment; on the other, a bench warrant and criminal prosecution awaited any potential return to the United States. He admitted to his parents that it 'does not seem wise to return immediately to the States, as ... some of the super-patriots might attack me and I see no reason of giving them the pleasure'.[1] There were a few grim weeks when the Malinas' horizon was clouded with the prospect of statelessness, unemployment and even prison. Then, in the middle of March 1953, came a turn of events so freakish and fortuitous that it changed their landscape forever.

In the months preceding Frank's departure from UNESCO he had been preoccupied by institutional matters, including the sudden resignation of Director General Jaime Torres Bodet. So he hadn't given much attention to what was going on at Aerojet. General Tire and Rubber, the company that Andy Haley had lined up to capitalise Aerojet back in 1943, had done rather well out of their initial investment. There was nothing like a darkening political climate at home and a hot war in Korea to boost the prospects for the missile business. General Tire's president, William O' Neil, a strong supporter of Senator Joseph McCarthy, had high hopes for Aerojet; he just didn't want the encumbrance of the company's founders as minority shareholders. He tried to buy them out with an offer of $350 a share but none were tempted. Then came the threat: he'd issue a dividend that would massively reduce the assets of the company and merge Aerojet with General Tire's other stock holdings at a very unfavourable ratio.[2] Von Kármán eventually accepted this offer for tax reasons, soon followed by Martin Summerfield and Fritz Zwicky, who also sold out after heavy pressure from the board.

Frank, too, was minded to sell, but the offer was so complicated that he struggled to understand the implications. And unlike the other founders, who were all in the US, Frank was difficult for General Tire to reach. One version of the story has it that O'Neil dispatched a representative to strong-arm Malina into selling, but that sunspots collapsed the French telephone system and the message to meet up couldn't get through.[3]

The whole thing came to a head the week that Stalin died. As the world was adjusting to the transition and what it meant, Frank was mulling over whether perhaps he should hold onto his Aerojet stock. A degree of sentiment may have been at work here, but his decision to keep the investment was not imbued with any great significance

– until, that is, the forces of predatory capitalism rewarded him with the most improbable wealth. Not only could he trouser the hefty dividend, he found himself sitting on a substantial holding of new Aerojet-General shares that soared in value, an episode summed up by a rueful von Kármán in a chapter of his memoir titled 'How I Lost "$12,000,000"'.[4]

Frank was rich overnight, yet the full extent of his windfall took years to unfold, as the Aerojet share price tracked the heightened tensions of the Cold War. His financial security was now founded on the military-industrial complex that he had disavowed upon leaving Caltech. This wasn't a contradiction he liked to talk about, but Andy Haley was less coy: 'any diminution in the "Cold War" would immediately be reflected in reduction in government contracts,' he warned Frank, 'which is really the only business we have'.[5]

Even after this bonanza, Frank didn't stop working. On the day after his resignation he started painting at 9 a.m., establishing a routine of creative work that he strictly observed (allowing him to heap scorn on artists who lacked this discipline).[6] The new financial security did take some of the pressure off. 'I am painting ... helping Doctor [von Kármán] a bit and enjoying family life,' he told Haley; that and keeping an eye on his Aerojet-General stock, which jumped from $20 in 1953, to $31 in 1954, $84 in March 1955 and $108 in April 1955.[7] He couldn't resist cashing in along the way, supporting his art and family into the foreseeable future.[8] 'As you no doubt know, the Aerojet shares have doubled again,' he boasted to von Kármán in 1957. 'Monte Carlo was never like this!'[9]

ONE IMMEDIATE BENEFIT of Frank's wealth was that it helped persuade the French government to give him continuing residency. But this didn't make a difference to his passport situation. Throughout the 1950s, he found himself stuck in France, unable to visit his ageing parents back in Texas or travel to England with Marjorie to see her ailing family in Yorkshire. Even in France he was vulnerable to the activities of the United States government behind the scenes. Their close friend Eric Roll, Deputy Head of the UK Delegation to NATO, was warned not to consort with Frank and Marjorie (Roll's wife Freda later admitted to Marjorie that her husband's recall to London was partly made to get them away from the Malinas).[10] A few other Paris friends dropped out, presumably under similar pressure.

Of all the members of Unit 122, Frank had been the least affected, yet the passport difficulty was a genuine hardship. The first practical attempt to overcome what he and Marjorie euphemistically referred to as 'The Problem' was to apply for residency in the United Kingdom. After all, Marjorie was a British citizen, and she knew of one girlfriend from her home village who had no trouble bringing her German husband home.

Frank was interviewed by the British authorities in March 1954, after which it all went quiet until August, when they heard the outcome: application denied. 'We cannot admit all the foreigners who would like to come here to live,' came the letter to Frank's British lawyer; the real difficulty, it acknowledged, was that 'Mr Malina does not possess a valid United States passport'. The Home Office were admirably blunt: 'we should only be creating difficulties for ourselves if we were to let Mr Malina come here without a passport'.[11] Neither Frank nor Marjorie could have known the irony of this rejection; that month, Prime Minister Winston Churchill's government was negotiating with the United States to acquire the very rocket that Frank had designed – the Corporal.

Churchill loved missiles, especially nuclear missiles. His 1952 Report of Defence Policy and Global Strategy made nuclear weapons the foundation of British defence.[12] The only problem was that Britain's home-grown rocket programme wasn't, at this stage, up to the job and Churchill didn't want to wait. So in 1954 he agreed a deal to buy 113 Corporal missiles plus launchers, ground handling, guidance and control equipment from the US.[13] Like many such deals with the Americans, the negotiations were delicate, and Churchill's government had no intention of jeopardising his much vaunted 'special relationship' by taking in a US fugitive.

The missile's architect may not have been welcome in Britain, but by the late 1950s his rocket was all but wearing a Union Jack. On breathless British Pathé newsreels it was deemed a harbinger of the Space Age. The Royal Artillery towed it around town centres as a recruitment tool. And at the very moment when the British state rejected Frank Malina's residency application, military planners were secretly sizing up Scotland's Outer Hebrides as a potential proving ground for his weapon.[14]

Soon enough the Corporal found its way to the heart of British popular culture. The toy manufacturers Dinky and Corgi rushed to produce die-cast replicas of the missile and its launcher to coincide with the first test firings in 1959. Dinky had a spring-loaded mechanism that could launch the Corporal across a child's bedroom, while Corgi offered a 'percussion warhead', sold separately, that could be 'easily and quickly fitted to your missile ... loaded with standard caps to give a really authentic explosion on impact'.[15] Parents would be left in little doubt that the toy was a correlate of Britain and America's new weapon of mass destruction.

All of this meant that the recently founded Campaign for Nuclear Disarmament (CND) would have its work cut out when Malina's little nuke was becoming fun for all the family – 76,000 of them from Corgi alone.[16] And that's not considering models emerging from US manufacturers, from Hawk's 'Glow-in-the-dark' Corporal ('sheds an eerie phosphorescent glow over the launch scene') to Monogram's entire US Missile Arsenal ('tell your dad about this sensational new hobby kit'), or more ephemeral versions such as Nabisco's 'First Family of the Nation's big Missiles' card series ('get the entire set by eating Nabisco Shredded Wheat regularly and trading with your friends').[17] Military hardware that lurked in a breakfast cereal or in a Christmas stocking was more than just a weapon; it turned the citizenry it purported to serve into a new kind of nuclear family. The vehicle for a twenty-kiloton warhead, more devastating than the one dropped on Hiroshima, became as ordinary and domestic as a Chevy or a Ford.

FRANK'S PASSPORT ISSUE TOOK years to sort out, which was an enduring puzzle to his lawyer, Andy Haley. The lawyer talked about Malina's 'Babylon[ian] captivity', yet always regarded it as an unaccountable bureaucratic error. Frank had never spelled it out to him. Haley hated communism ('Bolshevism is Ku Klux Klanism and Prohibition with a hell of a nastier background,' he wrote after one of their regular arguments).[18] He couldn't understand why Frank wasn't more forceful in denying the allegations.

All of this crystallised when the American Rocket Society, after heavy lobbying from Haley, decided to bestow a Fellowship on Malina (Louis Dunn notably declined to support the nomination).[19] The honour prompted a conservative journalist, Florence Fowler Lyons, to decry the recognition of this 'Castro-bearded' communist who had 'fled the country in 1947'.[20] Haley was outraged, and demanded that Frank file suit against both author and newspaper, requesting damages of $1 million.[21] Of course, Malina knew that there was no possibility of a libel action given that the substance of Lyons's complaint, his party membership, was essentially true (Lyons was a conservative activist who believed that UNESCO was using the public school system to brainwash American children with 'one world' communist propaganda). Frank acknowledged that 'some queer enemy of yours and mine is behind all this'; he simply ignored Haley's suggestion.[22] That was how he usually dealt with the allegation whenever it arose: he sidestepped and moved on.

The passport problem, however, put him in a situation where sidestepping wasn't possible. In June 1958, Malina heard news of another American artist who would transform his own life, as well as that of Linus Pauling, Paul Robeson and scores of others whose travel was deemed 'contrary to the best interests of the United States'. The painter Rockwell Kent sued Secretary of State John Foster Dulles, whose department had for many years made the issuance of passports an arm of US foreign policy.[23] The Supreme Court judgement in the Kent case was close (5–4), but on 16 June 1958 it established that 'the right to travel is a part of the "liberty" of which a citizen cannot be deprived without due process of law'.[24] After sixteen years of FBI surveillance and after more than seven without a US passport, the *Kent v. Dulles* decision meant that he could apply for a passport, confident that it couldn't legally be refused.

The problem was that Form DSP-11 still had The Question: 'Have you ever been a member of the Communist Party?' Frank asked Haley whether there was any alternative to 'the question we have for so long worried about'. But there was no way round it. Reading the accompanying instruction 'WRITE YES OR NO', Frank opted for the more cautious 'NOT TO MY KNOWLEDGE'. It was no less of a lie, but it was less of a demonstrable lie.

Even when Frank successfully reclaimed his passport, when the threat of consequences waned with the passing of the McCarthy era, it was many years before he could travel without worry. Whenever the Malinas visited Britain, for example, Marjorie and the boys got through fine but Frank was always separated and questioned. He was still granted entry but the special treatment made it clear that he remained on a blacklist.[25] And when Frank Malina Sr suffered a heart attack after Christmas 1958, Frank hesitated about returning to the US, despite his valid passport. His father died before he could get there.[26] It wasn't a captivity like Weinbaum's or even that of Tsien, but it was still distressing in its own way.

FRANK'S EXPERIENCE AT BRITISH ports was a bit like his experience with the American space and military establishment: he was recognised and eventually admitted, but not without a stamp of vague and lasting distrust. Of course, people gossiped. Decades later, Homer Joe Stewart didn't hide his misgivings about Frank when interviewed by a Caltech historian. 'There was official suspicion of something bad going on at Caltech,' he said of the WAC Corporal leak. 'I think Frank was hurt by it. Whether he deserved the hurt or not, I really don't know. That's a pretty awful feeling – that someone you know, a friend, may have done something that's really very bad.'[27]

Others took the view that Frank had been the innocent victim of an overzealous age, a position captured in a story told by Donald G. Agger, who had been a US representative to NATO in the 1950s.[28] On his first day back in Washington after a trip to Paris in 1963, Agger was interviewing a civilian employee of the Office of Naval Intelligence. In the course of this job interview, the candidate eyeballed a small package on Agger's desk addressed to Martin Summerfield. He noticed that the sender's name was Frank Malina. Perhaps the applicant had seen ONI files on Malina and Summerfield. In any case, his concern was enough that he interrupted Agger to warn him that 'the package might contain illicit material of some kind'. The package actually contained a packet of French butalbital suppositories that Martin had requested for Eileen Summerfield's regular migraines.[29] Knowing that Agger was travelling back to the US, Frank had asked him to mail it to the Summerfields from Washington to save on postage. Agger wearily explained all this, ripping open the paper to reveal the suppositories. The candidate wasn't hired.

What's interesting about this story is that Agger only told Frank many years later, and did so in order to bear witness to what he saw as unwarranted suspicion. 'I suppose that a case can be made for permitting sleeping dogs to continue sleeping,' Agger wrote. 'But I have also wondered whether our personal obligations to history do not require us to clarify what can be clarified.' It was an invitation, in other words, for Frank to open up about exactly what had happened to him, to write about his experience of McCarthyism just as he had written about the history of JPL. That was the last thing Frank wanted. He tried to draft a reply in longhand but the strain of his dissembling filled the page with vigorous crossings out and redrafting.[30]

Even as the spectre of McCarthysim receded, and Frank reemerged into tentative respectability in the 1960s and 1970s, this dilemma remained. How could he explain what had happened to him without detailing his past party membership? How could he explain the history of his rocketry work without mentioning its political context? And what was he to do with his voluminous archive of correspondence given how entangled his rocketry achievements were with references to his politics? At some point Frank or Marjorie did a little light engineering of this archive. Crucial letters that detailed his involvement with the party were removed, leaving interrupted threads of dialogue. All letters with the Weinbaums, for instance, have been destroyed up until 1946; and crucial letters with his parents when he joined the party have been weeded out. When arrangements were made to donate the collection to the Library of Congress, the Caltech archivist Judith Goodstein, who catalogued the material, reportedly told Tsien's biographer, Iris Chang, that 'all the sensitive material about Frank's early days as a member of the Communist Party ... had been *carefully removed* from the collection'.[31]

When asked about his political history, Frank chose evasion at every turn. He'd steer interviews away from politics. Even to friends like Agger he'd reply, 'there is no painful chapter for me to present … whatever speculations you have heard … are simply nonsense'.[32] Later in life, however, he had to contend with the fact that other writers would not be so careful. When an engineering colleague of Summerfield's at Princeton, Jerry Grey, publicly attributed Frank's security troubles to having been 'active in Communist activities', Malina drafted a furious letter accusing him of 'irresponsible and wanton lies', signing off 'your former friend and admirer'.[33] Frank didn't send the letter, of course. It wasn't a conversation he wanted to start.

As the Space Age took hold, first with Sputnik, then with the US satellite Explorer 1, all the way through to Apollo 11, Frank published several 'memoir' papers detailing his experiments in the Arroyo, his theory of long-duration burning in solid propellants, the discovery of hypergolic propellants, the success of JATOs, the WAC Corporal and the Bumper WAC Corporal. He took particular care to give credit to Jack Parsons, sometimes as a proxy for credit that more plausibly belonged to himself.

Malina's almost obsessive documentation of astronautical history was partly a defence against the emerging German-inflected accounts of American space exploration. It was irritating to see so much veneration given to unrepentant Nazis, two of whom – Wernher von Braun and Walter Dornberger – were busy writing their own version of space history for *Technology and Culture*, the academic journal of the Society for the History of Technology (SHOT).[34] NASA's first in-house historian, Eugene M. Emme, would regularly receive missives from Frank complaining that NASA or the Smithsonian had marginalised JPL. Indeed, there were times when he even blamed Emme for 'the suppression of the role of rocket research at Caltech', an omission, he said, that 'is incomprehensible to me'. Apparently 'only Goddard (the most famous unsuccessful inventor in America as someone unkindly called him) and the German V-2 made possible the evolution of space exploration … that's a sad denigration of a major contribution made in the USA'.[35]

Frank could just about tolerate the cult of Goddard, but he really drew the line at von Braun: 'All glory to the German V-2 … but why home-grown American rocket developments is [sic] so frequently minimised is beyond me,' he wrote to Emme.[36] Frank's complaints were often vituperative and counterproductive. But he had a point. The various institutions tasked with commemorating American space endeavour were strangely slow to take notice.

Even those who thought that Frank's complaints were well-founded did little to publicly redress the balance. In a break from working on Stanley Kubrick's film *2001: A Space Odyssey* at MGM studios, Arthur C. Clarke wrote to Frank that 'I quite agree with you that the early days of JPL are seldom recognised', yet his very next book, a review of space developments from the preceding twenty years, ignored Malina entirely and perpetuated the influence of the V-2.[37] Clarke, of course, worked closely with Wernher von Braun. So too did many of the key historians, such as Frederick Ordway III, who helped build the myth of the V-2's exclusive technical supremacy without asking the awkward questions about the thousands of prisoners who were worked to death in its construction. In *A History of Rocketry and Space Travel*, Ordway and von Braun did their best to minimise the WAC Corporal, claiming that 'Germans were dominant in every field of rockets and missiles in World War II'. The German V-2 team, they said, 'created modern military rocket technology … virtually all postwar missile developments were based, in varying degrees, on what went on in Germany'.[38] That, as Frank put it, was 'really a bit much'.[39]

After all, the WAC Corporal had led directly to the Corporal, the world's first nuclear missile, and to the Aerobee, the workhorse of the space sciences, which saw over one thousand launches, right up until the mid-1980s. When Wernher von Braun held aloft JPL's Explorer 1 satellite before the world's media in 1958, it had been successfully launched on Juno 1, which consisted of von Braun's Redstone rocket as a first stage but subsequently propelled by three stages of JPL's solid propellant Sergeants, direct descendants of Parsons and Malina's JATO work. The same solid propellant technology, developed by Aerojet, has been used in the front line of US nuclear security in the land-based Minuteman ICBM, in the submarine-based Polaris and in contemporary Trident II missiles. And even as Ordway and von Braun were writing, Aerojet were developing the propulsion system for Apollo's Service Module, a hypergolic rocket motor that descended from work by Malina and Summerfield. All these developments were made in America.

There was one institution, however, that did give Frank full recognition. On the afternoon of 31 October 1968, JPL held a celebration to honor Frank Malina, Ed Forman and Apollo Smith – the surviving members of the Suicide Squad – to mark the original methyl alcohol and gaseous oxygen test in the Arroyo Seco. A thirty-two-year anniversary wasn't a round number, but a few of the long-standing JPL engineers, including Director William Pickering, thought it important. Walt Powell, a colleague whose frustrations with Frank Malina once extended to smashing through his office door with a hatchet, organised a painstaking recreation of the test stand on which was mounted the original rocket motor.[40] In his speech at this event Frank gave warm tribute to Theodore von Kármán, who died in 1963, aged 81, and to Jack Parsons, who, Frank said, 'has not received ... his due for his pioneering work'. Louis Dunn was not present, nor was he mentioned, but his face glared down on proceedings from a display of former JPL staff photos.[41] He was, in fact, now working as a sporadic consultant for Aerojet.[42]

H. S. TSIEN HAD little sympathy for Frank's fixation with the past. When still stuck in California, he accused Frank of being hopelessly naive:

> Do you believe in history ... knowing it is being re-written all the time? Do you think there is justice and honesty in this part of the world? Do you expect to be famous and honored in the USA without being your own press-agent or without having a public relations man at your employ? Dear friend, let us not believe in fictions![43]

Tsien had every reason to be bitter. Having been under effective house arrest in Pasadena, denied access to classified research, he was eventually exchanged for American prisoners of war by ambassadors Alexis Johnson and Wang Ping-nan in August 1955. Colleagues reported that he wasn't keen to leave the United States, though his adopted country had effectively made personal and professional life impossible.[44]

'I do not plan to come back,' Tsien told reporters, standing on the dock of Los Angeles harbour on 17 September 1955. 'I have no reason to come back. I have thought about it for a long time. I plan to do my best to help Chinese people build up their nation to where they can live in dignity and happiness. I have been artificially delayed in this country from returning to my country. I suggest you ask your State Department why.'[45] Chinese Premier Zhou Enlai was delighted. In due course, Tsien developed such a successful space and ballistic missile programme that China became the third country to independently send humans into orbit. He is now hailed as a national hero; there is a museum in his honour at Jiao Tong University in Shanghai.

Although Frank did his best to maintain contact with Tsien, communication to and from China was often difficult. Years would go by between letters, yet even as their lives diverged there was always a degree of warmth between them. 'You said you were writing your memoirs,' Tsien teased Frank in 1971. 'What's got into you wanting to do such a thing, my old friend? Is that because you are getting on in years? I am never in such a state of mind. I have to advance as there is much to do. How could I spare time for memoirs? ... The world is getting better and better. Let us march forward, head up and chest out, to welcome the beautiful future!'[46]

In the half century following Tsien's return, the future turned out to be a bit less beautiful than perhaps he hoped. Missile technology that developed under his leadership found its way to other states with even less cordial relations to the United States. Naturally, none of this had anything to do with Tsien himself, but it's ironic that his Silkworm missiles and their derivatives have since been fired back at the American military, first by Iraq during the Gulf War of 1991 and, most recently, by Yemen's Shiite Houthi rebels in October 2016. The crew of the US Navy destroyer USS *Mason* who came under attack by

Houthis could hardly have imagined the strange thread that bound an incoming missile to a fateful deportation from the US in 1955, and, further back, to a small group of Caltech engineers meeting to talk about the Spanish Republic in 1938.

For all that the deportation decision was a strategic disaster, INS officials were actually correct that Tsien had been a member of the Communist Party in Pasadena. New sources suggest that when Tsien joined the Chinese Communist Party he submitted a biographical statement in which he detailed joining the American Communist Party in 1938.[47] His biographer, Iris Chang, claimed that 'an independent investigation conducted by the author revealed that it was unlikely that Tsien had ever joined the [American] Party', citing as evidence the fact that Liljan Malina did not remember seeing Tsien at a meeting.[48] This was a smokescreen. Although Liljan told Iris Chang that Tsien *wasn't* a member, she wrote the opposite in her private journal – that Tsien had indeed joined and had attended meetings regularly.[49] At first glance, this sort of behaviour looks much like J. Edgar Hoover's archetypal communists, 'trained in deceit and trickery', people who 'use every form of camouflage and dishonesty to advance their cause'.[50] But the real revelation is how difficult it was, even many years later, for American communists to shake off their prohibition of talking openly about who was a member.

Liljan was happy to reminisce with Chang about the old days, but she also wrote in her journal that 'after hours and hours of telephone conversations, I realised that she had absolutely <u>no</u> idea of what it meant to belong to the Communist Party at that time. [...] Those early 40s had another tempo, other vital necessities, another flavor altogether & this was especially so in the Caltech environment. I cannot in a few short hours delineate it for her.'[51]

ALTHOUGH FRANK KEPT UP correspondence and friendship with old Unit 122 comrades, some of those who testified against Sidney Weinbaum were not forgiven. Gus Albrecht, who had lined up as a prosecution witness against Sidney, was dropped from correspondence. 'Have you heard from Frank and his art?' Albrecht asked Liljan in one letter, adding 'I don't hear from him since I refused to go to jail for perjury for dear old Sidney'.[52]

Frank maintained contact with the Dubnoffs, but they found themselves socially ostracised by many in the party. Even though they seemingly cooperated with the FBI, they still went to much effort behind the scenes to protect others, notably Summerfield. No one could know that, however. Isolated by friends on the left, the Dubnoffs were more often reviled by those on the right. Belle, a teacher, was blacklisted from the public school system. Even when she started her own independent school for children with learning difficulties, her communist past regularly surfaced in ways that seemed to threaten the whole endeavour.[53]

As the strain on the Dubnoff family took its toll, Jacob and Belle both saw a psychotherapist. In fact, they saw the *same* psychotherapist: Dr James Jackson of Sherman Oaks, who wrote up their emotional lives in a clinically cold academic paper titled 'The concurrent psychotherapy of a latent schizophrenic and his wife'.[54] It's hard to imagine a psychotherapeutic betrayal more egregious than that of Phil Cohen leaking to both the party and the FBI, yet Jackson's treatment of the Dubnoffs was, if anything, even worse.

In an extraordinary turn of events, Jackson's own health deteriorated, and he spread ludicrous, anti-Semitic accusations that the Dubnoffs were 'Russian agents'. By this stage, Jacob had developed incurable cancer; he and Belle had divorced but she was still close to him, and together they suffered Jackson's claim that Jacob was Caltech's 'master of biological warfare', creating chemicals 'to turn unborn babies into monsters', that along with Charles Manson, Jacob was involved in the murder of Sharon Tate, and had staged the kidnapping of Patty Hearst 'to cover up the fact that the Hearst family are top echelon communists'.[55] Jacob did not live to see Jackson's libel conviction, for which he received a two-year jail sentence.[56]

It's hardly surprising that the Dubnoffs never talked about their life in the party with their own children. Their daughter Ena recalled that, much later, when Belle was herself dying, she still refused to reveal anything about this period. 'I want you to not know,' she said, as if ignorance was the best protection she could offer the next generation.[57]

IT'S NOT CLEAR THAT FRANK ever really understood the suffering that Sidney and Lina Weinbaum endured, nor how irreversibly affected their lives were by his imprisonment and her confinement. No sooner had Sidney been granted parole from McNeil Island than he invested everything in getting Lina released. This, he was told, would only be possible if he found a new home and a job – not easy when he was still blacklisted.[58] An old friend from the party gave him a job in a factory making children's clothes. Science was over, but at least he could be with Lina.

The joy of their reunion, however, was short-lived. The exact sequence of events isn't apparent beyond the fact that her emotional ill health required readmission to Stockton State Hospital; there's evidence to suggest that Lina was violent to Sidney and that at some stage she tried to strangle him.[59] She was in and out of Stockton for the rest of her life. The FBI just kept following her: they tracked her fruitless attempts to move back to the USSR; they spoke to the owner of a budget LA motel who had 'literally' thrown Lina out into the street 'because she wasn't in control of herself'. He told them that she 'received in the mail a copy of the "USSR Magazine" at which time she loudly extolled the virtues of Soviet Russia much to the dislike of the Hotel guests'. [60] Her inability to find her way home, wherever she thought that might be, was a constant sorrow; she worried about her brother in Russia and raged against the bureaucratic obstacles to her seeing him. At one point she felt that the Soviet Union was persecuting her by using impulses from the particle accelerator at Berkeley's Rad Lab.

Like the Dubnoffs, Sidney and Lina divorced. 'I love Sidney very much,' she wrote in a disordered letter to Linus Pauling, 'and Selina told me that Sidney never stopped loving me. It is a result (of tests of weapons) – the divorce, vulgar ... and so painful to us ... he sent me a summons after a quarrel! I am sure it was under the influence of the effects of the atomic nuclear tests.'[61] It seems that her beliefs may have been symptoms of psychosis, but it's remarkable that the abiding object of her concerns, the nuclear missile, was a spectral technology whose architects, J. Robert Oppenheimer and Frank Malina, were once so well known to her – more than this, that it was partly to protect these two scientists that Sidney had unwittingly staked his freedom and her sanity.

Selina Weinbaum graduated, married, got a PhD in zoology, and had kids of her own. She forged a career in San Francisco working in environmental protection. Sidney, too, eventually remarried and found a serenity that no one quite expected, least of all himself. 'I have a nice life now,' he told a researcher during an oral history interview in the 1980s. 'Just looking at my wife, I can spend hours looking at my wife,' he said, beaming across at Betty Weinbaum, with whom he lived for his remaining thirty-five years. 'I didn't expect I would have such a peaceful and pleasant life and so many friends.'[62] Frank Malina was counted among them. Sidney was never the most dutiful correspondent; he left it to his daughter to maintain the annual exchange of Christmas cards and family photos with the Malinas. But among Sidney's papers after his death, his family found a 1974 essay of Frank's entitled 'Reflections of an artist-engineer on the arts-science interface'. It was the sort of topic they used to love to argue about in the 1940s. It was signed: 'In old friendship, Frank J. Malina'.[63]

LIBERATED FROM THE PURSUIT of income, Frank lived a full life, which extended his contribution to astronautics if not to rocketry per se. One of his major legacies was the International Academy of Astronautics (IAA), a supranational organisation he founded in 1959 with von Kármán and Andy Haley that embodied Frank's convictions about the need for the international coordination of astronautics for peaceful purposes. As ever, Malina deferred to the Boss, who became the first president, with Malina as deputy automatically taking up the reins upon his death. Frank edited the IAA journal *Acta Astronautica* between 1961 and 1965, and it continues to be an important forum for new academic work across engineering, the life and social sciences, technology and economics.

It's instructive, however, that Frank never sought the presidency of the IAA for himself. Perhaps it was his experience of exhausting

UNESCO diplomacy, or perhaps he worried about his communist past. Perhaps he just liked to defer to von Kármán. But he certainly didn't exhibit the same appetite for political leadership that had once worried his colleagues back in 1940. His proximity to power without the burdens of ultimate responsibility still let him pursue important side projects, like the IAA's History of Astronautics Committee (another front on which he could defend the JPL legacy), and, in 1960, chairing the committee for the Lunar International Laboratory (LIL), which he established with Sir Bernard Lovell, director of the British Jodrell Bank Observatory.[64] His vision for a cooperative occupation of the moon both preceded the Apollo programme and was, in some ways, displaced by its overtly nationalistic overtone. Even so, the LIL's collaborative spirit would later find expression in the International Space Station.

Malina made such significant contributions to the political infrastructure of space exploration, pouring energy into these institutions, that it would be easy to think of him as having retreated from engineering. Not so: instead, he applied it to a new field. He had sketched, on and off, in both Texas and Pasadena. During his Pasadena years, this interest was expressed as an over-involvement in Liljan's painting, but it was only when he moved to Paris and settled down with Marjorie that he made serious art work of his own, experimenting first with portraiture and abstract compositions, then incorporating mixed media, with string, electric wire and wire mesh. Within months of his departure from UNESCO, he was being represented by an agent and was selling his work through the private Henri Tronche Gallery in Paris.[65] By 1954, one of his paintings was bought by the City of Paris for 40,000 francs.

Frank's work wasn't the usual fare of the Parisian scene. He had no 'nudes, flowers, landscapes, and dead fish'; rather he looked to the engineering world for inspiration.[66] By taking his background in technical design – all those diagrams he once drew for von Kármán's textbook – and applying it to astronomical objects, his work more than stood out from the art crowd. It only took a year for him to shift from thinking of himself as an engineer with art as a hobby to being primarily an artist with a sideline in rocketry.[67] His sons, Roger and Alan, grew up with the impression that painting was what all scientists did.[68]

To integrate these activities – what his friend C. P. Snow would dichotomise as the 'two cultures' of science and art – became the core of Malina's intellectual project. He didn't drop his positivism, and he still believed in a world that could be objectively known, described and transformed. But now he also aimed to convey a personal poetic vision. As with rocketry, he wanted to experiment with engineering as an instrument that could provide new perspectives on the place of Earth. He loved how, by rotating two layers of wire mesh the resulting 'moiré' effect produced a pleasing third image in the retinal perception of the viewer. Then one December, the boys were perturbed to see their father ripping the new electric lights from their Christmas tree in order to put them into a painting – a Eureka moment – which was soon followed by other motors and thermal interrupters, to produce a new kind of illuminated kinetic art.[69]

It opened up a new world. He made over 200 electric-kinetic paintings, patenting his method into the 'Lumidyne' system, in which new pieces conveyed a sinuous, celestial movement. One coloured pattern would spill organically into another, tracing ellipses and spirals with an aleatoric series of switches, meaning that no viewer could ever see the same composition twice.[70] The effect was mesmeric, though some gallery owners felt that if you had to plug it in, it wasn't art.

As ever, his eye was on the trajectory rather than the launch. It was less interesting to him to articulate his own creative vision of art and science than it was to create a vehicle for others to follow. This took the form of *Leonardo*, a proto-academic journal for professional artists to exchange technical dimensions of their work, extending his lifelong belief in the free dissemination of knowledge. It was not the most commercially promising publication – that wasn't the point – but Frank's Czech-British friend Robert Maxwell was willing to take the risk (his Pergamon Press already published *Acta Astronautica*). *Leonardo* absorbed Frank's energies until the end of his life, and was carried forward by his son Roger.

ONE OF MANY SOLICITED contributions for *Leonardo* came from a New York artist who these days went by the name Jan Wunderman. Her name had always been a malleable thing. As a child at home she was Lily, at high school she was Marie, and though Frank always called her Liljan, art school friends and party friends often called her Lee. She was still Liljan when she married Lester Wunderman, but, as that relationship ended in divorce, she generally became known as Jan. In the spring of 1972, Jan wrote to Frank out of the blue to say that as she and her new partner, Russ Gureasko, were to be in Paris – could they visit?[71] This meeting was always going to be a little awkward, but Jan hardly helped when the front gate opened and she mistook the smiling 'woman with greying hair, a very faded house dress and low shoes' for 'a domestic'.[72] It was, in fact, Marjorie Malina, who hosted her husband's ex-wife with grace and humour, serving lunch at the big table in their studio home.

They actually met up a few times throughout the 1970s, sometimes in Paris, sometimes in Princeton (hosted by Martin Summerfield, who had finally regained his security clearance and now led a notable engineering programme). After one Princeton cocktail party in May 1981, Jan observed that 'at 67, [Frank] still [has] a handsome face ... [but] habits of stubbornness persist. He <u>had</u> to be right. It seems to me he's more opinionated than ever, but pleasant and wryly humorous ... with Martin and Eileen ... it was a reunion of people who've known each other for a very long time.'[73]

A year or so before, Frank had asked Jan to write for *Leonardo* on her own art practice and experience, an essay that she titled 'My Background and Intent as a Painter of Nonfigurative or Abstract Pictures'.[74] It must have seemed like a good idea to both of them, but when Jan submitted her draft, it brought to the surface some old tensions. 'He was not capable of letting me say what I had to say,' Jan wrote in her journal. 'It had to conform to his thinking, and we battled over meanings and interpretations by mail for weeks. As a result, my article is frequently clumsy and removed from my meaning. It's a perfect example of Frank at his most infuriating.'[75]

Jan was not especially motivated to write about her own art work; what she had to say about her background was partly for

Frank's benefit. She tried, in her own oblique way, to pay tribute to the influence he had on her, even after they had separated: 'I said … that in my younger years I found myself too often in the right place at the wrong time.' Addressing Frank directly in the pages of her private journal, she wrote, 'I knew you would read it and I knew you would understand'.[76]

Maybe he did. But her 'right place, wrong time' line didn't survive his editorial pen in what turned out to be Frank Malina's last issue of *Leonardo*.

IT WAS NOVEMBER 1981 and Jan and Russ were at home in their Manhattan apartment, an old pre-war walk-up furnished with a mix of well-chosen antiques and high modernist design. It was cold out, and the wind was coming in from the north, stripping the last leaves from the trees outside. Streetlights seeped around the shutters as they sat in bed watching television.

The phone rang, a late hour to get a call. Russ felt for the remote to mute the noise as Jan picked up the receiver.

'Hello?'

It was Martin Summerfield sounding agitated. He immediately asked to speak to Russ. Jan handed over the phone and the room was quiet for a few moments.

'She's strong,' she heard Russ say, 'it doesn't have to come from me.'[77]

FRANK HAD BEEN AT HOME in Boulogne-Billancourt packing for a flight to a conference in Edinburgh.[78] He felt unwell, went to bed and was later found dead from a heart attack. Marjorie was already in Britain visiting relatives, and waiting for him to join her; Roger and Alan were working overseas. His death, aged 69, came without warning.

Martin and Jan held on to the phone for a long time, each using the other to adjust to the news. Jan remembered these two engineers as roommates in the days before she and Frank were married, and recalled how Martin had once been so annoyed by Frank's work ethic and early rising that he had smashed his alarm clock into pieces. 'Like squabbling brothers all these 45 years or more,' thought Jan.[79] She could tell that Martin was struggling for composure. For her part, she preferred not to cry in front of other people, not even Martin.

A few days later Jan found her journal. It was only a couple of weeks since she had last been writing about Frank – even after all these years, he occupied her thoughts to a surprising degree. She still kept a small photo of him tacked on a bulletin board above the desk in her studio.[80] In the days after his death she addressed him directly, as if in a letter.

> You've been much on my mind Frank. I'll miss knowing you're somewhere fighting for your ideals. I still think you were trying to do too much. You had a vision of the Renaissance man and you wanted to be known as one. Everything – science, art, politics, music, publishing, teacher – no one can say you didn't try. It was too soon for you to die.[81]

Her entries that week have a raw, unprocessed quality; there's anger in there, as well as grief and a defensiveness, tinged with regret at how events had unfolded around them.

> We never had a chance together you and I. Both too young, inexperienced, expecting all the wrong things from marriage. On top of all that we were caught in the maw of history, separated, frightened and shaken by a horrible war, all your thoughts and energies turned to experiments in jet propulsion. Kept so busy and so much away from me. This A.M. I was listening to that melody … Rondo and Capriccioso by Saint-Saëns, reminded of evenings with you.
>
> You taught me so much of music, of logic and literature and so much more. I respected you, admired you and it should've been enough to make our marriage work. But too much intruded.
>
> I wonder even now if you really loved me, and now it's too late to ask. It doesn't matter. It was a crazy seven years together.[82]

Roger Malina got the call out of the blue. Having done a PhD in astronomy, he was now working at the Space Sciences Laboratory in Berkeley and was passing through its front door on 9 November 1981, when hailed by a receptionist holding the telephone receiver. It gave Frank such satisfaction that, twice a year, Roger would travel down from Berkeley to White Sands Proving Ground to insert X-ray telescopes into the instrument bay of an Aerobee rocket – the direct technical descendant of the WAC Corporal.[83] Improved propellants and lighter materials made space so accessible these days. At an altitude of 400 kilometres – well above the protective atmosphere of Earth that absorbs X-radiation – the instrument's 'door' opened, showing Roger and the Space Astrophysics Group a new slice of the observable universe.

Frank J. Malina, the engineer who in 1936 dreamed of the rocket as a vehicle for investigating cosmic radiation, for understanding the wonder of the cosmos, had the knowledge that his own son was among those doing just that.

AFTERWORD

PART OF WHY I have been gripped by researching this story, and why I have spent too many years immersed in FBI files and private archives is that, as J. Edgar Hoover once put it, there's something addictive about secrets. They compel our curiosity. Most of the families of Unit 122 members that I spoke to did not know that their parents or grandparents had been involved in the Communist Party. Some had fragments of the story, others none at all. In many cases, revelations from FBI reports and personal journals have made a kind of retrospective sense, though there are a few for whom this history is difficult to reconcile with the people they knew.

Secrets run deep, especially in families. In the course of writing this book, I often found my mind returning to Ena Dubnoff's account of pleading with her mother Belle to talk about what happened to them back in the 1950s. No, she said, it's better that you don't know. Belle's subsequent silence is surely a measure of her fear. We tend to think of secrecy as being bound up with intimacy and trust, but sometimes concealment – keeping people out of secrets – arises out of a desire to protect loved ones from a knowledge that's impossible to bear. Knowing who was a party member made someone vulnerable, and Belle didn't want that for her own children, even if the times had changed.

Quite early in the research for this book, I encountered an unexpected paradox. At first, I thought that everyone involved in this story had passed on and that, unlike previous researchers, I had the disadvantage of coming too late to obtain first-hand accounts. That was true enough. But over the years, I realised that something else was happening: that the characters were no longer alive to withhold or obfuscate, and that the archival traces they could no longer suppress were often more voluble than their living presences would ever have permitted. It takes a lot of active work to maintain these boundaries. The grave can liberate secrets as well as entomb them.

Frank and Marjorie were careful with what they left behind. One or other of them weeded out letters that told the fuller story of Frank's party membership. Even after his death, Marjorie kept up appearances on his behalf. Jan Wunderman wrote in her journal that Tsien's biographer, Iris Chang, 'tells me Marjorie ... nervously waffles around many questions and seems to know nothing, not even that Frank had been denied re-entry into the States after he went to Paris! How odd! Is it possible he never told her any of it? So it seems.'[1] No, that wasn't the case, but Marjorie did maintain a disciplined front. She knew all about Frank's trouble – 'The Problem', she called it – but she hoped that a façade of ignorance would spare her some difficult questions. After Frank's resignation from UNESCO, she apologised to his parents that 'it does not seem to be a wise thing to go back to the States', yet forty years later she left notes for her sons saying that 'Frank never indicated that he was in political difficulties at all'.[2] None of this adds up.

Jan Wunderman did not talk to her own children about the party, but when researchers came to interview her in the 1990s, she opened up a little. She told Chang that Summerfield was 'apolitical' and that Tsien was not a member, neither of which was true. When the historian of science Benjamin Zibit asked Jan about Frank's experience with the FBI, she paused and said that 'this is the part I hate to go into. Frank has two sons in the United States. I don't know how much should be known.'[3] She thought herself lucky to have emerged from the scene comparatively unscathed. 'I have always found it strange,' she wrote to Chang, 'that I was never asked to testify especially after that spooky spy stuff that took place in Pasadena and N.Y., but I suppose I should be grateful.'[4] Spooky spy stuff? Is it a reference to the raid on her home in Pasadena and an FBI visit in Glen Cove? Or to William Perl? It's not clear. She seemed to think that her story would be forgotten, though she also noted in her journal that 'what will remain is the information with the FBI'.[5] She was right about that.

Martin Summerfield, as far as I can tell, avoided talking about his political past. Sidney and Betty Weinbaum said little, though his interview with a Caltech historian steered a delicate, if confusing, course between denial and candour. Tsien privately acknowledged his party membership but made no public admission. Richard Lewis, Frank Oppenheimer, Gus Albrecht and Jacob Dubnoff all made admissions directly to the FBI.

The continuing secrecy of Unit 122 members, even long after its dissolution, was arguably a protective reflex, a strategy with roots long into the history of labour activism in the United States. But it was also a bulwark against the shame that clung to party membership in the McCarthy era. From the perspective of the present, this avoidance of shame can feel excessive. To take a trivial example, after Frank died, the engineer Andrew Fejer sent Marjorie an audio recording of his memories of GALCIT life with Frank that contains the usual evasions about Unit 122 (we thought it was just a music group, we didn't know Dubnoff and Weinbaum were communists and so on). But then Fejer added that 'we really never did get to know ... [Liljan], not even to get superficially acquainted', despite the fact that Frank and Liljan's correspondence at the time is littered with references to their intimate socialising with the Fejers. It's flatly not true. They went to fundraising parties in aid of the Czech Red Cross, they met up in 1939 at Unit 122 meetings, and dropped in on each other regularly. Liljan complained about having to do Russian War Relief posters at short notice for the Fejers. Why would Fejer say otherwise? Perhaps he thought that Marjorie was a political innocent and that, with Liljan having been in the thick of it, some distance was needed. It's as if their secrecy became generative, unstoppable, going beyond the bounds of ordinary caution.

It's ironic, I suppose, that such a furtive culture existed hand in hand with a commitment to scientific openness and sharing. There's an irony, too, in the fact that much of this story was secret even to its principal protagonists. Frank Malina in 1946 may have fantasised about being a satellite 'perched above the world as an observer of the whole', but he witnessed comparatively few of the events and actions that shaped his own life.[6] Louis Dunn's betrayal was unknown to him. The same went for Jack Parsons and probably Phil Cohen. Frank had only a limited insight into the suffering of the Weinbaums, the Dubnoffs and the Brunners.

I can't pretend that this book gives a final definitive account, that all of the secrets are at last unveiled. There is so much we don't know. Like some of the characters, we can look back on the denials, evasions and accusations, and wonder: what really happened? Like them, our understanding is partial; we may harbour our own suspicions. The abiding secret, of course, is who leaked the WAC Corporal documents from JPL – and why? Nothing points definitively towards any particular culprit, though clearly this was a militarised scientific establishment that was politically ripe for espionage.

It's such a stark noun, espionage: it sounds so definite. What happened at JPL may never appear quite as clear. The leakers may have couched the sharing of the WAC Corporal documents as radical scientific openness or assisting a wartime ally – that, at least, was the dubious logic of the Rosenberg network. The JPL leak may have been careless and imprudent, or it may have been ruthless and strategic. Either way, there's something illusory about our desire for a singular, complete truth. Further information might add intrigue and nuance, but it's still likely to leave us in the same hall of mirrors.

THE ELISION OF FRANK MALINA from much space history does not have to be a conspiracy. It's enough that his political background be awkward – which of course it is (though weightier baggage wasn't such an encumbrance for Wernher von Braun). It's awkward because it's hard to know what to do with the fact that the party was both progressive *and* subversive, that it was both a grass-roots network of local activists and the centralised creation of Moscow, both an organic community of socialist ideas and a front for covert intelligence operations. There's no evidence that most of the people involved in campus communism had any knowledge of such operations, but it's reasonable to ask why they didn't. With Soviet operatives working in the West Coast aviation industry in the mid-1930s, it's hard to imagine why the NKGB (Soviet secret police, intelligence and counter-intelligence) or the GRU (Soviet military intelligence) wouldn't at some point turn their attention to GALCIT – even before the rocketry work started.[7]

Let's consider, too, that Unit 122 was established by Frank Oppenheimer, brother of the J. Robert Oppenheimer who was the highest-value target of the Soviet agencies. One NKGB memo reported of Oppie (code name: 'Chester') that 'his membership in the compatriot organization [US Communist Party] and friendly attitude toward our country provides grounds for counting on a positive result of his cultivation'. As it turned out, Oppie was neither easily approachable nor very sympathetic, but a key channel of NKGB access was identified as 'Ray', their code name for Frank Oppenheimer, who, they felt, was 'politically closer to us'. In other words, Soviet agencies must surely have been alert to the existence and potential opportunities of Unit 122 from the outset.[8]

So even though there were episodes of individual harassment and prejudice, the rationale for the FBI's surveillance and concern was understandable. I'm not sure that party members with this kind of proximity to classified intelligence should elicit the sort of sympathy that belongs to the thousands of ordinary Americans who were swept up by McCarthyism, many blacklisted, because they were, say, a school teacher who once attended an event organised by a front organisation. In that sense, it's not helpful to think about JPL rocketeers as 'victims' of an anti-communist witch-hunt. Their treatment wasn't always fair, but they were aware of the defence context in which they worked.

All of this makes for a messy history. It's not our business to tidy it up; and it's not necessary in order to recognise and honour the achievements of Malina and his colleagues. Does it matter that these people were members of the Communist Party? Yes, it does, but primarily because the shame of its association has concealed part of the means and the motivation by which our species left our planet. And while Malina's membership isn't an accurate gauge of his politics, far less his loyalty, it contextualises the surveillance and repression to which he was subject.

Credit for reaching space is always going to be diffuse – there are multiple branches of technical achievement. Even so, JPL's rocketeers constitute an exceptionally important strand of American space endeavour. The WAC Corporal represents the first moment that the promise of rocketry – attaining extreme altitudes – was successfully realised in the United States. With the BUMPER WAC Corporal, the US first entered the Space Age.

It's not that the story of JPL has been unknown until this book, far from it, but the extent to which the history of its engineering is profoundly entangled with political struggle has not been so apparent.[9] And it's a new context that inverts the previously V-2-dominated history of American astronautics. There's no doubt that von Braun and his team made an enormous contribution to our ascent into space, just as there's no doubt that many of them were once Nazis working in the service of fascist expansionism. Even if von Braun was not *primarily* driven by Nazi ideology, he was more than happy to use fascism as the vehicle for his rocketry ambitions.

The story of Frank Malina and his colleagues is the exact opposite. America's home-grown ascent into space was propelled by anti-fascism and anti-racism. Yes, they accepted military funding to help fight the Nazis – to 'give the fascists hell', as Frank put it – but their ultimate object was non-military civilian science. Frank was never much of a Marxist. Becoming a millionaire might have been an accident, but the same can't be said for setting up Aerojet in the first place. In 1947, Frank paid the cleaner of his Parisian apartment with stationery and canned goods that he lifted from UNESCO.[10] This is not to point the finger – contradiction, after all, is the condition of all politics – but simply to say that he was more a believer in the party as an effective, almost scientific instrument for radical change than he was a philosophical adherent of Marxism.

In hindsight, Frank's compact with both the military and the Communist Party looks like a questionable fulfilment of his original hopes. The military took over the GALCIT Rocket Research Project in its entirety, setting the terms for the research and establishing JPL as a predominantly classified institution. It shouldn't have surprised him that his propulsion success would soon be applied to the world's first nuclear missile, but he still seemed shocked to see his vehicle for instrumental exploration turned towards remote-controlled mass destruction. And although he absented himself from its remaining development, the totem of the guided nuclear missile was founded on his own work. No, the Corporal missile didn't kill anyone, not directly, but as the philosopher of science Ian Hacking has argued, the damage caused by this kind of weapons research was as much to the world of ideas.[11] The casualty was the child digging in a cereal box for an analogue missile; it was a culture bent to the thrall of nuclear deterrence.

Yet even if Malina's civilian ideal of rocketry initially foundered, there's something to be recovered from his belief that transforming the Earth is as important as transcending it. Admittedly, it's a strange turn of events that made Frank Malina the first space millionaire and, outside the USSR at least, the first space communist. But now that twenty-first-century capitalism has slipped the bonds of gravity, scenting new opportunities in space tourism, asteroid mining and lunar territorial claims, the spirit of international cooperation and regulation that he fought for at UNESCO and the IAA has never been more important.

Today, the future of space exploration is arguably being advanced less by traditional state structures than by a handful of billionaires. Doubtless they'll find encouragement from the many popular histories of American space exploration, which are riddled with conservative tropes of heroic masculinity, isolated genius and rugged individualism. The fuller story of American space endeavour, however, looks rather different. The recently recovered role of the women computers in NASA, including at JPL in the 1950s, is a good example: so many white male historians couldn't see the contribution of black women because they apparently didn't have the right stuff.[12]

Malina's eccentric film proposal with Jack Parsons tells us quite plainly what worried him about the rockets they were making: that their technology would first fall into the hands of corporate capital, then be annexed by the far right. In a crude sense both of these fears were well founded.

The remarkable thing about their movie synopsis is how, to a truly improbable degree, it anticipated in 1936 the storms that would engulf them fifteen years later: Malina's character, the communist Jan Kavan, is denied a passport; Belvedere, the mystic, dies in an explosion; Lin Lao, the Chinese rocketeer, is forced back to China. In one respect, however, their story takes a different trajectory from history. Their synopsis ends with the rocketeers destroying all trace of their knowledge of how to build a successful rocket – just in case it falls into the hands of the Nazis. In reality, of course, their rocket didn't need to be captured by Nazis – they brought their Vengeance Weapon with them and made it American.

THE USUAL HONOUR GIVEN to pioneers of space flight is that the International Astronautical Union names an extraterrestrial feature in recognition of their achievement. Von Braun has a crater on the Moon named after him, as does Goddard, and even Jack Parsons. Theodore von Kármán's memorialisation has been particularly extensive, with craters on both the Moon and Mars, as well as the Kármán line. If you count Von Kármán Avenue in Orange County, California, he has eponymous features on three celestial bodies – which is decent going, even for someone who claimed to be the most important scientist of all time bar Einstein and Newton. Frank Malina, however, is not well known, nor, at the time of writing, have his legacies – theoretical, practical or institutional – received meaningful commemoration. There is no Malina crater.

Back in the mid-1970s, an 'International Space Hall of Fame' was set up in Alamogordo, New Mexico, to mark personal achievement and leadership in space. The first year saw it take in thirty-five scientists and engineers, including Neil Armstrong and Theodore von Kármán, as well as the V-2 engineers Wernher von Braun, Klaus Riedel and Walter Thiel. 'The snub of JPL is only too obvious,' complained Frank, noting that their failure to recognise Jack Parsons was 'a poor joke to play on the USA'.[13] Years passed, and while other V-2 engineers were installed, Malina remained out in the cold. Only in

1990, nearly a decade after his death, and after the final collapse of the Berlin Wall, was Malina posthumously inducted. Recognition has come slowly.

He did, however, get generous credit from one of the many illustrious visitors to Frank and Marjorie's house in Boulogne-Billancourt. In the year before Frank's death, the acclaimed sci-fi writer Ray Bradbury and Frank got talking about late 1930s Los Angeles. They didn't actually know each other in those days, but they shared a memory of the optimism of the time. Frank talked about Jack Parsons, and about why their work on solid propellants was important. After the evening had concluded and Bradbury had returned to his hotel, he was jolted into a sudden memory of having attended a meeting of the Los Angeles Science Fiction Society. It was in 1939 or 1940, back when he was a cub reporter, and he could visualise the young scientist who came to talk about rockets. After the talk, Bradbury got over his shyness at being a mere newsboy and pestered the 'proper' scientist with a barrage of questions. Forty years later, Bradbury realised that this charismatic man was none other than Jack Parsons. 'I was stunned and touched,' Bradbury wrote to Frank, 'to think our lives almost touched, so long ago.'

In closing this letter, he added:

> I'm glad I'm in Paris, and I'm glad that the years have cycled around and here we are, you with a long and glorious history behind you, one that will have changed the destiny of people on Earth forever, and this is not said idly. A million years from tonight, when future historians speak of the most important years in the history of the thinking beasts, your name will be there with von Kármán's and all the rest. What a glad knowledge to have of yourself.[14]

But as long as the name of Frank Malina remains stuck in the reputational mire of McCarthyism, it will not ring out on some distant horizon. It's a name that should be raised up, actively remembered, if only to prove that HUAC and J. Edgar Hoover do not determine the shape of our history. We can live with this untidy past.

For his part, Frank always had more faith in engineers than in historians. As he put it in one of his regular missives to NASA's Eugene Emme, 'one must resign oneself to the sacrifice of truth and accuracy in popular … histories', on the grounds that historians generally lacked the technical qualifications to write about engineering.[15] Ouch.

Even so, a million years from tonight, I think his name should indeed be there – with von Kármán's, and all the rest.

FURTHER READING

Many readers will know that this isn't the only work to deal with the JPL story. The first such book, Clayton Koppes' *JPL and the American Space Program* remains an outstanding institutional history. Benjamin Zibit's doctoral thesis, *The Guggenheim Aeronautical Laboratory at Caltech and the Creation of the Modern Rocket Motor (1936–1946): how the dynamics of rocket theory became reality* (City University of New York, 1999) greatly added to my understanding of the scientific context. I often found original sources and interviews in the late Iris Chang's *Thread of the Silkworm*, in George Pendle's *Strange Angel*, in M. G. Lord's *Astroturf* and in Ewen Chadronnet's *Mojave Epiphanie*. W. Patrick McCray's forthcoming *Wiring Art Work: Artists, Engineers, and Collaborations Between Creative Cultures Since 1960* will have much more to say about Frank Malina's artistic life. And if that's not enough, dive into the FBI's vault and have a look at Malina's available files for yourself: https://vault.fbi.gov.

NOTES

Principal archives

CAL   Caltech Archives, Pasadena
FBI   Federal Bureau of Investigation, files accessed via FOIA
FJM   Frank J. Malina Collection, microfiche edition, Caltech Archives
            [original material in the Library of Congress, control no. 84061793]
GLC   Gil Lewis private collection
GYC   Gerald Yorke Collection, Warburg Institute, London
ICP   Iris Chang Papers, UC Santa Barbara Special Collections
JPL   Jet Propulsion Laboratory Archives
KWC   Karen Wunderman private collection
LPP   Linus Pauling Papers, Oregon State University
MFC   Malina Family private collection
MWC   Marc Wunderman private collection
NA   National Archives, UK
NARA   US National Archives and Records Administration
TVK   Theodore Von Kármán Papers, Caltech Archives
WPP   William Pickering Papers, Pasadena History Museum

Prologue

1. Details of Katja Liepmann's escape from the Nazis, as told over dinner, are detailed in a letter from Frank Malina to his parents, 20 November 1939, Box 21, Folder 10, FJM.
2. Letter from Frank Malina to Liljan Malina, 23 April 1945, MFC.

3. The address, given simply as Laurel Canyon in Jan Wunderman's journal, 16 August 1984, is confirmed as North Orange Grove Avenue in Helen Blair's testimony to HUAC: United States Congress House Committee on Un-American Activities, *The Southern California District of the Communist Party: structure, objectives and leadership*, afternoon session, Tuesday 25 February 1959, p. 264.

4. All of the details of this raid are from Jan Wunderman journal, 16 August 1984, KWC.

5. 'extreme altitudes': Robert H. Goddard, 'A Method of Reaching Extreme Altitudes', *Nature*, vol. 127, 1920, pp. 809–11. The primary objective of early rocketry was to exceed the altitude of contemporary balloon technology – around 100,000 feet. See also J.D. Hunley, 'The Enigma of Robert H. Goddard' *Technology and Culture*, vol. 36, no. 2, 1995, pp. 327–50.

6. George Pendle, *Strange Angel: The Otherworldly Life of Rocket Scientist John Whiteside Parsons*, Orland, FL, Harcourt Inc, 2005; John Carter, *Sex and Rockets: the Occult World of Jack Parsons*, Los Angeles, Feral House, 2004. See also Iris Chang, *Thread of the Silkworm*, New York, Basic Books, 1995; M. G. Lord, *Astroturf: the Private Life of Rocket Science*, New York, Walker and Company, 2005.

7. Walter A. McDougall, *The Heavens and the Earth: A Political History of the Space Age*, New York, Basic Books, 1985.

8. Dwight Eisenhower, Radio and Television Address to the American People on Science in National Security, 7 November 1957, https://www.presidency.ucsb.edu/, Legacy PID no. 10946.

9. William Safire, *The New Language of Politics: An Anecdotal Dictionary of Catchwords, Slogans, and Political Usage*, New York, Random House, 1968, p. 47.

10. Michael J. Neufeld, *The Rocket and The Reich: Peenemünde and the Coming of the Ballistic Missile Era*, New York, Simon and Schuster, 1995, p. 264.

11. See Yves Béon, *Planet Dora: A Memoir of the Holocaust and the Birth of the Space Age*, edited by Michael J. Neufeld, Boulder, Westview Press, 1997.

12. The question of whether von Braun directly witnessed executions or saw piles of bodies in tunnels cannot be fully answered. It's a question that perhaps stands in for his general lack of truthfulness about the extent to which he knew about the horrors of Dora and the Mittelwerk. See: Michael J. Neufeld, *Wernher von Braun: Dreamer of Space, Engineer of War*, New York, Alfred A. Knopf, 2007, p. 165; Michael J. Neufeld, '"Smash the Myth of the Fascist Rocket Baron": East German Attacks on Wernher von Braun in the 1960s', in Alexander C.T. Geppert (ed.) *Imagining Outer Space: European Astroculture in the Twentieth Century*, London, Palgrave Macmillan, 2012, pp. 106–26 (p. 122).

13. Quoted in Michael J. Neufeld, *The Rocket and The Reich*, p. 258.
14. For a detailed discussion of this see Michael J. Neufeld, 'Wernher Von Braun, the SS, and Concentration Camp Labor: Questions of Moral, Political, and Criminal Responsibility', *German Studies Review*, vol. 25 no. 1, 2002, pp. 57–78; also Michael J. Neufeld, *Wernher von Braun*, p. 177.
15. Arno Mayer interview, *This American Life* episode #595, Deep End of the Pool, 26 August 2016, 52 minutes, http://www.thisamericanlife.org/radio-archives/episode/595/deep-end-of-the-pool accessed, 20 September 2016.
16. On von Braun's anti-communism, see Michael J. Neufeld, 'Wernher Von Braun, the SS, and Concentration Camp Labor', p. 58.
17. John D. Anderson Jr, *Hypersonic and High Temperature Gasdynamics*, 2nd edition, American Institute of Aeronautics and Astronautics, Reston, Virginia, 2006, p. 2.
18. Letter from Frank Malina to Frank H. Winter, Smithsonian Inst., 16 March 1979, Box 6, Folder 34, FJM.
19. Letter from Frank Malina to Andy Haley, 26 June 1958, MFC.
20. Passport application, MFC.

Chapter 1: California

1. Letter from Frank Malina to parents, 14 September 1934, Box 21, Folder 5, FJM.
2. Ibid.
3. Ibid.
4. Benjamin Zibit interview of Jan Wunderman, 22 April 1996, Frank Malina file, CAL.
5. Frank J. Malina, oral history by James H. Wilson, 8 June 1975, JPL Oral History Program, JPL, p. 24.
6. Letter from Frank Malina to Liljan Malina, 6 September 1944, MFC.
7. Letter from Frank Malina to parents, 14 September 1934, Box 21, Folder 5, FJM.
8. Ibid.
9. 25 cents an hour, from Frank J. Malina, 'The Beginning of Rocketry and JPL' in F. E. C. Culick (ed.) *Guggenheim Aeronautical Laboratory at the California Institute of Technology: the First Fifty Years*, San Francisco, San Francisco Press, 1983, pp. 65–73 (p. 65); Howard Hughes test, letter from Frank Malina to parents, 22 September 1935, Box 21, Folder 6, FJM. By March 1936, Malina was earning 80 cents an hour overtime: letter to parents, 6 March 1936, Box 21, Folder 6, FJM.
10. 'Don't worry about me working very much. I am by nature lazy': letter to Liljan, 5 September 1938; '[I] only wish I was not so lazy and buckled down more often': letter to Liljan, 14 February 1939, MFC.

11. Letter from Frank Malina to parents, 4 January 1936, Box 21, Folder 6, FJM.

12. Theodore von Kármán (with Lee Edison), *The Wind and Beyond: Pioneer in Aviation and Pathfinder in Space*, Boston, Little, Brown and Co., 1967, p. 176, p. 4.

13. Technically, Robert Millikan was Chair of Executive Council, the nearest equivalent to President at that time.

14. Judith R. Goodstein, *Millikan's School: A history of the California Institute of Technology*, New York, W. W. Norton, 1991, p. 99.

15. See Egon Krause and Ulrich Kalkmann, 'Von Kármán between Aachen and Pasadena', *Progress in Aerospace Sciences*, vol 59, no. 1, 2013, pp. 2–12.

16. Benjamin Zibit, *The Guggenheim Aeronautics Laboratory at Caltech and the Creation of the Modern Rocket Motor: How the Dynamics of Rocket Theory Became Reality*, unpublished PhD thesis, City University of New York, 1999, p. 74; von Kármán quote from *The Wind and Beyond*, p. 235.

17. Homer J. Stewart interviews by John L. Greenberg (Oct–Nov 1982) and by Shirley K. Cohen (3 November 1993), Oral History Project, CAL, p. 20.

18. Andrew Fejer interview with Iris Chang 12/7/91, tape 29, Box 12, Folder 26, ICP.

19. Cited in Frank J. Malina, 'On the GALCIT Rocket Research Project, 1936–38' in *Smithsonian Annals of Flight no. 10: First Steps Toward Space* [Proceedings of the First and Second History Symposia of the International Academy of Astronautics], Belgrade, Yugoslavia, vol. 26. 1967, pp. 113–127, p. 126; other sources mention a similar article in the *Pasadena Post*: Frank Malina interview with R. Cargill Hall, 29 October 1968, JPL [copy at ICP].

20. Frank J. Malina, 'Memoir on the GALCIT Rocket Research Project', prepared for the First International Symposium on the History of Astronautics, IAA, Belgrade 25–26 September, 1967, http://www.olats.org/OLATS/pionniers/memoir1.shtml, accessed 30 June 2015. This text is similar but not identical to Malina, 'On the GALCIT Rocket Research Project'.

21. Theodore von Kármán, *The Wind and Beyond*, p. 268.

22. John Carter, *Sex and Rockets*, p. 56.

23. George Pendle, *Strange Angel*, p. 27.

24. Ibid., p. 31.

25. John Whiteside Parsons, *Analysis of a Master of the Temple of the Critical Nodes in the Experience of his Material Vehicle*, unpublished mss, *c.* 1948, GYC.

26. George Pendle, *Strange Angel*, p. 45.

27. The phrase 'powder monkeys' comes from Pendle's interview with Marjorie Zisch, *Strange Angel*, p. 47; $20 note, ibid., p. 45.

28. 1928 move to rockets in John Carter, *Sex and Rockets*, p. 4.

29. John Whiteside Parsons, *Analysis of a Master of the Temple*, GYC.

30. George Pendle, *Strange Angel*, p. 64.
31. Ibid., p. 87.
32. Letter from Frank Malina to parents, 9 November 1935, Box 21, Folder 6, FJM.
33. Letter from Frank Malina to parents, 2 February 1936, Box 21, Folder 6, FJM.
34. Ibid.
35. Letter from Frank Malina to parents, 9 November 1935, Box 21, Folder 6, FJM.
36. Letter from Malina to parents, 14 February 1936, Box 21, Folder 6, FJM.
37. Frank J. Malina, 'Memoir on the Galcit Rocket Research Project', unpaginated.
38. Frank J. Malina, 'On the Galcit Rocket Research Project, 1936–38', p. 114.
39. Ibid., p. 114.
40. Frank J. Malina interview by Mary Terrall, 14 December 1978, Oral History Project, CAL, p. 33.
41. George Pendle, *Strange Angel*, p. 90.
42. Benjamin Zibit, *The Guggenheim Aeronautics Laboratory at Caltech and the Creation of the Modern Rocket Motor*, p. 84.
43. Letter from Frank Malina to parents, 29 June 1936, Box 21, Folder 6, FJM.

Chapter 2: The Suicide Squad

1. The complaint about too little time is in the letter from Frank Malina to parents, 18 July 1936, Box 21, Folder 6, FJM.
2. Letter from Frank Malina to parents, 6 March 1936, Box 21, Folder 6, FJM.
3. Letter from Frank Malina to parents, 21 November 1936, Box 21, Folder 6, FJM.
4. Ibid.
5. Robert H. Goddard, 'A Method of Reaching Extreme Altitudes'.
6. Letter from Frank Malina to parents, 16 March 1936, Box 21, Folder 6, FJM.
7. Robert Millikan was Chair of Executive Council, the nearest equivalent to President at that time.
8. Obituary: Clark Blanchard Millikan (1903–1966), *Engineering and Science*, January 1966, pp. 20–22.
9. Letter from Frank Malina to parents, 4 January 1936, Box 21, Folder 6, FJM.
10. Letter from Frank Malina to parents, 5 June 1936, Box 21, Folder 6, FJM.
11. Frank J. Malina interviewed by Mary Terrall, 14 December 1978, CAL, p. 5.
12. Ibid., p. 5.
13. Millikan's 'Kármán's folly' remark mentioned in a letter from Martin Summerfield to William H. Pickering, 10 March 1981, copy in Box 6, Folder 25, FJM.

14. Theodore von Kármán, *The Wind and Beyond*, p. 176.
15. These details compiled from ibid., p. 176; Benjamin Zibit interview of Jan Wunderman, 22 April 1996, Frank Malina file, CAL, p. 4.
16. 'most peculiar types of people': Frank Malina's eulogy for Theodore von Kármán, included in a letter from Malina to Andy Haley, 19 June 1963, Box 2, Folder 10, FJM.
17. David A. Clary, *Rocket Man: Robert H. Goddard and the Birth of the Space Age*, New York, Hyperion, 2004, p. 178.
18. Ibid., p. 178.
19. Frank Malina, 'Memoir on the Galcit Rocket Research Project', unpaginated.
20. Letter from Frank Malina to Milton Lehman, 14 August 1956, Box 6, Folder 1, FJM.
21. Iris Chang, *Thread of the Silkworm*, p. 1.
22. Frank Malina, 'Memoir on the Galcit Rocket Research Project', unpaginated.
23. Theodore von Kármán, *The Wind and Beyond*, p. 242.
24. Letter from Frank Malina to parents, 29 June 1936, Box 21, Folder 6, FJM.
25. Letter from Frank Malina to parents, 6 March 1936, Box 21, Folder 6, FJM.
26. Letter from Frank Malina to parents, 29 June 1936, Box 21, Folder 6, FJM.
27. Letter from Frank Malina to parents, 4 October 1936, Box 21, Folder 6, FJM.
28. George Pendle, *Strange Angel*, p. 105.
29. Untitled MGM manuscript, MFC.
30. Letter from Frank Malina to Liljan Malina, 23 November 1938, MFC.
31. Letter from Frank Malina to Liljan Malina, 16 January 1939, MFC.
32. Letter from Frank Malina to parents, 22 May 1937, Box 21, Folder 7, FJM; on the film, see H. Mark Glancy, 'Hollywood and Britain: MGM and the British "Quota" Legislation' in Jeffrey Richards (ed.), *The Unknown 1930s: an Alternative History of the British Cinema, 1929–1939*, IB Tauris, 2001 [1998], pp. 57–74 (p. 67); Michael Sragow, *Victor Fleming: An American Movie Master*, Lexington, University Press of Kentucky, 2013, p. 404.
33. Letter from Frank Malina to parents, probably 1935, quoted in *The American Rocketeer*, documentary film dir. Blaine Baggett, Pasadena, JPL, 2011, 8 min. 40 sec.
34. Clayton R. Koppes, *JPL and the American Space Program: A History of the Jet Propulsion Laboratory*, New Haven, Yale University Press, 1982, p. 7.
35. Frank Malina, 'Memoir on the Galcit Rocket Research Project', unpaginated.
36. Letter from Frank Malina to parents, 1 November 1936, Box 21, Folder 6, FJM.
37. Letter from Frank Malina to parents, 7 November 1936, Box 21, Folder 6, FJM.
38. Ibid.

39. Ibid.
40. Benjamin Zibit, *The Guggenheim Aeronautics Laboratory at Caltech and the Creation of the Modern Rocket Motor*, p. 203.
41. Frank J. Malina, 'The Beginning of Rocketry and JPL', p. 68. Other sources say $100.
42. Frank J. Malina, 'The Rocket Pioneers: Memoirs of the Infant Days of Rocketry and Caltech', *Engineering and Science*, vol. 31, 1968, pp. 9–32 (p. 30).
43. Letter to parents, 22 May 1937, Box 21, Folder 6, FJM.
44. Frank J. Malina, 'The Rocket Pioneers', p. 12.
45. Frank J. Malina interviewed by Mary Terrall, 14 December 1978, CAL, p. 4.
46. John D. Clark, *Ignition! An Informal History of Liquid Rocket Propellants*, New Brunswick, Rutgers University Press, 1972, p. 11.
47. Clayton Koppes, *JPL and the American Space Program*, p. 14.
48. For a brilliant discussion of this, see: Carol Cohn, 'Sex and death in the rational world of defense intellectuals', *Signs: Journal of Women in Culture and Society*, vol. 12, 1987, pp. 687–718.
49. Frank J. Malina, oral history, James H. Wilson, 8 June 1973, JPL Oral History Program, JPL, pp. 32–3.
50. An example: 'I hope you will see to it that my baby … gets the recognition it deserves': letter from Frank Malina to Fred Durant III, Smithsonian Inst., 30 January 1965, Box 5, Folder 17, FJM.

Chapter 3: Facts and Fancies
1. 'satin ribbon' etc.: undated entry, Jan Wunderman journal, KWC.
2. 'warm dark brown eyes', ibid.
3. Letter from Frank Malina to parents, 15 July 1938, Box 21, Folder 8, FJM.
4. Jan Wunderman interview, *The American Rocketeer*, JPL.
5. 'He was smitten': undated entry, Jan Wunderman journal, KWC.
6. 'rumbunctious Irish boy', ibid.
7. 'plans of marrying us off', ibid.
8. 'Benny Goodman', 'Artie Shaw', 'melt in each other's arms', ibid.
9. 'wild dreams of space rocketry': undated entry, Jan Wunderman journal; 'friendly, funny', ibid.
10. 'Barber of Seville', letter from Frank Malina to parents, 25 July 1938, Box 21, Folder 8, FJM.
11. Letter from Frank Malina to Liljan Malina, 5 September 1938, MFC.
12. The 'situation could have continued': undated entry, Jan Wunderman journal, KWC.
13. Malina, 'Memoir on the GALCIT Rocket Research Project', unpaginated.
14. Frank Malina oral history, James H. Wilson, 8 June 1973, JPL.

15. Letter from Frank Malina to Liljan Malina, 5 September 1938, MFC.
16. Letter from Frank Malina to Liljan Malina, 30 September, 1945, MFC.
17. Draft letter from FJM to Liljan, undated and possibly unsent but from contents written between end of August and 24 September, 1938, MFC.
18. Letter from Frank Malina to parents, 26 September 1938, Box 21, Folder 9, FJM.
19. Letter from Frank Malina to Liljan, 12 September 1938, MFC.
20. This paper was eventually published as: F. J. Malina, 'Characteristics of the rocket motor unit based on a theory of perfect gases', *Journal of the Franklin Institute*, vol. 230, October 1940, pp. 433–54; on the prize, see Frank H. Winter, 'The Birth and Early Rise of "Astronautics": the REP-Hirsch Astronautical Prize', 1928–1940, *Quest*, vol. 41, no. 1, 2007, pp. 35–43.
21. Letter from FJM to Liljan, 16 November 1938, MFC.
22. Letter from FJM to Liljan, 20 October 1938, MFC.
23. 'forceful', 'decisive': undated entry, Jan Wunderman journal, KWC.
24. Undated draft letter [*c.* Sept 1938] from Frank Malina to Liljan Malina, MFC.
25. 'I felt terribly young': undated entry, Jan Wunderman journal, KWC.
26. Undated draft letter [*c.* Sept 1938] from Frank Malina to Liljan Malina, MFC.
27. 'way of being forceful': undated entry, Jan Wunderman journal, KWC.
28. 'no surprises', ibid.
29. 'no birth control', ibid.
30. Letter from Frank Malina to his parents, 10 October 1938, Box 21, Folder 9, FMC.
31. Letter from Frank Malina to his parents, 24 October 1938, Box 21, Folder 9; letter from Frank Malina to his parents, 7 November 1938, Box 21, Folder 9, FJM.
32. Letter from Frank Malina to his parents, 3 October 1938, Box 21, Folder 9, FJM.
33. Letter from Frank Malina to his parents, 24 October 1938, Box 21, Folder 9, FJM.
34. Letter from Frank Malina to his parents, 19 November 1938, Box 21, Folder 9, FJM.
35. Frank J. Malina, 'On the GALCIT Rocket Research Project, 1936–1938' *Smithsonian Annals of Flight*, vol. 10, 1974, pp. 113–27.
36. Letter from Frank Malina to his parents, 10 December 1938, Box 21, Folder 9, FJM; Frank J. Malina, 'On the GALCIT Rocket Research Project', p. 125.
37. Frank J. Malina and A. M. O. Smith, 'Flight Analysis of a Sounding Rocket', *Journal of Aeronautical Sciences*, vol. 5, 1938, pp. 199–202, p. 202.

38. F. J. Malina, 'The Rocket Motor and its Application as an Auxiliary to the Power Plants of Conventional Aircraft', GALCIT Rocket Research Project Report, no. 2, 24 August 1938 (unpublished), JPL.
39. Theodore von Kármán, *The Wind and Beyond*, p. 243.
40. Benjamin Zibit, *The Guggenheim Aeronautics Laboratory at Caltech and the Creation of the Modern Rocket Motor*, p. 315, fn. 428.
41. Frank J. Malina, 'The Jet Propulsion Laboratory: its origins and first decade of work', *Spaceflight*, vol. 6, no. 6, 1964, pp. 216–23, p. 220.
42. Letter from Frank Malina to his parents, 13 March 1939, Box 21, Folder 9, FJM.
43. Letter from Liljan Malina undated but from sequence in folder probably late 1938, MFC.
44. Letter from Frank Malina to his parents, 10 December 1938, Box 21, Folder 9, FJM.
45. Letter from Frank Malina to his parents, 13 March 1939, Box 21, Folder 9, FJM.
46. Jan Wunderman journal, undated entry, KWC.
47. Letter from Frank Malina to his parents, 8 July 1939, Box 21, Folder 9, FJM; the Raymond Fault from Francis H. Clauser interview by Rachel Prud'homme, 25 March 1983, Oral History Project, CAL, p. 64.
48. Letter from Frank Malina to his parents, 29 October 1938, Caltech fiche, Box 21, Folder 9.
49. Jan Wunderman journal, undated entry, KWC.
50. Ibid.
51. Theodore von Kármán and Maurice A. Biot, *Mathematical Methods in Engineering: An Introduction to the Mathematical Treatment of Engineering Problems*, New York, McGraw-Hill, 1940.
52. 'since we would have nothing (?) to do on our honeymoon he would send the proofs to me': letter from Frank Malina to his parents, 20 May 1939, Box 21, Folder 9, FJM.
53. Jan Wunderman journal, undated entry, KWC.
54. Letter from Liljan to Frank Malina, undated but *c.* December 1938, MFC.
55. Letter from Frank Malina to parents, 27 June 1939, Box 21, Folder 9, FJM.
56. Jan Wunderman journal, undated, KWC.
57. Jan Wunderman journal, 16 August 1984, KWC.

Chapter 4: Unit 122
1. These authors listed in Jan Wunderman journal, 16 August 1984, KWC.
2. Ibid.
3. Benjamin Zibit interview of Jan Wunderman, 22 April 1996, CAL.

4. Draft letter from Frank Malina to Liljan Malina, undated and possibly unsent, *c.* Aug–Sept 1938, MFC.
5. Letter from Liljan Malina to Frank Malina, 6 February 1939, MFC.
6. Letter from Liljan to Frank Malina, 20 March 1939, MFC.
7. Jan Wunderman journal, 16 August 1984, KWC.
8. Jan Wunderman journal, 10 November 1981, KWC.
9. Benjamin Zibit interview of Jan Wunderman, 22 April 1996, CAL, p. 3.
10. See Leonard H. Caveny, 'Martin Summerfield and His Princeton University Propulsion and Combustion Laboratory', 47th AIAA/ASME/SAE/ASEE Joint Propulsion Conference & Exhibit, 2011, pp. 1–70 p. 2, Summerfield's reticence in talking about family mentioned in letter from Frank Malina to Liljan Malina, 30 January 1943, MFC.
11. Coffee and cake detail from Benjamin Zibit interview of Jan Wunderman, 22 April 1996, CAL.
12. Jan Wunderman journal, undated entry between 19 October 1987 and 10 November 1987, KWC.
13. Sidney Weinbaum interview by Mary Terrall, Pasadena, California, 15, 20 and 25 August 1985, Oral History Project, CAL p. 5.
14. Mary Jo Nye, 'Mine, Thine, and Ours: Collaboration and Co-Authorship in the Material Culture of the Mid-Twentieth Century Chemical Laboratory', *Ambix*, vol. 61, no. 3, 2014, pp. 211–35.
15. Address from personnel security questionnaire completed 8 April 1949, Weinbaum FBI file.
16. Author redacted, report name redacted, details of interview with Parsons on 13 March 1950, report dated 14 April 1950, Parsons FBI file.
17. Forman's testimony to the FBI in: author redacted, 'Internal Security–R, Fraud Against the Government', PX 46–306, dated 1 June 1953, Malina FBI file; see also Arthur Wittenburg, 'Dr Sidney Weinbaum' LA 65–4846, 27 April 1950, p. 11, Weinbaum FBI.
18. Jan Wunderman journal, 16 August 1984, KWC.
19. Jan Wunderman journal, 16 August 1984, KWC.
20. This detail provided by an anonymous witness to Army Intelligence, a report from which was later quoted by the FBI: Arthur Wittenburg, 'Dr Sidney Weinbaum' LA 65–4846, 5 April 1949, p. 18, Weinbaum FBI file.
21. Ibid.
22. Sidney Weinbaum interview by Mary Terrall, CAL, p. 30.
23. Hans Liepmann, letter to *Engineering and Science*, Winter 1992, p. 40.
24. Benjamin Zibit interview of Jan Wunderman, 22 April 1996, CAL.
25. Frank Oppenheimer interview by Judith R. Goodstein, San Francisco, California, 16 November 1984, Oral History Project, CAL, p. 14.

26. Ibid., p. 15; see also: K. C. Cole, *Something Incredibly Wonderful Happens: Frank Oppenheimer and the World He Made Up*, New York, Houghton Mifflon Harcourt, 2009, p. 47–8.

27. Verification that the Dubnoffs lived in the Soviet Union comes from author interview with Ena Dubnoff, 9 June 2016. That Dubnoff recruited Weinbaum is reported by FBI informant T-2, Arthur Wittenburg, 'Dr Hsue-Shen Tsien wa [with alias]. John M. Decker', 29 August, 1949, LA 65–4857, p. 11, Tsien FBI file, ICP. Date of Weinbaum's membership from memo from SAC, Los Angeles to Director FBI, 'Communist Infiltration of the Jet Propulsion Laboratory, Caltech, Espionage–R', 26 January 1949, p. 4, Weinbaum FBI file.

28. Arthur Wittenburg, 'Dr Sidney Weinbaum, with aliases...,' LA 65–4846, 1 September 1949, p. 4, Weinbaum FBI.

29. Ibid., Frank Oppenheimer interview by Judith R. Goodstein, CAL, p. 10.

30. Kathryn S. Olmsted, *Right Out of California: The 1930s and the Big Business Roots of Modern Conservatism*, New York, The New Press, 2015, p. 110f.

31. Frank Oppenheimer interview by Judith R. Goodstein, CAL, p. 15.

32. See Robert W. Cherny, 'Prelude to the Popular Front: The Communist Party in California, 1931–35', *American Communist History*, vol. 1, no.1, 2002, pp. 5–42.

33. Leon Trotsky, *Literature and Revolution*, Chicago, Haymarket Books, 2005, p. 62.

34. David Caute, *The Fellow Travellers: a Postscript to the Enlightenment*, New York, Macmillan, 1973, p. 163, p. 4.

35. David Caute, p. 163.

36. Testimony of William Ward Kimple in letter from J. Edgar Hoover to Jack Neal, State Department, 'Frank J. Malina', 17 October 1947, Malina FBI file.

37. Ibid.; letter from FJM to parents, 19 November 1938, Box 21, Folder 9, FJM.

38. Letter from Frank Malina to parents, 7 November 1936, Box 21, Folder 6, FJM.

39. Evidence of Weinbaum's pro-Olson campaigning from an anonymous FBI informant, Arthur C. Wittenburg, 'Dr Sidney Weinbaum, Espionage–R', 15 June 1949, LA file 65–4846, Weinbaum FBI file.

40. Robert W. Cherny, 'The Communist Party in California, 1935–1940: From the Political Margins to the Mainstream and Back', *American Communist History*, vol. 9, no. 1, 2010, p. 3–33, p. 26.

41. Malina's CP paperwork is complicated. One version of the Malina FBI file used by the scholar James L. Johnson contains an enclosure that resembles his CP application form. There are at least four different versions of this FBI file in circulation, all with subtly different redactions. In January 2016, an

application for a renewed and full redaction was lodged by the author but not cleared at the time of publication. It's likely that many of the original files have since been destroyed. See also James L. Johnson, 'Rockets and the Red Scare: Frank Malina and the American Missile Development, 1936–1954', *Quest: the History of Spaceflight*, vol. 19, no. 1, 2012, pp. 30–36 (p. 32).

42. Testimony of Jack Parsons, interviewed 11 January 1950, file no. 65–4857, 24 February 1950, Tsien FBI file.
43. Unidentified informant in author redacted, 'Frank J. Malina wa. Frank Parma', file LA100–8196, 19 December 1950 Frank Malina FBI file.
44. Maths tuition: Arthur Wittenburg, 'Dr Hsue-Shen Tsien wa. John M. Decker', 24 February, 1950, LA 65–4857, p. 2, Tsien FBI file, ICP; 'not mathematically talented', Frank Malina oral history with James H. Wilson, 8 June 1973, JPL Archives Oral History Program, p. 23.
45. Ibid., Wittenburg, p. 2.
46. Arthur Wittenburg, 'Dr Hsue-Shen Tsien wa. John M. Decker', 24 February 1950, Tsien FBI.
47. Parsons quoted in report: Arthur Wittenburg, 'Sidney Weinbaum w.a.s., Espionage–R, Perjury', LA 65–4846, 14 April 1950, p. 20, Weinbaum FBI file.
48. Kimple's data doesn't explicitly mention Summerfield, only an identity called 'Fred Kane' whose real name is marked 'secret'. In the 250 or so pages of Summerfield's FBI file, it is clear that agents assume that Kane is Summerfield but demonstrating that in court would not have been easy. Kimple (as informant T-2) quoted in report: Philip J. Reilly, 'Martin J. Summerfield, Espionage–R', LA 65–4852, p. 6, Martin Summerfield FBI file.
49. Harvey Klehr, John Earl Haynes and Fridrikh Igorevich Firsov, *The Secret World of American Communism*, New Haven, Yale University Press, 1995, p. 34.
50. Ibid., p. 14, p. 34.
51. This is detailed in Kimple's testimony to House Committee on Un-American Activities, *Investigation of Communist Activities in the Los Angeles, Calif., Area*, Part 2, 30 June 1955, United States Congress.
52. Scott Allen McLellan, *Policing the Red Scare: The Los Angeles Police Department's Red Squad and the Repression of Labor Activism in Los Angeles, 1900–1940*, unpublished PhD dissertation, UC Irvine, 2010, p. 45.
53. Frank Donner, *Protectors of Privilege: Red Squads and Police Repression in Urban America*, Los Angeles: University of California Press, 1993, p. 60.
54. Ibid., p. 60.
55. Ibid., p. 60.
56. Ibid., p. 59.
57. For detail see Dorothy Healey and Maurice Isserman, *California Red: A Life in the American Communist Party*, New York, Oxford University Press, 1993, p. 40.

58. William Lawrence Ryan and Sam Summerlin, *The China Cloud: America's Tragic Blunder and China's Rise to Nuclear Power*, New York, Little, Brown, 1968, p. 57.

59. The Californian communist Dorothy Ray Healey once remarked that 'the FBI did a much better job of keeping track of our members than we did': Dorothy Ray Healey and Maurice Isserman, *California Red*, p. 73.

60. Robert W. Cherny, 'The Communist Party in California, 1935–1940', p. 25.

61. Frank Oppenheimer, oral history with Mary Terrall, 16 November 1984, CAL, p. 16.

62. Sidney Weinbaum interview by Mary Terrall, CAL, p. 56.

63. Testimony of FBI informant Gus Albrecht in: Arthur C. Wittenburg, 'Dr Sidney Weinbaum, with aliases, Espionage–R', 65–4846, 1 September 1949, Weinbaum FBI file.

64. Ibid.

65. Testimony of FBI informant Richard Rosanoff in Arthur C. Wittenburg, 'Sidney Weinbaum w.a.s., Espionage–R, Perjury', LA 65–4846, 14 April 1950, p. 17.

66. Francis Clauser Caltech oral history, by Peter Westwick, 13 April 2011, CAL, p. 61.

67. The Fejers shared with Malina an admiration for the liberal interwar Czech presidencies of Tomáš Masaryk and Edvard Beneš: Andrew Fejer, audio letter to Marjorie Malina, 1983, MFC.

68. Letter from Caroline Malina to Frank Malina, 23 November 1938, MFC.

69. William Ryan and Sam Summerlin, *The China Cloud*, p. 53.

70. Letter from Richard N. Lewis to Edward S. Lesnick, 17 September 1989, GLC.

71. Details of this influence in letter from Richard N. Lewis to Edward S. Lesnick, 3 December 1989, GLC.

72. Jan Wunderman journal, 16 August 1984, KWC.

73. Ibid.

74. Three scraps of Caltech Calendar among Malina's correspondence, MFC.

75. For a discussion of Marx as often peripheral to the interests and motivations of communist adherents, see David Caute, *The Fellow Travellers*, p. 25f.

76. Jan Wunderman journal, undated, probably April 1983, KWC.

77. Arthur C. Wittenburg, 'Dr Sidney Weinbaum, Espionage–R', 27 April 1950, LA file 65–4846, Weinbaum FBI file, p. 8.

78. Andrew Fejer, audio letter to Marjorie Malina, 1983, MFC.

79. Rachmaninoff: Weinbaum oral history with Mary Terrall, p. 56. Evidence of Weinbaum playing Prokofiev in letter from Lina Weinbaum to Frank Malina, 30 January 1948, MFC; the mutual friend of Prokofiev was Jenny Marling (also known as Evgina Afinogenov) American communist wife

of playwright Alexander Afinogenov, for which see memo from SAC Los Angeles to Director FBI, 'Communist Infiltration of Jet Propulsion Laboratory, Caltech',26 January 1949, p. 4, Weinbaum FBI file; see also Simon Morrison, *The Love and Wars of Lina Prokofiev*, London, Vintage, 2014, p. 160f.

80. Jan Wunderman journal, undated, probably April 1983, KWC.
81. Harvey Klehr, John Earl Haynes and Fridrikh Igorevich Firsov, *The Secret World of American Communism*, p. 38.
82. Testimony of FBI informant Richard N Lewis report: redacted author, 'Frank J. Malina wa, Internal Security–R,' 25 January 1952, p. 10, Malina FBI file.
83. Benjamin Zibit interview of Jan Wunderman, 22 April 1996, CAL, p. 7.

Chapter 5: War Games

1. Ellen Schrecker, *Many Are the Crimes: McCarthyism in America*, New York, Little, Brown and Co., 1998, p. 8.
2. Report by Arthur C. Wittenburg, 'Dr Sidney Weinbaum was., Espionage–R', 30 November 1949, p. 7, Weinbaum FBI file.
3. Letter from R. B. Hood to J. Edgar Hoover, 1 March 1949, FBI File no. 65–4846, Los Angeles, p. 3, Weinbaum FBI file.
4. Letter from Frank Malina to parents, 14 August 1939, Box 21, Folder 6, FJM.
5. 'Secretary of the Student Board', letter from Frank Malina to his parents, 2 October 1939, Box 21, Folder 10, FJM.
6. Letter from Frank Malina to his parents, 26 May 1940, Box 21, Folder 10, FJM.
7. Weinbaum details Pasadena Young Dems led by CP members: Sidney Weinbaum oral history with Mary Terrall, Caltech Archives, August 1985, p. 35; see also Dorothy Ray Healey and Maurice Isserman, *California Red*, p. 77.
8. Theodore von Kármán, *The Wind and Beyond*, p. 206.
9. Letter from Frank Malina to parents, 26 September 1938, Box 21, Folder 9, FJM; the work was eventually published as: F. J. Malina, 'Recent developments in the dynamics of wind-erosion', *Eos Trans. American Geophysical Union*, vol. 22, no. 2, 1941, pp. 262–87.
10. Letter from Frank Malina to Liljan Malina, 7 June 1939, MFC.
11. Letter from Frank Malina to Liljan Malina, 9 June 1939, MFC.
12. Letter from Frank Malina to his parents, 24 October 1938, Box 21, Folder 9, FJM.
13. Letter from Frank Malina to his parents, 22 January 1939, Box 21, Folder 9, FJM.
14. F. J. Malina, 'Recent developments in the dynamics of wind-erosion', p. 263.

15. C. J. Whitfield, 'Wind erosion, dunes, etc', Roundtable Discussion on the Role of Hydraulic Laboratories in Geophysical Research at the National Bureau of Standards, 13 September 1939, MFC.
16. Letter from Frank Malina to Liljan Malina, 1 October 1941, MFC. See also 'Obituary: George K. Morikawa', *Seattle Times*, 25 April 2006.
17. Letter from Frank Malina to Liljan Malina, 1 October 1941, MFC.
18. Ibid.
19. 'conservative Republican': Frank J. Malina JPL oral history, interviewed by James H. Wilson on 8 June 1973, p. 42, JPL.
20. All of this is detailed in the Homer J. Stewart interview by John L. Greenberg, Pasadena, Oct–Nov 1982, Session 3, Tape 4 Side 1, p. 69, CAL.
21. Report by R. H. Hallerberg, 7 May 1940, file no. 65–265, von Kármán FBI file, p. 7.
22. Millikan's friendship with Goetz detailed in Lee DuBridge oral history with Judith Goodstein, 19 February 1981, Caltech Archives, p. 7.
23. Report by W. E. Dettweiller, 'Theodore von Kármán, with aliases – Espionage G' 3 June 1942, file no. 65–265, p. 1, von Kármán FBI file.
24. Homer J. Stewart interview with John L. Greenberg, p. 67.
25. Ibid., p. 69.
26. Dorothy Ray Healey and Maurice Isserman, *California Red*, p. 82; John Holmes, 'American Jewish Communism and Garment Unionism in the 1920s', *American Communist History*, vol. 6, no. 2, 2007, pp. 171–95.
27. Maurice Isserman, *Which Side Are You On? The American Communist Party During the Second World War*, Chicago, University of Illinois Press, 1993, p. 37.
28. Ibid., p. 37.
29. Benjamin Zibit interview of Jan Wunderman, 22 April 1996, CAL, p. 11.
30. Jan Wunderman journal, 16 August 1984, KWC.
31. 'theoretical acrobatics', Jan Wunderman journal, undated, KWC.
32. Benjamin Zibit interview of Jan Wunderman, 22 April 1996, CAL, p. 8; Stravinksy and Shostakovich were favourites, see Andrew Fejer interview with Iris Chang Box 12, Folder 25, ICP.
33. Letter from Frank Malina to parents, 3 September 1939, Box 21, Folder 10, FJM.
34. Letter from Frank Malina to parents, 6 May 1940, Box 21, Folder 10, FJM.
35. Letter from Frank Malina to parents, 14 May 1939, Box 21, Folder 9, FJM.
36. Letter from Frank Malina to parents, 28 January 1941, Box 21, Folder 11, FJM.
37. 'wouldn't trust Mr Churchill', letter from Frank Malina to parents, 16 December 1940, Box 21, Folder 11, FJM; on slogans: Robert W. Cherny, 'The Communist Party in California, 1935–1940: From the Political Margins to

the Mainstream and Back', *American Communist History*, vol. 9, no. 1, 2010, pp. 3–33 (p. 30).

38. Carol Jean Newman, quoted in Dorothy Ray Healey and Maurice Isserman, *California Red*, p. 86.

39. Arthur C. Wittenburg, 'Sidney Weinbaum, w.a.s., Espionage–R, Perjury, LA file LA-65–4846, 14 April 1950, p. 16, Weinbaum FBI.

40. Jan Wunderman journal, 30 January 1986, KWC.

41. See Jan Wunderman journal, several entries; see also letter from Frank Malina to parents, 20 August 1941, Box 21, Folder 11, FJM.

42. Jan Wunderman journal, 29 December 1982, KWC.

43. Letter from Liljan Malina to Frank Malina, undated but June 1940, MFC.

44. Jan Wunderman journal, 29 December 1982, KWC.

45. See: 'Obituary: James Burford, 79; Unionist, Organizer', *Los Angeles Times*, 21 July 1990. A California Legislature enquiry into Un-American Activities found that Burford was the principal architect of communist influence in Young Dems: *Report of Joint Fact Finding Committee on Un-American Activities in California*, p. 84; 'moving power' quote from Jan Wunderman journal, 29 December 1982, KWC.

46. *Report of Joint Fact Finding Committee on Un-American Activities in California*, p. 69.

47. Jan Wunderman journal, 29 December 1982, KWC.

48. Burford wrote to her, 'I've loved you since your hair brushed my face that beautiful evening on the boat to Sausalito. When you stood close to me even when I did not dare to touch you first', undated letter from Jim Burford to Liljan Malina, MWC.

49. Jan Wunderman journal, 29 December 1982, KWC.

50. Letter from Jim Burford to Liljan Malina, 27 February 1946, MWC.

51. Jan Wunderman journal, 29 December 1982, KWC.

52. Liljan describes Bowling as a photographer. The 1940 Census for James R. Bowling says 'draftsman'. Correspondence with Bowling's niece Leanne Hines Hill confirmed he ran a photography business.

53. 'satisfying folly': Jan Wunderman journal, 11 November 1980; 'the most important thing in my life': Jan Wunderman journal, 28 May 1996, KWC.

54. Jan Wunderman journal, 11 November 1980, KWC.

55. 'the Bowling episode': letter from Frank Malina to Liljan Malina, 29 November 1945, MFC.

56. Jan Wunderman journal, 6 May 1981, KWC.

57. 'the Bowling episode': letter from Frank Malina to Liljan Malina, 29 November 1945, MFC.

Chapter 6: Give the Fascists Hell

1. Descriptions of sweater, smoking, from Jan Wunderman journal, 7 April 1983, KWC.

2. Benjamin Zibit interview of Jan Wunderman, 22 April 1996, CAL, p. 4.

3. Parsons and Forman pictured in: 'Rocket Works Like a Machine Gun', *Popular Science*, November 1938, p. 78; H. S. Tsien and F. J. Malina, 'Flight analysis of a sounding rocket with special reference to propulsion by successive impulses', *Journal of the Aeronautical Sciences*, vol. 6 no. 2, 1938, pp. 50–58. To be fair, Parsons and Forman also published on this topic, albeit in the less prestigious journal of the American Rocket Society: John W. Parsons and Edward S. Forman, 'Experiments with Powder Motors for Rocket Propulsion by Successive Impulses', *Astronautics*, vol. 9, no. 43, August 1939, pp. 4–11.

4. Benjamin Zibit, *The Guggenheim Aeronautics Laboratory at Caltech and the Creation of the Modern Rocket Motor*, p. 338.

5. P. Thomas Carroll, 'Historical Origins of the Sergeant Missile Powerplant', in Kristan R. Lattu and R. Cargill Hall, eds., *History of Rocketry and Astronautics: Proceedings of the Seventh and Eighth History Symposia of the International Academy of Astronautics*, AAS History Series, vol. 8, San Diego: Univelt, Inc., for the American Astronautical Society, 1989, pp. 121–46 (p. 123).

6. F. J. Malina, 'The US Army Air Corps Jet Propulsion Research Project, GALCIT project no. 1, 1939–1946: A memoir', *Essays on the History of Rocketry and Astronautics, Vol. 2*, 1977, pp. 153–201 (p. 158).

7. Theodore von Kármán, *The Wind and Beyond*, p. 245.

8. 'exotic furniture' etc. from Jan Wunderman journals, 10 November 1981, KWC.

9. Benjamin Zibit, *The Guggenheim Aeronautics Laboratory at Caltech and the Creation of the Modern Rocket Motor*, p. 343.

10. Homer J. Stewart interview with John L. Greenberg. Pasadena, California, Oct–Nov 1982, Oral History Project, CAL, p. 21.

11. See letter from Frank Malina to parents, 21 October 1939, FJM Collection, Box 21, Folder 10.

12. Letter from Frank Malina to parents, 13 February 1939, Box 21, Folder 9, FJM.

13. Malina calls the theory by this name in a letter to P. Thomas Carroll, 23 May 1972, Box 5, Folder 13, FJM.

14. 'Goddard furious': Clayton Koppes, *JPL and the American Space Program*, p. 10; 'demons': David Clary, *Rocket Man: Robert H. Goddard and the Birth of the Space Age*, New York, Hyperion, 2004, p. 202.

15. F. J. Malina, 'The US Army Air Corps Jet Propulsion Research Project, GALCIT Project no. 1, 1939–1946: A memoir', *Essays on the History of Rocketry and Astronautics, Vol. 2*, 1977, pp. 153–201 (p. 158).

16. Letter from Martin Summerfield to G. Edward Pendray, 10 July 1965, copy in Box 6, Folder 25, FJM.

17. See Clayton Koppes, *JPL and the American Space Program*, p. 11.

18. Letter from Frank Malina to parents, 30 December 1940, Box 21, Folder 11, FJM.

19. Le Page's stationery glue: Homer A. Boushey, 'A Brief History of the First US JATO Flight Tests of August 1941: a memoir', *Proceedings of the Eighteenth and Nineteenth History Symposia of the International Academy of Astronautics*, Lausanne, Switzerland, AAS History Series, vol. 14, 1993, pp. 127–37 (p. 129).

20. Fred Miller, 'First JATO: ammonium nitrate, cornstarch, black powder and glue', American Rocket Society News, *Journal of Jet Propulsion*, vol. 26, no. 1, 1956, pp. 51–2 (p. 51).

21. Homer A. Boushey, 'A Brief History of the First US JATO Flight Tests of August 1941', p. 131.

22. Ibid.; F. J. Malina, 'The US Army Air Corps Jet Propulsion Research Project, GALCIT Project no. 1, 1939–1946'.

23. Footage of these tests is included in *The American Rocketeer*, dir. Blaine Baggett, 34 min., JPL DVD, 56 sec.; quote from Theodore von Kármán *The Wind and Beyond*, p. 250.

24. All of this from Theodore von Kármán, *The Wind and Beyond*, p. 250.

25. For a longer discussion of Barbara Canright and the JPL computers see Nathalia Holt, *Rise of the Rocket Girls: The Women who Propelled Us, from Missiles to the Moon to Mars*, New York, Little, Brown, 2016.

26. Involvement of Millikan and Stewart, Theodore von Kármán, *The Wind and Beyond*, p. 250.

27. The poster is visible and legible in JPL film footage, *The American Rocketeer*, dir. Blaine Baggett, 35 min.

28. Homer A. Boushey, 'A Brief History of the First US JATO Flight Tests of August 1941', p. 134.

29. Ibid., p. 134; Fred Miller, 'First JATO: ammonium nitrate, cornstarch, black powder and glue', p. 51.

30. Frank J. Malina, oral history by James H. Wilson, 8 June 1975, JPL Oral History Program, JPL, p. 38.

31. Malina, F. J., 'The US Army Air Corps Jet Propulsion Research Project, GALCIT Project no. 1, 1939–1946', p. 172.

32. These are mentioned in a letter from Frank Malina to P. Thomas Carroll, 24 February 1971, Box 5, Folder 13, FJM; on Miller's claim, see letter from William E. Zisch to Frank Malina, 22 March 1961, Box 6, Folder 2, FJM.

33. John Carter, *Sex and Rockets*, p. 72.

34. Letter from Frank Malina to parents, 3 September 1940, Box 21 Folder 11, FJM.

35. Report: author redacted, 'Frank J. Malina, Internal Security–R, Custodial Detention', 28 November 1942, file no. 11–8196, p. 1, Malina FBI file.

36. Letter from Frank Malina to parents, 15 June 1942, Box 22, Folder 1, FJM.

37. Clayton Koppes, *JPL and the American Space Program*, p. 17.

38. This dialogue from Nathalia Holt, *The Rise of the Rocket Girls*, p. 26. Holt's source is: Walter B. Powell, 'Comments on Malina Memoirs', 6 August 1970, four pages, handwritten, JPL. Upon request, these notes were not available to this author having apparently been misfiled.

39. John Carter, *Sex and Rockets*, p. 76.

40. Nathalia Holt conflates the axe and necktie episodes, though Clayton Koppes sees them as separate and unrelated; Clayton Koppes, *JPL and the American Space Program*, p. 16.

41. This is detailed in a letter from JPL 'computer' Melba Nead to Kyky Chapman, 5 November 1991, JPL Archives.

42. Ibid.

43. John Carter, *Sex and Rockets*, p. 97.

44. Clayton Koppes, *JPL and the American Space Program*, p. 15.

45. Theodore von Kármán, *The Wind and Beyond*, p. 252.

46. Clayton Koppes, *JPL and the American Space Program*, p. 15.

47. Theodore von Kármán, *The Wind and Beyond*, p. 251.

48. Ibid., p. 252.

49. See F.J. Malina, 'The US Army Air Corps Jet Propulsion Research Project, GALCIT Project no. 1, 1939–1946', p. 167.

50. Leonard Caveny, 'Martin Summerfield and His Princeton University Propulsion and Combustion Laboratory', 47th AIAA/ASME/SAE/ASEE Joint Propulsion Conference & Exhibit, 2011, available from arc.aiaa.org, p. 9.

51. Powell quoted in Clayton Koppes, *JPL and the American Space Program*, p. 14.

52. Letter from Frank Malina to parents, 22 March 1942, Box 22, Folder 1, FJM.

53. Letter from Frank Malina to parents, 29 March 1942, Box 22, Folder 1, FJM.

54. Theodore von Kármán, *The Wind and Beyond*, p. 254.

55. Letter from Frank Malina to parents, 22 March 1942, Box 21, Folder 11, FJM.

56. This number, albeit without a clear methodology for its calculation, is mentioned in Frank H. Winter and George S. James, 'Highlights of 50 years of Aerojet, a pioneering American rocket company, 1942–1992', *Acta Astronautica*, vol. 35, nos. 9–11, 1995, pp. 677–98, (p. 680).
57. Theodore von Kármán, *The Wind and Beyond*, p. 254.

Chapter 7: The Way of The Beast

1. Crowley hated Christians 'as socialists hate soap': Ronald Hutton, *The Triumph of the Moon: A History of Modern Pagan Witchcraft*, New York, Oxford University Press, 1999, p. 187; Crowley 'vermin socialists': Tobias Churton, *Aleister Crowley: The Biography*, London, Watkins Publishing, 2011, p. 140.
2. Martin P. Starr, *The Unknown God: Wilfred T. Smith and the Thelemites*, Bolingbrook, Teitan Press, 2003, p. 260, p. 257.
3. Parsons wrote about how his 'repressed homosexual component' was capable of 'serious disorder': 'Analysis by a Master of the Temple of the critical Modes in the Experience of his Material Vehicle', undated, *c.* 1948, Folder DD2, GYC.
4. Francis King, *The Magical World of Aleister Crowley*, London, Arrow Books, 1978, p. 173; disrobing tended to be in private rather than in public performances; Martin P. Starr, *The Unknown God*, p. 260, fn. 28.
5. Letter from Aleister Crowley to Helen Parsons Smith, 1943, quoted in Starr, *The Unknown God*, p. 260, fn. 29.
6. Hugh B. Urban, *Magia Sexualis: Sex, Magic, and Liberation in Modern Western Esotericism*, University of California Press, 2006, p. 122.
7. Lawrence Sutin, *Do What Thou Wilt: A Life of Aleister Crowley*, New York, St Martin's Press, 2000, p. 114.
8. Starr, *The Unknown God*, p. 257.
9. 'Semper Fidelis', George Pendle, *Strange Angel*, p. 204.
10. Martin P. Starr, *The Unknown God*, p. 381.
11. Francis X. King, *The Secret Rituals of the OTO*, Samuel Weiser, New York, 1973, p. 28.
12. Diary of Jane Wolfe, edited by Phyllis Seckler, published as Jane Wolfe, 'The Sword', *In The Continuum* [this is the journal of Seckler's College of Thelema, not the OTO], vol. 3, no. 7 (1985), p. 34.
13. Martin P. Starr, *The Unknown God*, p. 247.
14. Quoted in ibid., p. 182.
15. Jane Wolfe, 'The Sword', *In The Continuum*, vol. 3, no. 6, 1984, p. 38.
16. Martin P. Starr, *The Unknown God*, p. 263.
17. Forman, Canrights and Miller are mentioned in the Appendix of OTO members and their degree work in ibid., pp. 365–6.

18. Benjamin Zibit interview of Jan Wunderman, 22 April 1996, CAL, p. 9.
19. Martin P. Starr, *The Unknown God*, p. 257.
20. Benjamin Zibit interview of Jan Wunderman, 22 April 1996, CAL, p. 9. Martin P. Starr, *The Unknown God*, p. 257.
21. The Kahl-Seckler-Smith tensions are detailed in Jane Wolfe's diary for 13 June 1941; see Jane Wolfe, 'The Sword', *In The Continuum*, vol. 3, no. 6, 1984, p. 36.
22. Benjamin Zibit interview of Jan Wunderman, 22 April 1996, CAL, p. 10.
23. Martin P. Starr, *The Unknown God*, pp. 254–5.
24. George Pendle, *Strange Angel*, p. 203.
25. Martin P. Starr, *The Unknown God*, p. 256.
26. George Pendle, *Strange Angel*, p. 204.
27. Martin P. Starr, *The Unknown God*, p. 271.
28. Ibid., p. 256. Helen Parsons Smith gave the same detail to Martin P. Starr, pers. comm, 15 May 2016. Within the OTO, Betty's childhood sexual experience was interpreted as precociousness. As Jane Wolfe put it in a letter to Crowley, Betty 'stepped across the threshold of womanhood at the age of 10: vital experiences began at 12'; letter from Jane Wolfe to Aleister Crowley, 26 November 1943, quoted in Jane Wolfe, Hollywood, 'The Sword', *In The Continuum*, vol. 4, no. 3, 1988, p. 33. Variance regarding Betty's precise age here must not obscure the reality of abuse.
29. John Whiteside Parsons, 'Analysis by a Master of the Temple of the critical Modes in the Experience of his Material Vehicle', undated, *c.* 1948, Folder DD2, GYC.
30. Hugh B. Urban, *Magia Sexualis*, pp. 129–30.
31. John Carter, *Sex and Rockets*, p. 88.
32. Aleister Crowley, *Hymn to Pan*, Chicago, Renshaw Press, 1919.
33. See in particular Nathalia Holt, *Rise of the Rocket Girls*.
34. Patricia Canright Smith, 'She Would Love To See China', (em) *Review of Text and Image*, issue one, Fall 2012, pagination unknown, republished http://www.patriciacanrightsmith.com/she-would-love-to-see-china/ accessed 25 May 2017; interview with Patricia Canright Smith, Friday 12 May 2017.
35. The Canrights' daughter Patricia sees her mother's participation in the OTO as complex, as both having her own agency but within the misogynist culture of the Lodge; pers. comm. with author, 6 August 2018.
36. Jane Wolfe, 'Hollywood, The Sword', *In The Continuum*, vol. 3, no. 7, 1985, p. 38.
37. 'detestable institution': Crowley quote from George Pendle, *Strange Angel*, p. 204.

38. Helen Parsons Smith diary, 16 August 1941, quoted in Martin P. Starr, *The Unknown God*, p. 271.
39. Jane Wolfe, 'Hollywood, The Sword', *In The Continuum*, vol. 3, no. 7, 1985, p. 42.
40. Martin P. Starr, *The Unknown God*, p. 304.
41. Jane Wolfe, 'Hollywood, The Sword', *In The Continuum*, vol. 3, no. 7, 1985, p. 33.
42. Pendle cites correspondence with Phyllis Seckler for the night watchman job: George Pendle, *Strange Angel*, p. 322. Malina recalls Seckler as working as a mechanic with Parsons and Fred Miller: Frank Malina oral history by James H. Wilson, 8 June 1973, JPL CL-08_2255, JPL, p. 25.
43. Jane Wolfe, 'Hollywood, The Sword', *In The Continuum*, vol. 3, no. 7 (1985), p. 33.
44. Frank Malina oral history by James H. Wilson, p. 27.
45. Ibid., p. 27.
46. Ibid., p. 27.
47. Ibid., p. 27.
48. Letter from Frank Malina to parents, 4 November 1941, Box 22, Folder 1, FJM.
49. Martin P. Starr, *The Unknown God*, p. 274.
50. Frank Malina oral history by James H. Wilson, p. 25.
51. Ibid., p. 26.
52. Ibid., p. 27.
53. Jan Wunderman journal, undated but probably May 1996, KWC.

Chapter 8: Enterprise

1. See Daniel S. Greenberg, *The Politics of American Science*, London, Penguin, 1967, p. 137.
2. Frank J. Malina, 'The US Army Air Corps Jet Propulsion Research Project, GALCIT Project no. 1, 1939–1946: A memoir', p. 194.
3. Frank J. Malina interview with Mary Terrall, Pasadena, California, 14 December 1978, Oral History Project, CAL, p. 10.
4. Theodore von Kármán, *The Wind and Beyond*, p. 256.
5. Ibid.
6. Frank J. Malina, 'The US Army Air Corps Jet Propulsion Research Project, GALCIT Project no. 1, 1939–1946: A memoir', p. 194.
7. Letter from Frank Malina to parents, 7 December 1941, Box 22, Folder 1, FJM.
8. Letter from Frank Malina to parents, 14 December 1941, Box 22, Folder 1, FJM.
9. Jerry Grey, *Enterprise*, New York, Morrow, 1979, p. 117.

10. Letter from Frank Malina to parents, 22 March 1942, Box 22, Folder 1, FJM Collection.
11. Charles H. Ehresman, *Aerojet Engineering Corporation First Manufacturing Plant Pasadena, California*, American Institute of Aeronautics and Astronautics pamphlet, Reston, Virginia, undated, p. 3.
12. Theodore von Kármán, *The Wind and Beyond*, p. 259.
13. Letter from Frank Malina to Liljan Malina, 26 October 1942, MFC.
14. Letter from Frank Malina to parents, 3 September 1942, Box 22, Folder 3, FJM.
15. Reaction Motors Ltd had already been established by the American Rocket Society.
16. Letter from Frank Malina to parents, 20 November 1942, Box 22, Folder 2, FJM.
17. Ibid.
18. On the Parsons and Forman move to Aerojet, Malina only visiting the company 'maybe once a week': see Frank Malina oral history by Mary Terrall. Pasadena, California, 14 December 1978, CAL, p. 13.
19. Letter from Frank Malina to parents, 25 November 1941, Box 22, Folder 1, FJM.
20. Letter from Frank Malina to parents, 12 September 1941, Box 22, Folder 1, FJM.
21. Frank Malina oral history by James H. Wilson, 8 June 1973, JPL CL-08_2255, JPL, p. 46.
22. Letter from Lina Weinbaum to Frank Malina, 5 October 1948, MFC; of Dunn's devotion, details from an author interview with Linda Dunn Payne, 2 May 2017.
23. This comment is given by informant T-7 in a report, redacted author, 'Frank J. Malina, wa. Frank Parma, Internal Security–R', LA-100–8196, 26 July 1949, p. 7, Malina FBI file. The identity of T-7 is given in a report from Special Agent R. B. Hood to FBI Director, 4 June 1949, Martin Summerfield FBI file.
24. 'Frank J. Malina, with alias Dr Frank J. Malina', 28 November 1942, Internal Security–R Custodial Detention, Frank Malina FBI file, p. 4.
25. See J. Edgar Hoover, 'Alien Enemy Control', *Iowa Law Review*, vol. 29, 1943, p. 396.
26. redacted author, 'Frank J. Malina, wa. Frank Parma, Internal Security–R', LA-100–8196, 26 July 1949, p. 7 , Malina FBI.
27. On Parsons' drug use, see George Pendle, *Strange Angel*, p. 216.
28. Quote from letter from Aleister Crowley to Jane Wolfe, 4 May 1943 reproduced in: Jane Wolfe, Pasadena, *In The Continuum*, vol. 4, no. 1, 1987, p. 37; concern about drug use from the same source, p. 37.

29. Letter from McMurtry to Parsons, quoted in Pendle, *Strange Angel*, p. 215; amphetamine use, p. 216.

30. Martin P. Starr, *The Unknown God*, p. 288.

31. Jane Wolfe, 'Hollywood, The Sword', *In The Continuum*, vol. 3, no. 7, 1985, p. 40.

32. Jane Wolfe, 'Pasadena', *In The Continuum*, vol. 3, no. 8, 1985, p. 33.

33. 'Jack said he would try very hard to get the money needed' from Jane Wolfe, 'Pasadena', *In The Continuum*, vol. 3, no. 8, 1985, p. 33.

34. Martin P. Starr, *The Unknown God*, p. 272.

35. Much later, after the effective breakdown of Agape Lodge, 1003 was known as 'the Parsonage'. I have stuck to 'Grim Gables' in part because it better describes the character of what took place there.

36. All these details from Martin P. Starr, *The Unknown God*, p. 273.

37. Jane Wolfe, 'Pasadena', *In The Continuum*, vol. 3, no. 9, 1986, p. 33.

38. All wildlife details from Jane Wolfe, 'Pasadena' *In The Continuum*, vol. 3, no. 9, 1986, p. 36, p. 37.

39. Jane wrote to tell on him: Jane Wolfe, 'Pasadena', *In The Continuum*, vol. 3, no. 9, 1986, p. 41; 'white-livered lunatic': letter from Aleister Crowley to Jane Wolfe, in Jane Wolfe, 'Pasadena', *In The Continuum*, vol. 4, no. 1, 1987, p. 37.

40. Aleister Crowley, *Liber Oz*.

41. Date from Martin P. Starr, *The Unknown God*, p. 285; report from 1943 detailed in author redacted, 'John Whiteside Parsons, aka Jack Parsons', File no. 65–5131, 7 December 1950, p. 9, Parsons FBI file.

42. Ibid., p. 8.

43. Jane Wolfe and Karl Germer, quoted in Jane Wolfe, 'Pasadena', *In The Continuum*, vol. 3, no. 9, 1986, p. 44.

44. Martin P. Starr, *The Unknown God*, p. 302; Letter from Aleister Crowley to Jack Parsons, 19 October 1943, quoted in Jane Wolfe, 'Hollywood, The Sword', *In The Continuum*, vol. 4, no. 2, 1987, p. 35.

45. Aleister Crowley, *Liber CXXXII*, NS15, GYC.

46. Ibid., emphasis in the original; see also Martin P. Starr, *The Unknown God*, p. 295.

47. Ibid., p. 296.

48. Letter from Aleister Crowley to Jack Parsons, 19 October 1943, quoted in Jane Wolfe, 'Hollywood, The Sword', *In The Continuum*, vol. 4, no. 2, 1987, p. 31.

49. Letter from Aleister Crowley to Jane Wolfe, 8 November 1943, quoted in ibid., p. 37.

50. Frank J. Malina, 'America's First Long-Range-Missile and Space Exploration Program', p. 343.

51. Theodore von Kármán, *The Wind and Beyond*, p. 264.

52. Letter from Frank Malina to parents, 17 August 1943, Box 22, Folder 2, FJM.

53. Frank J. Malina, 'America's First Long-Range-Missile and Space Exploration Program', p. 343.

54. Theodore von Kármán, 'Memorandum on the Possibilities of Long-Range Rocket Projectiles', JPL-GALCIT, Memo no. 1, JPL-1, 20 November 1943, JPL.

55. Clayton Koppes, *JPL and the American Space Program*, p. 19.

56. Frank J. Malina, 'America's First Long-Range-Missile and Space Exploration Program', p. 344.

57. Ibid., p. 345.

58. Clayton Koppes, *JPL and the American Space Program*, p. 20.

59. Theodore von Kármán, *The Wind and Beyond*, p. 267.

60. It seems as if this move was incremental rather than abrupt; Parsons is still listed as a GALCIT Jet Propulsion Project employee as late as May 1943: Benjamin Zibit, *The Guggenheim Aeronautics Laboratory at Caltech and the Creation of the Modern Rocket Motor*, p. 420.

61. Parsons is not listed in JPL's first organisational chart of personnel dated 9 January 1945, Box 74, Folder 4, Theodore von Kármán papers, CAL.

62. Jerry Grey, *Enterprise*, p. 119.

63. Theodore von Kármán, *The Wind and Beyond*, p. 260.

64. Ibid., p. 260.

65. Large orders: Iris Chang, *Thread of the Silkworm*, p. 97 (Chang cites JPL historian John Bluth for this information); worst capital position: Frank Malina oral history with R. Cargill Hall, 29 October 1968, ICP.

66. Letter from Frank Malina to parents, 9 April 1943, MFC.

67. Frank H. Winter and George S. James, 'Highlights of 50 Years of Aerojet, a Pioneering American Rocket Company', p. 681.

68. Interview with Apollo M. O. Smith by Jennifer K. Stine, Caltech Oral History, CAL, p. 17.

69. Letter from Frank Malina to Andy Haley, 9 March 1955, Box 2, Folder 6, FJM; see also Frank J. Malina, 'The US Army Air Corps Jet Propulsion Research Project, GALCIT Project no. 1: A memoir', p. 194.

70. Theodore von Kármán, *The Wind and Beyond*, p. 316.

71. Andrew G. Haley, *Rocketry and Space Exploration*, New York, Van Nostrand, 1958, p. 158.

72. Ibid.

73. This is evident in a reply from Crowley to Jack quoted in Jane Wolfe, 'Hollywood, The Sword', *In The Continuum*, vol. 4, no. 2, 1987, p. 30; 'lack of stability' from letter from Jane Wolfe to Aleister Crowley, 26 November

1943, in Jane Wolfe, 'Hollywood' *In The Continuum*, vol. 4, no. 3, 1988, p. 33.

74. Frank Malina oral history with James H. Wilson, JPL Archives Oral History Program, 8 June 1973, JPL CL-08–2255, p. 13, JPL.

75. Letter from Frank Malina to Liljan Malina, undated but sequence suggests December 1943, MFC.

76. Letter from Frank Malina to parents, 6 January 1944, Box 22, Folder 3, FJM.

77. Letter from Frank Malina to Liljan, undated but sequence suggests December 1943, MFC.

78. Jane Wolfe, 'Hollywood, The Sword', *In The Continuum*, vol. 3, no. 6, 1984, p. 34.

79. Letter from Liljan Malina to Frank Malina, undated but sequence suggests 1943, MFC.

80. Letter from Frank Malina to Liljan Malina, 12 December 1943, MFC.

81. Jan Wunderman journals, undated, between 1992 and 1998, KWC.

82. Charles Bartley interview with John Bluth, 3 and 4 October 1994, JPL Archives Oral History Program, pp. 14–15, JPL. Jack Parsons is not recorded in the minutes of JPL meetings. Until more recent times (when JPL has used the 1936 Halloween test as a charismatic, if inaccurate, myth of origin, it traced its genesis to 1944) long after Parsons had gone. This is detailed in the programme 'In commemoration of the first test of a rocket motor by Caltech students and associates, October 31, 1936', JPL, California Institute of Technology, 31 October 1968, Box 4, Folder 1, FJM.

83. Theodore von Kármán, *The Wind and Beyond*, p. 317.

84. George Pendle, *Strange Angel*, p. 240; these figures should be treated with caution, not least because they appear not to reflect the extent of their original investment. A letter from Frank Malina to his parent claims Frank invested $2,000 overall: 19 December 1944, Box 22, Folder 4, FJM.

85. Theodore von Kármán, *The Wind and Beyond*, p. 317.

86. Charles Bartley interview with John Bluth, JPL, p. 15.

87. J. D. Hunley, *The Development of Propulsion Technology for US Space-Launch Vehicles, 1926–1991*, College Station, Texas A&M University Press, 2013, p. 294.

88. Untitled notebook of Frank Malina, undated, 4"×6", MFC.

Chapter 9: The Quest for Space

1. Maurice Isserman, *Which Side Were You On? the American Communist Party During the Second World War*, Chicago, University of Illinois Press, 1993.

2. Frank Oppenheimer interview by Judith R. Goodstein, CAL, p. 16.

3. K. D. Cole, *Something Incredibly Wonderful Happens*, p. 57.

4. Sidney Weinbaum oral history with Mary Terrall, Caltech Archives, August 1985, p.44.

5. Letter from Frank Malina to parents, 11 May 1944, Box 22, Folder 3, FJM.

6. Letter from Eli Epstein to Liljan Malina, 2 December 1944, MWC.

7. Letter from Liljan to Frank Malina, 26 October 1944, MFC.

8. Letter from Liljan to Frank Malina, 15 September 1944, MFC.

9. Letter from Frank Malina to parents, 11 May 1944, Box 22, Folder 3, FJM.

10. Letter from Frank Malina to parents, 22 April 1944, Box 22, Folder 3, FJM.

11. Letter from Frank Malina to Liljan Malina, 19 April 1944, MFC.

12. 'I spoke to Frank about wanting a divorce. His answer? "Not now, can't you see how tied up I am with the war effort? We'll talk about this when the war is over."' Jan Wunderman journal, undated, probably after 2003, KWC.

13. 'whims' is used by Burford quoting Liljan's original letter, now lost. See letter from Jim Burford to Liljan Malina, 4 May 1944, MWC.

14. Ibid.

15. Some of Liljan's Mexico paintings were exhibited at the George Gastines Galleries: no author, 'Three Women from Southland Exhibit Oils', *Los Angeles Times*, 3 June 1945, p. 26.

16. Art school friend mentioned in letter from Frank Malina to parents, 26 July 1944, MFC; quote from letter from Liljan to FJM, 15 September 1944, MFC; Mary Tracy's full name in a letter from Jan Wunderman to Fabrice Lapelletrie, 27 July 2009, Fabrice Lapelletrie private collection.

17. Letter from Liljan Malina to Frank Malina, dated only August 1944, MFC.

18. Letter from Liljan Malina to Frank Malina, 15 September 1944, MFC.

19. Letter from Liljan Malina to Frank Malina, dated only August 1944, MFC.

20. Letter from Liljan Malina to Frank Malina, 15 September 1944, MFC.

21. Ibid.

22. Letter from Frank Malina to Liljan Malina, 27 August 1944, MFC.

23. Letter from Frank Malina to Liljan Malina, 6 September 1944, MFC.

24. Ibid.

25. Ibid.

26. Michael J. Neufeld, *Von Braun*, p. 185.

27. Letter from Frank Malina to his parents, 14 November 1944, Box 22, Folder 3, FJM.

28. Letter from Frank Malina to Liljan Malina, 30 October 1944, MFC.

29. Steven J. Zaloga, *V-2 Ballistic Missile 1942–52*, Oxford, Osprey Publishing, 2003, p. 23.

30. For more on this see Frank J. Malina interviewed by James H. Wilson, 8 June 1973, p. 16f, JPL.

31. Frank J. Malina, 'America's First Long-Range-Missile and Space Exploration Program', p. 351.

32. Letter from Frank Malina to Liljan Malina, 22 October, 1944, MFC.
33. Letter from Frank Malina to Liljan Malina 8 November 1944, MFC.
34. Letter from Frank Malina to Liljan Malina, 4 December 1945, MFC.
35. Malina, Frank J., 'America's First Long-Range-Missile and Space Exploration Program', p. 352.
36. Letter from Frank Malina to Liljan Malina, 27 August 1944, MFC.
37. Theodore von Kármán, *The Wind and Beyond*, p. 265.
38. Frank J. Malina, 'America's First Long-Range-Missile and Space Exploration Program', p. 353.
39. See Douglas J. Mudgway, *William Pickering: America's Deep Space Pioneer* Washington DC, NASA History Division, 2008, p. 45.
40. 'over a weekend': Homer J. Stewart interview with John L. Greenberg, CAL, p. 70; the report was F. J. Malina and H. J. Stewart, 'Considerations of the Feasibility of Developing a 100,000ft Altitude Rocket (The WAC Corporal)', JPL-GALCIT memo 4–4, January 1945, unpublished, JPL.
41. See for instance William Pickering, Oral History by Shirley Cohen, 22 and 29 April 2003, p. 12, CAL.
42. When von Kármán left for Washington he asked Tsien to join him as a member of the Scientific Advisory Group, which meant that Stewart – with Malina's approval – was promoted to section head. See Iris Chang, *Thread of the Silkworm*, p. 109.
43. Homer J. Stewart interview with John L. Greenberg, CAL, p. 74.
44. Ibid.
45. Memo from Special Agent in Charge, Washington Field Office to FBI Director, 20 February 1953, ' Frank J. Malina, Fraud Against the Government', Malina FBI.
46. Frank J. Malina, 'America's First Long-Range-Missile and Space Exploration Program', p. 360.
47. See Benjamin Zibit, *The Guggenheim Aeronautics Laboratory at Caltech and the Creation of the Modern Rocket Motor*, p. 455.
48. Letter from Frank Malina to parents, 13 May 1945, Box 22, Folder 4, FJM.
49. Letter from Frank Malina to parents, 28 May 1945, Box 22, Folder 4, FJM.
50. See Jan Wunderman journal, 16 August 1984, KWC.
51. Letter from Liljan Malina to Frank Malina, 2 April 1945, MFC.
52. Letter from Frank Malina to Liljan Malina, 13 April 1945, MFC.
53. Letter from Frank Malina to parents, 31 March 1940, Box 21, Folder 10, FJM.
54. Letter from Frank Malina to Liljan Malina, 23 April 1945, MFC.
55. Ibid.
56. See also letter from Frank Malina to parents, 13 May 1945, Box 22, Folder 4, FJM.

57. Theodore von Kármán, *The Wind and Beyond*, p. 279.

58. Iris Chang, *Thread of the Silkworm*, p. 116.

59. Leonard H. Caveny, 'Martin Summerfield and His Princeton University Propulsion and Combustion Laboratory', p. 20.

60. Letter from Frank Malina to Liljan Malina, 23 September 1945, Malina Family Archive.

61. Letter from Frank Malina to Liljan Malina, 30 September 1945, Malina Family Archive.

62. Ibid.

63. None of the written accounts mention a countdown as the prelude to launch, but film footage of the launch shows hand signals counting down from 'five', *The American Rocketeer*, DVD, JPL.

64. Benjamin Zibit, *The Guggenheim Aeronautics Laboratory at Caltech and the Creation of the Modern Rocket Motor*, p. 475.

65. Frank J. Malina, 'America's First Long-Range-Missile and Space Exploration Program', p. 365.

66. Laundry evident in letter from Liljan Malina, undated but probably September 1945, MFC.

67. See Fraser MacDonald, 'Instruments of Science and War: Frank Malina and the Object of Rocketry' in Fraser MacDonald and Charles W. J. Withers, eds, *Geography, Technology and Instruments of Exploration*, Farnham, Ashgate Publishing Ltd, 2015, pp. 219–40.

68. On historical definitions of space see: Jonathan C. McDowell, 'The edge of space: Revisiting the Karman Line', *Acta Astronautica*, vol. 151, 2018, pp. 668–77 (p. 669). For practical purposes, the US Air Force acknowledged 80 km/50 miles as the edge of space, the point at which 'astronaut wings' were conferred on pilots. The Kármán line was not established until 1963.

69. Letter from Frank Malina to Liljan Malina, 11 October 1945, MFC.

70. Letter from Frank Malina to parents, 29 September 1945, Box 22, Folder 5, FJM.

71. Letter details conversations with von Kármán the previous week: letter from Frank Malina to parents, 23 August 1945, Box 22, Folder 5, FJM.

72. Jan Wunderman journals, 11 November 1980, KWC.

73. This phrase from Jan Wunderman journals, 16 August 1984, KWC.

74. 'My mother did not want me back in their house': Jan Wunderman journal, undated but after 2003, KWC.

75. The Liepmanns had been to dinner with Frank and Liljan, detailed in letters from Frank Malina to parents, 20 November 1939, 30 October 1939, Box 21, Folder 10, FJM Collection; the matter of Hans' assessment of Malina is suggested in William Fowler Oral History by John Greenberg and Carol

Bugé, 3 May 1983–31 May 1984, 3 October 1986, Oral History Project, CAL, p. 179.

76. The character differences between Hans and Katja Liepmann, and their eventual divorce, are noted in William Sears, *Stories of a Twentieth-Century Life*, Stanford, Parabolic Press, 1994, p, 79.

77. The available version of Malina's FBI file, of which only parts are unredacted, shows very little FBI activity or interest from 1943 to 1947. Liljan's account of this raid is, however, exactly coincident with US Army Intelligence concern with the leak of the WAC Corporal document. These security files have not been found.

78. The FBI files on von Kármán that relate to the investigation of wind tunnel data are copied to Army Counter Intelligence Corps. All attempts to trace these CIC files at the National Archives and Records Adminstration have been unsuccessful.

79. Details of the WAC Corporal plans leak are retrospectively detailed in numerous reports across the FBI files for Malina, Summerfield, Tsien and Weinbaum.

80. Homer J. Stewart interview with John L. Greenberg, CAL, p. 69.

Chapter 10: The Shrinks

1. Letter from Frank Malina to Liljan Malina, 29 November 1945, MFC.
2. Jan Wunderman journal, undated, 1992–98, Karen Wunderman Private Archive.
3. Letter from Frank Malina to Liljan Malina, 29 November 1945, MFC.
4. Letter from Frank Malina to Liljan Malina, 29 November 1945, MFC.
5. Paul Epstein interview with Alice Epstein, Pasadena, California. 22 November 1965 CAL, p. 74; Russell Jacoby, *The Repression of Psychoanalysis: Otto Fenichel and the Freudians*, Chicago, University of Chicago Press, 1986, p. 184, no. 33. The Weinbaums subscribed to the *International Journal of Psychonalaysis*: see letter from Lina Weinbaum to Frank Malina, 3 February 1946, Box 3, Folder 35, FJM. Lina also wrote to Frank about the writings of Havelock Ellis: see Lina to Frank Malina, undated, Box 3, Folder 35, FJM. Epstein's relationship with Weinbaum evident in the former's support for the latter's legal fund. On von Kármán and psychoanalysis see *The Wind and Beyond*, p. 93.
6. Letter from Frank Malina to Liljan Malina, 10 December 1945, MFC; Frank often referred to the psychoanalyst as a 'psychologist'.
7. Letter from Frank Malina to Liljan Malina, 14 December 1945, MFC.
8. For instance, Frank's friend and attorney, Jack Frankel, his stepdaughter Saki Dikran, her partner Sidney, as well as some of Liljan's friends in the army.

9. See Benjamin Harris, 'The Benjamin Rush Society and Marxist psychiatry in the United States, 1944–1951', *History of Psychiatry*, vol. 6, no. 23, 1995, pp. 309–31 (p. 310).

10. Joseph B. Furst, 'What Psychoanalysis Can Do', *New Masses*, 22 January 1946, pp. 14–18; Joseph B. Furst, 'Psychoanalysis Today', *New Masses*, 30 October 1945, pp. 13–15; Joseph Wortis, 'Freudianism and the Psychoanalytic Tradition', *New Masses*, 2 October 1945, p. 10; Francis Bartlett, 'Recent Trends in Psychoanalysis', *Science and Society*, vol. IX, Summer 1945, pp. 214–31.

11. Letter from Frank Malina to Liljan Malina, 29 November 1945, MFC.

12. Sterling Hayden, *Wanderer*, New York, Bantam Books, 1963, p. 365; 'short, pudgy and bespectacled' from Bernard Gordon, *The Gordon File: A Screenwriter Recalls Twenty Years of FBI Surveillance*, Austin, University of Texas Press, 2004, p.92; 'dentist', Victor Navasky, *Naming Names*, New York, Viking Press, 1980, p. 133.

13. Sterling Hayden, *Wanderer*, p. 324

14. For a detailed description, see Bernard Gordon, *The Gordon File*, p. 90f.

15. Sterling Hayden, *Wanderer*, p. 364.

16. Victor Navasky, *Naming Names*, p. 133; p. 138.

17. Victor Navasky, *Naming Names*, p. 131.

18. Sigmund Freud, *Civilization and Its Discontents*, trans. by David McLintock, London, Penguin Books, 2002, p. 103.

19. Stilwell Committee mention in letter from Frank Malina to parents, 2 January 1946, MFC.

20. 'cold sweats': Frank J. Malina interview with Mary Terrall, Pasadena, California, 14 December 1978, CAL. p. 14.

21. Letter from Frank Malina to Liljan Malina, 21 January 1945, MFC.

22. Letter from Frank Malina to Liljan Malina, 2 February 1946, MFC.

23. Ibid.

24. Benjamin Harris, 'The Benjamin Rush Society and Marxist psychiatry in the United States', p. 314.

25. Bernard Gordon, *The Gordon File*, p. 90.

26. This is detailed in the letter from FJM to Liljan, 2 February 1946, MFC.

27. Letter from Frank Malina to Liljan Malina, 2 February 1946, MFC.

28. Letter from Frank Malina to Liljan Malina, 2 February 1946, MFC.

29. Letter from Liljan Malina to Frank Malina, 20 February 1946, MFC.

30. See: Marvin E. Gettleman, '"No Varsity Teams": New York's Jefferson School of Social Science, 1943–1956', *Science & Society*, vol. 66, no. 3, 2002, pp. 336–59.

31. Malina's car registration was recorded outside a party meeting at 6121 Wilshire Blvd, 13 May 1946, detailed in: author redacted, 'Frank Malina, wa

Frank Parma, Internal Security–R', LA 100–8196, 26 July 1949, p. 4, Malina FBI; 'textile designer' letter from Frank Malina to his parents, 17 January 1946, MFC.

32. Citizens United to Abolish the Wood-Rankin Committee (1946) [display advertisement], *New York Times*, 14 March 1946, p. 18. Details of Liljan's anti-Rankin work are found in letter to Frank Malina, 25 March 1946, MFC; see also Stacy Spaulding, 'Off the Blacklist, But Still a Target: The Anti-Communist Attacks on Lisa Sergio', *Journalism Studies*, vol. 10, no. 6, 2009, pp.789–804, (p. 794).

33. Letters from Liljan Malina to Frank Malina, 1 April 1946; Liljan Malina to Frank Malina, 8 April 1946, MFC.

34. Letter from Frank Malina to Liljan, 10 December 1945, MFC.

35. Details of ICC meeting in letter from Frank Malina to Liljan Malina, 17 January 1946, MFC.

36. Letter from Frank Malina to Liljan Malina, 25 February 1946, MFC.

37. Winston Churchill, 'The Sinews of Peace' in Mark A. Kishlansky, ed., *Sources of World History*, New York, Harper Collins, 1995, pp. 298–302.

38. Parsons described psychoanalysis as 'a confusion of conformity and cure', letter from Jack Parsons to Marjorie Cameron Parsons, 27 January 1950, DD7, GYC.

39. A letter from Wilfred Smith to Aleister Crowley mentions Agape Lodge member Roy Mellinger's extensive knowledge of psychoanalysis, 27 February 1942, NS15, GYC.

40. Letter from Jane Wolfe to Karl Germer, 13 March 1946 as quoted in Jane Wolfe, 'Hollywood' *In The Continuum*, vol. 4, no. 6, 1989, pp. 31–42; it is possible but uncertain that this couple were Dick and Barbie Canright, the JPL computers.

41. Russell Miller, *Bare-Faced Messiah*, London, Michael Joseph, 1987, p. 118.

42. Anthony Boucher [aka H. H. Holmes], *Rocket to the Morgue*, New York, Duell, Sloan and Pearce, 1942.

43. Russell Miller, *Bare-Faced Messiah*, p. 118.

44. Quoted in Hugh B. Urban, 'The Occult Roots of Scientology? L. Ron Hubbard, Aleister Crowley, and the Origins of a Controversial New Religion', *Nova Religio: The Journal of Alternative and Emergent Religions*, vol. 15, no.3, 2012, pp. 91–116, (p. 94).

45. Letter from Aleister Crowley to Jane Wolfe, reproduced in Jane Wolfe, 'Pasadena', *In The Continuum*, vol. 4, no. 1, 1987, p. 40.

46. Nieson Himmel, Hubbard's roommate at 1003, quoted in Miller, p. 118.

47. Quoted in John Symonds, *The King of the Shadow Realm*, London, Duckworth, 1989, pp. 562–63.

48. On Forman, see George Pendle, *Strange Angel*, p. 257; 'troublesome spirits': Jane Wolfe, 'Hollywood', *In The Continuum*, vol. 4, no. 4, 1988, p. 42.

49. Robert Cornog, quoted in George Pendle, *Strange Angel*, p. 256.

50. Lawrence Wright, *Going Clear: Scientology, Hollywood, and the prison of belief*, Los Angeles, Silvertail, 2013, p. 53.

51. Letter from Jane Wolfe to Karl Germer, undated January 1946, quoted in Jane Wolfe, 'Hollywood', *In The Continuum*, vol. 4, no. 5, 1989, p. 42.

52. Distinction between office and flesh in: letter from Karl Germer to Jane Wolfe, quoted in Jane Wolfe, 'Hollywood', *In The Continuum*, vol. 4, no. 10, 1991, p. 40.

53. Letter from Jack Parsons to Aleister Crowley, 21 January 1946, OS DD19, GYC.

54. Jack Parsons, *The Book of Babalon*, DD4, GYC.

55. Ibid.

56. Letter quoted in Lawrence Sutin, *Do What Thou Wilt*, p. 413.

57. Jack Parsons, *The Book of Babalon*, DD4, GYC.

58. Parsons had not been admitted to all the OTO's precursor degree work – his initiation into IX° was, as he detailed to Crowley, through the unsanctioned collaboration with the banned Wilfred Smith: see Russell Miller, *Bare-Faced Messiah*, p. 126.

59. Hugh B. Urban, *Magia Sexualis: Sex, Magic, and Liberation in Modern Western Esotericism*, University of California Press, 2006, p. 136.

60. Jack Parsons, *The Book of Babalon*, DD4, GYC.

61. For a discussion of this see Hugh B. Urban, 'The Occult Roots of Modern Scientology', pp. 100–101.

62. Letter from Aleister Crowley to Karl Germer, 19 April 1946, quoted in Suton, *Do What Thou Wilt*, p. 414; other sources say 'louts' rather than 'goats'.

63. Letter from Frank Malina to Liljan Malina, 21 March 1946, MFC.

64. Letter from Liljan to Frank Malina, 1 April 1946, MFC.

65. See letter from Frank Malina to Liljan Malina, 18 February 1946, MFC.

66. Letter from Liljan to Frank Malina, 4 March 1946, MFC.

67. Letter from Frank Malina to Liljan Malina, 21 March 1946, MFC.

68. 'I can't say I understood or laughed at the baby book you gave me before I left': letter from Liljan Malina to Frank Malina, 21 June 1946, MFC; 'I now know I can produce babies': letter from Frank Malina to Liljan Malina, 23 December 1946, MWC.

69. Letter from Liljan Malina to Frank Malina, 1 April 1946, MFC.

70. Frank J. Malina, 'America's First Long-Range-Missile and Space Exploration Program', p. 370.

71. Letter from Frank Malina to Liljan Malina, 24 June 1946, MFC.

72. Janet Malcolm, *Psychoanalysis: the impossible profession*, London, Granta, 2004 [1981], p. 102.

73. Letter from Frank Malina to Liljan Malina, 6 January 1947, MWC.

74. Victor Navasky, *Naming Names*, p. 134.

75. Ibid., p. 140.

76. Ibid., p. 138.

77. Ibid., p. 142.

78. Bernard Gordon, *The Gordon File*, p. 92.

79. Dorothy Ray Healey and Maurice Isserman, *California Red*, p. 130.

80. Russell Miller, *Bare-Faced Messiah*, p. 127.

81. Ibid., p. 128.

82. Ibid., p. 128.

83. Letter from Jack Parsons to Crowley quoted in Russell Miller, *Bare-Faced Messiah*, p. 129.

84. *John W. Parsons v. Lafayette Ron Hubbard and Sara Elizabeth Northrup*, Circuit Court, 11th Judicial District of Florida in and for Dade County, Florida in Chancery no. 101634, 16 July 1946, Chancery book 777, p. 1.

85. George Pendle, *Strange Angel*, p. 270. Although no documentary source is cited for this claim, it aligns with Martin P. Starr's evidence regarding Parsons' child rape of Betty: Martin P. Starr, *The Unknown God*, p. 256.

86. 'Analysis by a Master of the Temple of the critical Modes in the Experience of his Material Vehicle': undated, *c.* 1948, Folder DD2, GYC.

87. Letter from Jack Parsons to Karl Germer, 19 June 1949, Folder DD8, GYC.

88. L. Ron Hubbard, *Dianetics: the modern science of mental health*, London, Derricke Ridgway, 1951, p. xii.

89. Laura Hirshbein, 'L. Ron Hubbard's science fiction quest against psychiatry', *Medical Humanities*, vol. 42, 2016, pp. 10–14; W. Vaughn McCall, 'Psychiatry and psychology in the writings of L. Ron Hubbard', *Journal of Religion and Health*, vol. 46, no. 3, 2007, pp. 437–47.

90. 'Wife of Dianetics Founder Asks Divorce on Cruelty Grounds: claims systematic torture', *Los Angeles Evening Herald Express*, Monday 23 April 1951, p. 1; Lawrence Wright, *Going Clear*, p. 87; clique quote of L. Ron Hubbard in Stephen A. Kent and Terra A. Manca, 'A war over mental health professionalism: Scientology versus psychiatry', *Mental Health, Religion & Culture*, vol. 17, no. 1, 2014, pp. 1–23 (p. 9).

91. On Hubbard as *bricoleur*, see Hugh B. Urban, 'The Occult Roots of Scientology'.

92. Alexander Mitchell, 'The odd beginnings of Ron Hubbard's career', *Sunday Times*, 5 October 1969.

93. Hugh B. Urban, *The Church of Scientology: A History of a New Religion*, Princeton, Princeton University Press, 2013, p. 41.

94. Quotations from the statement from the Church of Scientology quoted in George Pendle, *Strange Angel*, p. 330.
95. Sterling Hayden, *Wanderer*, p. 371.
96. *Dr. Strangelove or: How I Learned to Stop Worrying and Love the Bomb*, dir. Stanley Kubrick, Columbia Pictures, 1964.
97. Sterling Hayden, *Wanderer*, p. 378.
98. Ibid., p. 372.
99. Ibid., p. 378.

Chapter 11: The Problem of Escape
1. Clayton Koppes, *JPL and the American Space Program*, p. 24.
2. Letter from Frank Malina to parents, 28 November 1945, MFC.
3. Ibid.
4. This is outlined in a letter from JPL historian Clayton Koppes to Frank Malina, 5 August 1980, MFC.
5. Koppes, p. 27.
6. Letter from Frank Malina to Liljan, 3 October 1945, MFC.
7. Frank J. Malina interview with Mary Terrall, Pasadena, California, 14 December 1978, Oral History Project, CAL, p. 9.
8. Clayton Koppes, *JPL and the American Space Program*, p. 27; see F. J. Malina, 'Memorandum on the Future of Jet Propulsion Research at the California Institute of Technology', November 1945, CL#12–0693, JPL.
9. Homer J. Stewart, Supplemental Oral History with Shirley K. Cohen, CAL, p. 15.
10. Letter from Frank Malina to Liljan Malina, 21 March 1946, MFC.
11. Clayton Koppes, *JPL and the American Space Program*, p. 32.
12. David H. DeVorkin, *Science With a Vengeance: How the Military Created the US Space Sciences After World War II*, New York, Springer-Verlag, 1992, p. 68.
13. Michael J. Neufeld, *Von Braun*, p. 217.
14. David DeVorkin, *Science with a Vengeance*, p. 174.
15. 'my brain [is] beaten to a pulp': letter from Frank Malina to Liljan, 23 December 1946, MWC.
16. A paradox famously developed in Donald A. MacKenzie, *Inventing Accuracy: a historical sociology of nuclear missile guidance*, Cambridge, The MIT Press, 1993.
17. Frank J. Malina, 'America's First Long-Range Missile and Space Exploration Program', p. 374.
18. Frank J. Malina, 'Is the Sky The Limit? Ordnance Rockets Demolish the Barriers of Space', *Army Ordnance*, vol. 31, no. 157, pp. 45–8 (Jul./Aug.

1946), p. 47; correspondence with the journal editor R. E. Lewis suggests a February submission.

19. William Pickering 'Chapter XXII: Origin of the Jet Propulsion Laboratory', copy of unknown publication, Box 3, Folder 7, WPP, p. 214.

20. James Bragg, *Development of the Corporal: the empryo of the Army missile program*, Army Ballistic Missile Agency, Redstone Arsenal, Alabama, 1961, p. 76; see also Stanley O. Starr, 'The Launch of Bumper 8 from the Cape: The End of an Era and the Beginning of Another', AAS History Series, vol. 32, 2010, pp. 75–97 (p. 78).

21. Arthur C. Clarke, 'V-2 for Ionosphere Research?' *Wireless World*, vol. LI, no. 3, Feburary 1945, p. 58; Arthur C. Clarke, 'Extra-Terrestrial Relays', *Wireless World*, vol. LI, no. 10, October 1945, pp. 305–8.

22. For Malina's account of presenting to the board see: Frank J. Malina, 'America's First Long-Range Missile and Space Exploration Program', p. 374; Stilwell's diary mentioned in Barbara W. Tuchman, *Stilwell and the American Experience in China, 1911–45*, New York, Macmillan, 1970, p. 523.

23. R. Cargill Hall, 'Early US satellite proposals', *Technology and Culture*, vol. 4, no. 4, 1963, pp. 410–34 (p. 413); this $15,000 contract was number a(5)-7913.

24. The name Jimmy is from William L. Ryan and Sam Summerlin, *The China Cloud*, p. 146.

25. Ibid., p. 146.

26. Homer J. Stewart oral history interview with John L. Greenberg, Pasadena, California, October–November 1982, CAL, p. 79; not talking about politics, Ryan and Summerlin, *The China Cloud*, p. 146.

27. W. Z. Chien, *Preliminary Calculations on the Performance of a High Altitude Test Vehicle*, 3 January 1946, JPL Memorandum, no. 8–1; see also W. H. Pickering and J. H. Wilson, 'Countdown to space exploration: A memoir of the Jet Propulsion Laboratory, 1944–1958' in R. Cargill Hall, ed., *History of Rocketry and Astronautics*, American Astronautical Society History Series, 1972, vol. 7, part 2, pp. 385–421 (p. 410).

28. 'Professor Dr Chien Wei-Zang' (Qian Wei-Chang), *Facta Universitatis, Mechanics, Automatic Control and Robotics*, vol. 2, no. 8, 1998, pp. 789–90 (p. 789).

29. Sidney Weinbaum interview by Mary Terrall, Pasadena, California, 15, 20 and 25 August 1985, CAL, p. 50.

30. Letter from Frank Malina to parents, 14 June 1946, MFC.

31. 385 employees, J. D. Hunley, *The Development of Propulsion Technology for US Space-Launch Vehicles, 1926–1991* (vol. 17), College Station, Texas A&M University Press, 2013, p. 19.

32. Clayton Koppes, *JPL and the American Space Program*, p. 31.

33. Frank's complaints about powdered egg and smoked herring, shortage of bread and beer, letter from Frank Malina to parents, 10 July 1946, Box 22, Folder 5, FJM.

34. Letter from Frank Malina to parents, 19 July 1946, Box 22, Folder 5, FJM.

35. Frank J. Malina, 'America's First Long-Range Missile and Space Exploration Program', p. 375.

36. UNESCO, *Preparatory Commission of UNESCO, Inventory of Archives, 1945–1946*, http://www.unesco.org/archives/files/ago3fa00001e.pdf, accessed 20 October 2017, p. 26.

37. Robert E. Filner, 'The Social Relations of Science Movement (SRS) and JBS Haldane', *Science & Society*, vol. 41, no. 3, 1977, pp. 303–16; William McGucken, *Scientists, Society, and State: The Social Relations of Science Movement in Great Britain, 1931–1947*, Columbus, Ohio State University Press, 1984.

38. Simon Winchester, *Book, Bomb and Compass: Joseph Needham and the Great Secrets of China*, London, Penguin, 2009, p. 168.

39. Patrick Petitjean, 'The joint establishment of the World Federation of Scientific Workers and of UNESCO after World War II', *Minerva*, vol. 46, no. 2, 2008, pp. 247–70 (p. 257).

40. Frank J. Malina interview with Mary Terrall, 14 December 1978. Oral History Project, CAL, p. 15; Frank Malina first mentions UNESCO opportunity in letter to his sister Caroline, 27 July 1946, MFC.

41. Letter from Joseph Needham to Frank Malina, 2 September 1946, MFC; letter from Joseph Needham to Lee DuBridge, undated copy, MFC.

42. See letter from Frank Malina to his parents, 17 September 1946, MFC.

43. See letter from Frank Malina to Liljan Malina, 20 June 1946, MWC.

44. T. H. Huxley quote from James P. Sewell, *UNESCO and World Politics*, Princeton, Princeton University Press, 1975, p. 86.

45. Huxley had once enthused about the place of science in the Soviet experiment, see David Caute, *The Fellow Travellers*, p. 110; Julian Huxley, *A Scientist among the Soviets*, London, Chatto & Windus, 1932.

46. John Toye and Richard Toye, 'One World, Two Cultures? Alfred Zimmern, Julian Huxley and the Ideological Origins of UNESCO', *History*, vol. 95, no. 319, 2010, pp. 308–31 (p. 330).

47. The conditionality of the job on Huxley's appointment is spelled out in a letter from Frank Malina to Liljan Malina, 5 December 1946, MWC.

48. Ibid., p. 328.

49. Frank A. Ninkovich, *The Diplomacy of Ideas: US Foreign Policy and Cultural Relations, 1938–1950*, Cambridge, Cambridge University Press, 1981, p. 98.

50. Letter from Frank Malina to parents, 4 December 1946, Box 22, Folder 6, FJM.

51. Kathryn S. Olmsted, *Red Spy Queen: A Biography of Elizabeth Bentley*, Chapel Hill, University of North Carolina Press, 2002.

52. Letter from Lina Weinbaum to Frank Malina, 5 October 1948, Box 3, Folder 35, FJM.

53. Ellen Schrecker, *Many Are the Crimes*, p. 169.

54. Amy Knight, *How the Cold War Began: the Igor Gouzenko Affair and the Hunt for Soviet Spies*, New York, Basic Books, 2007, p. 32; 'pregnant': quote in Christopher Andrew, *The Defence of the Realm: The Authorized History of MI5*, London, Allen Lane, 2009, p.339.

55. Ellen Schrecker, *Many Are the Crimes*, p. 170; Brian Cathcart, 'May, Alan Nunn (1911–2003)', *Oxford Dictionary of National Biography*, Oxford, Oxford University Press, Jan 2007; online ed., May 2008.

56. Chapman Pincher, *Treachery: Betrayals, Blunders and Cover-Ups: Six Decades of Espionage*, Edinburgh, Mainstream, 2011, p. 27; William A. Tyrer, 'The Unresolved Mystery of ELLI', *International Journal of Intelligence and CounterIntelligence*, vol. 29, no. 4, 2016, pp. 785–808.

57. Ellen Schrecker, *Many Are The Crimes*, p. 157.

58. Ellen Schrecker, *The Age of McCarthyism: a brief history with documents*, Boston, St Martin's Press, 1994, p. 15.

59. Ellen Schrecker, *Many Are The Crimes*, p. xvi.

60. Hoover quoted in Schrecker, *Many Are The Crimes*, p. 161.

61. Letter from Bessie Jacobs to Frank Malina, 18 December 1946 (enclosed with Jack Frankel correspondence), Box 1, Folder 29, FJM.

62. Ibid.

63. Mae Dena Huettig, *Economic Control of the Motion Picture Industry: A Study in Industrial Organization*, University of Pennsylvania Press, 1946; Wyatt D. Phillips, '"A Maze of Intricate Relationships": Mae D. Huettig and Early Forays into Film Industry Studies', *Film History: An International Journal*, vol. 27, no. 1, 2015, pp. 135–63.

64. See John Earl Haynes and Harvey Klehr, *Venona: Decoding Soviet Espionage in America*, New Haven, Yale University Press, 1999, p. 227.

65. Philip J. Reilly, 'Martin J. Summerfield, Espionage–R', 65–4852, 18 July 1949, p. 3, Summerfield FBI.

66. Letter from Mae Churchill to Frank Malina, 4 July 1947, Box 1, Folder 17, FJM.

67. Letter from Frank Malina to Liljan Malina, 3 June 1946, MWC.

68. Joseph Needham, *History Is On Our Side*, London, Allen and Unwin, 1946; discussion of sending the book to Marion O'Gorman, and his additional recommendation of John Strachey, *Theory and Practice of Socialism*, in a letter from Frank Malina to Liljan Malina, 17 January 1947, MWC.

69. Letter from Frank Malina to Liljan Malina, 5 December 1946, MWC.

70. Letter from Frank Malina to Liljan Malina, 6 January 1947, MWC.

71. Letter from Frank Malina to Liljan Malina, 3 July 1946, MWC.

72. Letter from Frank Malina to Liljan Malina, 6 February 1947, MWC.

73. Letter from Frank Malina to Liljan Malina, 23 December 1946, MWC.

74. Telegram reproduced in letter from Frank Malina to Liljan Malina, 24 December 1946, MWC.

75. Letter from Frank Malina to Liljan Malina, 17 January 1947 [letter is mistakenly dated 1946], MWC.

76. Ibid.

77. Julian Huxley, *UNESCO: Its Purpose and Its Philosophy*, Preparatory Commission of the United Nations Educational, Scientific and Cultural Organization, London, 1946.

78. Letter from Frank Malina to Liljan Malina, 9 February 1947, MWC.

79. Letter from Frank Malina to Liljan Malina, 6 February 1947, MWC.

80. Ibid.

81. Quoted in Frank A. Ninkovich, *The Diplomacy of Ideas,* p. 100; original emphasis.

82. Ibid.

83. Memo from General Hoyt Vandenberg to Truman, 15 February 1947, quoted in Simon Winchester, *Book, Bomb and Compass: Joseph Needham and the Great Secrets of China*, London, Viking, 2008, p. 172.

84. Aant Elzinga, 'UNESCO and the Politics of International Cooperation in the Realm of Science' in Patrick Petitjean, ed., *Les Sciences Coloniales: Figures et Institutions*, vol. 2, 1996, pp. 163–202 (p. 168); Patrick Petitjean, 'The joint establishment of the World Federation of Scientific Workers and of UNESCO after World War II', *Minerva*, vol. 46, no. 2, 2008, pp. 247–70, (p. 249).

85. Note in file by A. S. Halford, Foreign Office, 28 February 1947, MI5 file on Joseph Needham, KV/2/3055, NA.

86. Ibid.

87. Letter from R. H. Hollis to J. A. Cimperman, 7 March 1947, MI5 file on Joseph Needham, KV/2/3055, NA.

88. Ibid.

89. Peter Wright with Paul Greengrass, *Spycatcher: The Candid Autobiography of a Senior Intelligence Officer,* New York, Viking USA, 1987.

90. Chapman Pincher, *Treachery*; Christopher Andrew, *The Defence of the Realm: The Authorized History of MI5*, London, Allen Lane, 2009. For a critique of Andrew, see Paul Monk, 'Christopher Andrew and the Strange Case of Roger Hollis', *Quadrant*, vol. 54, no. 4, 2010, p. 37. A further investigation tilts away from Hollis as ELLI and points to incompetence rather than conspiracy: William A. Tyrer, 'The Unresolved Mystery of ELLI',

International Journal of Intelligence and CounterIntelligence, vol. 29, no. 4, 2016, pp. 785–808.

91. See letter from Frank Malina to Liljan Malina, 11 February 1947, MWC.
92. Details in letter from Frank Malina to Liljan Malina, 20 February 1947, MWC.
93. All details of these visits from: 'Notes made during trip on U.S. East Coast before going to join UNESCO Secretariat in Paris, 5 March–1 April 1947', MFC.
94. Ellen Schrecker, *The Age of McCarthyism*, p. 37.
95. Jessica Wang, 'Science, Security, and the Cold War: The case of EU Condon', *Isis,* vol. 83, no. 2, 1992, pp. 238–69 (p. 246); William Odlin Jr, 'Condon Duped Into Sponsoring Commie-Front's Outfit's Dinner', *Washington Times-Herald*, 23 March 1947, Section A-2, p. 1.
96. Jessica Wang, 'Science, Security, and the Cold War', p. 249.
97. Letter from Frank Malina to parents, 2 April 1947, Box 22, Folder 6, FJM.
98. All notes regarding meetings from 'Notes made during trip on US East Coast', MFC.
99. The PCA had formed after a split in the Independent Citizens Committee for the Arts Sciences and Professions (ICCASP), which Frank had joined in 1946 and to which Einstein, too, had lent his support. ICCASP's strong constituency of communists upset some liberal supporters, who resigned in protest. ICCASP in turn merged with the CIO's National Citizens Political Action Committee (NCPAC) to form the PCA. For a discussion of ICCASP, NCPAC and the PCA, see Andrew Hemingway, *Artists on the Left: American Artists and the Communist Movement, 1926–1956*, New Haven, Yale University Press, 2002, p. 195f.
100. For evidence of Einstein's support for Wallace, see Richard Crockatt, *Einstein and Twentieth-Century Politics: 'A Salutary Moral Influence'*, New York, Oxford University Press, 2016, p. 163.
101. John C. Culver and John Hyde, *American Dreamer: a life of Henry A. Wallace*, New York, W. W. Norton, 2000, p. 438.

Chapter 12: These Terrible Times

1. FBI report Arthur Wittenburg, 'Dr Sidney Weinbaum Espionage–R', 5 April 1949, LA 65–4846, p. 39. Weinbaum FBI.
2. Ibid., p. 27.
3. Letter from Lina Weinbaum to Frank Malina, 17 February 1947, Box 3, Folder 35, FJM.
4. See FBI report Arthur Wittenburg, 'Dr Sidney Weinbaum Espionage–R', 15 June 1949, LA 65–4846, p. 11, Weinbaum FBI.

5. Letter from Lina Weinbaum to Frank Malina, 13 September 1946, Box 3, Folder 35, FJM.

6. Letter from Sidney Weinbaum to Frank Malina, 17 February 1947, Box 3, Folder 35, FJM; 'keep your eyes and ears open' is also attributed to Sidney as a phrase that expresses the duty that members owe to the party, see the testimony of Richard Rosanoff, in *United States of America v. Sidney Weinbaum*, Southern District of California Central division, Los Angeles, California, 31 August 1950, reporter's transcript, Record Group 21, Criminal Case Files, 21383–21408, Folder 21408, Box no. 1261.

7. Sidney Weinbaum interview with Mary Terrall, 15, 20 and 25 August 1985, Oral History Project, CAL, p. 50.

8. Frank E. Marble interview with Shirley K. Cohen, Pasadena, January–March 1994, 21 April 1995, CAL, p. 42; Andrew Fejer, audio recording, 9 min., April 1993, MFC.

9. Letter from Lina Weinbaum to Frank Malina, 24 September, 1947, Box, 3, Folder 35, FJM.

10. Quoted in Clayton Koppes, *JPL and the American Space Program*, p. 35.

11. Sidney Weinbaum and H. L. Wheeler Jr, 'Heat Transfer in Sweat-Cooled Porous Metals', *Journal of Applied Physics*, vol. 20, no. 1, 1949, pp. 113–22.

12. Letter from Lina Weinbaum to Frank Malina, 4 July 1948, Box 3, Folder 35, FJM.

13. 'relieved of his duties': Louis Dunn quoted in report by Philip J. Reilly, 'Martin J. Summerfield, Espionage–R', 16 May 1949, p. 8, Summerfield FBI file.

14. Letter from Dorothy Lewis to Frank Malina, 7 August 1946, Box 2, Folder 27, FJM.

15. Letter from Lina Weinbaum to Frank Malina, 24 September 1947, Box 3, Folder 35, FJM.

16. Letter from Martin Summerfield to Frank Malina, 29 July 1947, MFC.

17. The suspension was detailed in the author's interview with David Altman, 12 September 2016; the exact date in 1947 is unknown.

18. Friendships detailed in an FBI report: David E. Todd, 'Sidney Weinbaum – Espionage–R', 6 March 1949, p. 3, Weinbaum FBI.

19. Martin Kamen, *Radiant Science, Dark Politics: a Memoir of the Nuclear Age*, Berkeley, University of California Press, 1985, p. 164.

20. Ibid., p. 181.

21. An entire chapter is devoted to theories relating to the death of Jean Tatlock in Kai Bird and Martin J. Sherwin, *American Prometheus: the Triumph and Tragedy of J. Robert Oppenheimer*, New York, Knopf, 2005.

22. Ibid., p. 174.

23. David E. Todd, 'Sidney Weinbaum – Espionage–R', 6 March 1949, p. 3, Weinbaum FBI.

24. John Earl Haynes and Harvey Klehr, *Venona: Decoding Soviet Espionage in America*, New Haven, Yale University Press, 1999, p. 325.

25. Letter from Lee Carlton, *Engineering and Science*, Winter 1992, p. 39.

26. Recent evidence from declassified Soviet archives supports the idea that Kamen was not involved in espionage. See Gregg Herken, 'Target *Enormoz*: Soviet Nuclear Espionage on the West Coast of the United States, 1942–1950', *Journal of Cold War Studies*, vol. 11, no. 3, 2009, pp. 68–90 (p. 74).

27. Author interview with David Altman, 12 September 2016.

28. Letter from Frank Malina to parents, 17 April 1947, Box 22, Folder 6, FJM.

29. Letter from Frank Malina to parents, 24 April 1947, Box 22, Folder 6, FJM.

30. Letter from Frank Malina to parents, 20 July 1947, Box 22, Folder 6, FJM.

31. Letter from J. Edgar Hoover to Jack Neal, State Department, 17 October 1947, Malina FBI.

32. Simon Winchester, *Book, Bomb and Compass*, p. 173.

33. Aant Elzinga, 'UNESCO and the Politics of International Cooperation in the Realm of Science', p. 167.

34. Letter from Frank Malina to parents, 14 September 1947, Folder 22, Box 7, FJM.

35. Letter from Gus Albrecht to Frank Malina, 30 July 1947, MWC.

36. Letter from Frank Malina to his parents, 18 August 1947, Box 22, Folder 7, FJM.

37. Joël Kotek, 'Youth organizations as a Battlefield in the Cold War', *Intelligence and National Security*, vol. 18, no. 2, 2003, pp. 168–91.

38. For details of the prize including Oberth's award see Frank H. Winter, 'The Birth and Early Rise of "Astronautics": the REP-Hirsch Astronautical Prize, 1928–1940', *Quest*, vol. 14, no. 1, 2007, pp. 35–43. The timing of Malina's discovery of the award is at odds with that reported by Winter, p. 39.

39. Details of the REP-Hirsch prize in a letter from Frank Malina to his parents, 17 August 1947, Box 22, Folder 8, FJM.

40. Julian Huxley, *Memories vol II*, London, Allen and Unwin, p. 21.

41. Max Weber's *Sociology* is detailed in Malina's list of books read in January 1948, MFC.

42. Letter from Frank Malina to his parents, 25 January 1948, Box 22, Folder 8, FJM.

43. 'eye to eye': letter from Frank Malina to his parents, 13 November 1948, Box 22, Folder 8, FJM.

44. 'getting acquainted': letter from Frank Malina to his parents, 17 July 1948, Box 22, Folder 8, FJM; 'Italian revolutionary' letter from Frank Malina to his parents, 10 August 1948, Box 22, Folder 8, FJM.

45. Letter from Frank Malina to parents, 19 September 1948, Box 22, Folder 8, FJM.

46. Letter from Frank Malina to parents, 10 October 1948, Box 22, Folder 8, FJM.

47. Letter from Lina Weinbaum to Frank Malina, 5 October 1948, Box 3, Folder 35, FJM.

48. Letter from Selina Weinbaum to Frank Malina, 15 July 1948, Box 3, Folder 35, FJM. Lest Selina be thought too young for the journey, she claimed to have a travelling companion, Eve Borsook, daughter of Caltech biologist Henry Borsook, though she seems not to have raised this possibility with Eve herself, pers. comm., Eve Borsook to author, 6 May 2014.

49. The friendship with Gibling is detailed in the Sidney Weinbaum interview with Mary Terrall, 15, 20 and 25 August 1985, Oral History Project, CAL, p. 58.

50. Simon Morrison, *The Love and Wars of Lina Prokofiev*, London, Vintage, 2013, pp. 174–5.

51. Ibid., p. 238.

52. Letter from Lina Weinbaum to Frank Malina, 20 November 1948, Box 3, Folder 35, FJM.

53. Details of the FBI surveillance of Marling in report: Arthur Wittenberg, 'Sidney Weinbaum, Espionage–R', 5 April 1949, pp. 15–16, Weinbaum FBI.

54. All this detail from Simon Morrison, *Love and Wars*, p. 239.

55. Ibid., p. 239.

56. Letter from Lina Weinbaum to Frank Malina, 20 November 1948, Box 3, Folder 35, FJM.

Chapter 13: Definite Leftists

1. All these quotes are from a memo: SAC Los Angeles to Director FBI, 'COMMUNIST INFILTRATION OF JET PROPULSION LABORATORY, CALTECH, ESPIONAGE–R', 26 January 1949, Weinbaum FBI. The identity of the JPL informant who expresses concern about JPL security is redacted in the numerous files connected with Malina, Tsien and Weinbaum. Fortunately, it is revealed as Louis Dunn in an unredacted FBI file recently obtained for Martin Summerfield – see letter from Special Agent R. B. Hood, SAC Los Angeles to Director FBI, LA 65–4852, 4 June 1949, Summerfield FBI (FOIPA request: 1324818–000, released to author 28 March 2016). Patty Line's friendship with Frank is detailed in a letter from Frank Malina to Theodore von Kármán, 11 July 1959, included in letters to Haley, Box 2, Folder 8, FJM.

2. 'National Affairs – Eggs in the Dust' *Time*, 13 September 1948, original pagination unknown.

3. Ralph Dighton, 'Spaceman Looks Ahead: Man to Explore Mars in 10–15 years', *Scene, Independent Star News*, Sunday 18 January 1959, p. 3.

4. SAC Los Angeles to Director FBI, 'COMMUNIST INFILTRATION OF JET PROPULSION LABORATORY, CALTECH, ESPIONAGE–R', 26 January 1949, Weinbaum FBI; Hood's investigations into Brecht and Mann in Alexander Stephan, *Communazis: FBI Surveillance of German Emigré Writers*, New Haven, Yale University Press, 2000, p. 49.

5. David Kaiser, 'The atomic secret in red hands? American suspicions of theoretical physicists during the early Cold War', *Representations*, vol. 90, no. 1, 2005, pp. 28–60, p. 42.

6. Report by Arthur C. Wittenburg, 'Dr Sidney Weinbeim, Espionage–R', 5 April 1949, p. 25, Weinbaum FBI.

7. Ibid.; my emphasis.

8. See letter from J. Edgar Hoover to Attorney General, 4 February 1949, Sidney Weinbaum FBI.

9. Arthur C. Wittenburg, 'Dr Sidney Weinbeim, Espionage–R', 5 April 1949, p. 3, Weinbaum FBI.

10. Ibid., p. 18.

11. Letter from R. B. Hood to FBI Director, 1 March 1949, Weinbaum FBI.

12. Details of collaboration between Pauling and Sherman in: Mary Jo Nye, 'Mine, Thine, and Ours: Collaboration and Co-Authorship in the Material Culture of the Mid-Twentieth Century Chemical Laboratory', *Ambix*, vol. 61, no. 3, 2014, pp. 211–35.

13. Details of these flights taken from Stanley O. Starr, 'The Launch of Bumper 8 from the Cape: The End of an Era and the Beginning of Another', p. 10; Gregory P. Kennedy, *The Rockets and Missiles of White Sands Proving Grounds, 1945–1958*, Atglen, PA, Schiffer Military History, 2009, p. 51.

14. Roger Launius, 'Hypersonic Flight: Evolution from X-15 to Space Shuttle', AIAA International Air and Space Symposium and Exposition: The Next 100 Years, Dayton, Ohio, 2003, https://doi.org/10.2514/6.2003–2716

15. Willy Ley, 'Development of the Spaceship', in W. Ley, W. von Braun, and H. Haber, eds, *The Complete Book of Satellites and Outer Space*, New York, MACO Magazine Corporation, 1953, pp. 4–15 (p. 12).

16. Memo from SAC to FBI Director, 'Re Sidney J. Weinbaum, Espionage-R', 16 March 1949, Weinbaum FBI.

17. Letter from R. B. Hood to FBI Director, 'Re Sidney J. Weinbaum, Espionage-R', 16 March 1949, Weinbaum FBI.

18. Letter from Louis Dunn to Frank Malina, 22 January 1947, MFC.

19. Report by Special Agent Arthur Wittenberg, 'Dr Sidney Weinbaum, Espionage–R', 5 April 1949, p. 32, Weinbaum FBI.

20. Ibid.

21. Letter from R. B. Hood to FBI Director, 18 April 1949, Weinbaum FBI.

22. Report by Philip J. Reilly, 'Martin J. Summerfield – Espionage–R', 16 April 1949, Summerfield FBI.

23. Weinbaum felt that the timing of the PSQ was about the statute of limitations; see oral history Sidney Weinbaum interview with Mary Terrall, 15, 20 and 25 August 1985, CAL, p. 51. The FBI file suggests it might also have been about the precise wording of the PSQ; see R. B. Hood to FBI Director 'Sidney J. Weinbaum, Espionage–R', 15 April 1949, Weinbaum FBI file.

24. Ibid., Hood note.

25. Letter from R. B. Hood to Director FBI, 11 April 1949, Summerfield FBI.

26. George Pendle, *Strange Angel*, p. 283. See also report by NJB, 'John Whiteside Parsons, aka Jack Parsons, Espionage–IS', 22 November 1950, p. 2, Parsons FBI.

27. Report by Hollis H. Bowers, 'Huse-Shen Tsien, J. M. Decker – Espionage–R' file no. 65–5154, 24 May 1949, Tsien FBI.

28. Letter from R. B. Hood to FBI Director, 1 March 1949, 'Sidney J. Weinbaum, Espionage–R', Weinbaum FBI.

29. That Dunn called him is mentioned in Sidney Weinbaum interview with Mary Terrall, 15, 20 and 25 August 1985, CAL, p. 52.

30. All details and dialogue from Arthur C. Wittenburg, 'Dr Sidney Weinbaum, Espionage–R', 15 June 1949, LA file 65–4846, Weinbaum FBI.

31. This detail from Weinbaum quoted in 'Never a Red, Weinbaum tells jury' *Los Angeles Herald and Express*, 6 September 1950, p. 18.

32. Ibid.

33. Arthur C. Wittenburg, 'Dr Sidney Weinbaum, Espionage–R', 15 June, 1949, LA file 65–4846, p. 20. Weinbaum FBI.

34. These letters are mention in the Sidney Weinbaum interview with Mary Terrall, CAL, p. 40.

35. Letter from Sidney Weinbaum to Frank Malina, April 28 [1949] (filed among letters from 1947), Box 3, Folder 35, FJM.

36. Malina left Paris for Britain on 14 May and left London for Texas on 20 May; dates detailed in a report by redacted author, 'Frank J. Malina, Internal Security–R', 26 July 1949, file LA-100–8196, p. 3, Malina FBI.

37. Sidney Weinbaum interview with Mary Terrall, CAL, p. 40.

38. Ibid., p. 40.

39. Ibid., p. 40.

40. US Congress, House of Representatives, Committee on Un-American Affairs, *Hearings Regarding Communist Infiltration of Radiation Laboratory and Atomic Bomb Project at the University of California, Berkeley, California,*

vol. 1–3, 81st Cong., 1st Sess., 22, 26 April, 25 May, 10, 14 June 1949, Washington DC, Government Printing Office, 1949–1950.

41. Stetson Kennedy, *The Klan Unmasked* [second edition, with a new introduction by David Pilgrim and a new author's note], Tuscaloosa, University of Alabama Press, 2011, p. 2.

42. Russell Olwell, 'Physical Isolation and Marginalization in Physics: David Bohm's Cold War Exile', *Isis*, vol. 90, pp. 738–56 (p. 741).

43. *Hearings Regarding Communist Infiltration of Radiation Laboratory*, p. 321.

44. Ibid., p. 320.

Chapter 14: Degradation Ceremonies

1. These dates detailed in a report by redacted author, 'Frank J. Malina, Internal Security–R', 26 July 1949, file LA-100–8196, Malina FBI.

2. Letter from Mae Churchill to Frank Malina, 2 May 1949, Box 1, Folder 18, FJM.

3. All these details from Arthur Wittenburg, 'Report on Sidney Weinbaum, Espionage–R', 7 September 1949, p. 17, Weinbaum FBI.

4. Ibid., p. 6.

5. The date is detailed in a report by redacted author, 'Frank J. Malina, Internal Security–R', 26 July 1949, file LA-100–8196, p. 11, Malina FBI.

6. Conversation about a job in Paris mentioned in letter from Jack Parsons to Frank Malina, 15 June 1949, MFC.

7. George Pendle, *Strange Angel*, p. 288.

8. Jack Parsons, 'The Book of the Antichrist and The Manifesto of the AntiChrist', NS 110 Springback folder, GYC.

9. Ibid.

10. Foreword to *The Book of Babalon*, Folder DD4, Old Series, GYC.

11. Ibid.

12. John Whiteside Parsons, 'The Oath of the AntiChrist', DD3 Old Series, GYC.

13. See Jack Parsons, 'The Book of the Antichrist and The Manifesto of the AntiChrist', NS 110 Springback folder, GYC.

14. Letter from Jack Parsons to Karl Germer, 19 June 1949, DD8, GYC.

15. Frank J. Malina oral history with James H. Wilson, 8 June 1975, JPL, p. 25.

16. Letter from Frank Malina to parents, 16 June 1949, MFC.

17. All dialogue reproduced from US Congress, House of Representatives, Committee on Un-American Affairs, *Hearings Regarding Communist Infiltration of Radiation Laboratory and Atomic Bomb Project at the University of California, Berkeley, California*, vol. 1–3, 81st Cong., 1st Sess., 22, 26 April, 25 May, 10, 14 June 1949, Washington DC, Government Printing Office, 1949–1950.

18. Harold Garfinkel, 'Conditions of successful degradation ceremonies', *American Journal of Sociology*, vol. 61, no. 5, 1956, pp. 420–24; see also Victor Navasky, *Naming Names*, p. 314f.

19. Iric Nathanson, 'The Oppenheimer Affair: Red Scare in Minnesota', *Minnesota History*, vol. 60, no. 5, 2007, pp. 172–86 (p. 174).

20. Ibid.

21. Ibid.

22. Size of ranch in interview of Frank Oppenheimer by Charles Weiner on 21 May 1973, Niels Bohr Library & Archives, American Institute of Physics, College Park, MD, USA.

23. All details and Oppenheimer quote from K. C. Cole, *Something Incredibly Wonderful Happens*, pp. 99–102.

24. Sidney Weinbaum interview with Mary Terrall, CAL, p. 53.

25. Thomas Hager, *Force of Nature: the life of Linus Pauling*, New York, Simon and Schuster, 1995, p. 222.

26. For a longer discussion of the professional relationship between Pauling and Weinbaum at Caltech, see Thomas Hager, *Force of Nature*, p. 660.

27. Sidney Weinbaum interview with Mary Terrall, CAL, p. 53f.

28. Date of the board in: 'Dr Weinbaum's Trial Begins on Perjury Charge', *Los Angeles Times*, 30 August 1950, p. 6.

29. 'the heat was awful that day, 102°F': letter from Lina Weinbaum to Frank Malina, 5 November 1949, Box 3, Folder 35, FJM; records from Weather Underground, weather history for KLAX, www.wunderground.com.

30. Ibid.

31. Sidney learned on 3 December, see letter to from Sidney Weinbaum to Linus Pauling, 3 December 1949, Box 433, Folder 5, LPP.

32. Frank Malina interview with Mary Terrall, 14 December 1978, CAL, p. 16.

33. Twenty-one paintings in three weeks detailed in letter from Marjorie Malina to Carolyn Mercer, 20 November 1949, Box 22, Folder 8, FJM.

34. Marjorie Malina to Malina family, 21 November 1949, Box 22, Folder 8, FJM.

35. Letter from Andy Haley to Frank Malina, 1 July 1949, MFC.

36. Unsent draft of letter from Frank Malina to Andy Haley, undated but probably July 1949, MFC.

37. 'Two More Arrested by FBI On Charges They Divulged Secrets to the Soviet Union', *New York Tribune*, Saturday 17 June 1950, p. 1.

38. Letter from Mrs Janet Knight Cheney to J. Edgar Hoover, 26 June 1950, Sidney Weinbaum FBI.

39. Sidney Weinbaum interview with Mary Terrall, CAL, p. 55.

40. 'Caltech Man's Red Label Hit By Wife', *The Pasadena Independent*, Sunday 18 June 1950, p. 2.

41. Memo from A. H. Belmont to D. M. Ladd, 'Hsue-Shen Tsien', 29 June 1950, p. 4, Tsien FBI.

42. Ibid., p. 3.

43. Dialogue from Earnest C. Watson interview with Larry Shirley, Pasadena, California, 20 January 1969, Oral History Project, CAL, p. 3.

44. Iris Chang, *Thread of the Silkworm*, p. 152.

45. Ryan and Summerlin, *The China Cloud*, p. 87.

46. Frank E. Marble interview with Shirley K. Cohen, Pasadena, California, Jan–March 1994, 21 April 1995, Oral History Project, CAL, p. 44.

47. Iris Chang, *Thread of the Silkworm*, p. 160.

48. Iris Chang, *Thread of the Silkworm*, p. 157.

49. Quote from Milton Viorst interview with Iris Chang, ICP.

50. Details in *United States of America v. Sidney Weinbaum*, Southern District of California Central division, Los Angeles, California, 31 August 1950, reporter's transcript. Record Group 21, Criminal Case Files, 21383–21408, Folder 21408, Box no. 1261; quote in letter from David Harker to Linus Pauling, 6 February 1951, quoted in Thomas Hager, *Force of Nature*, p. 660.

51. Letter from Henri Levy to Linus Pauling, 2 September 1950, Box 455, Folder 5, LPP.

52. 'terrified': in memo to Director FBI, 7 July 1950, 'Frank Friedman Oppenheimer – Security Matter–C', Weinbaum FBI.

53. One example is Oppenheimer testifying that he didn't know if Weinbaum was a party member when it's plain from his own subsequent oral histories that they were the core of the unit. See, for instance, Frank Oppenheimer interview with Judith R. Goodstein, San Francisco, California, 16 November 1984, Oral History Project, CAL, p. 9.

54. 'Three Balk Over Testifying at Weinbaum Trial', *Los Angeles Times*, 25 August 1950, p. 1.

55. David Caute, *The Great Fear*, New York, Simon & Schuster, 1978. p. 152.

56. Sidney Weinbaum interview with Mary Terrall, CAL, p. 57; 'deception' is based on several FBI interviews with the Berkuses in which their knowledge of Weinbaum is made plain.

57. Author interview with Alice Brunner, 21 May 2016; Wendel Brunner, pers. comm., 6 September 2018.

58. Letter to the editor from Lee Carleton, *Science & Engineering*, p. 39.

59. 'Court Outcry by Weinbaum: accused ex-Caltech scientist throws trial into uproar', *Los Angeles Examiner*, 7 September 1950, pagination unknown; 'Arguments Ended in Weinbaum Case', *Los Angeles Times*, 7 September 1950, pagination unknown.

60. 'Weinbaum Ordered Jailed for Four Years', *Los Angeles Times*, 12 September 1950, pagination unknown.

Chapter 15: The Coffee Can

1. In April 1949, agents wrote that Parsons had been interviewed on several previous occasions, see: Arthur Wittenburg, 'Dr Sidney Weinbaum, Espionage–R', LA 65–4846, 5 April 1949, Weinbaum FBI. What is currently available in Parsons' FBI file details interviews from 1950. Evidence of Parsons' testimony from 1942 is in: author redacted, 'Frank J. Malina with alias Dr Frank J. Malina', 100–8196, 28 November 1942, Malina FBI.

2. Martin P. Starr, *The Unknown God*, p. 284.

3. This note is from a 'War Department' file – most likely CIC – and is cited in Martin P. Starr, *The Unknown God*, p. 284, fn. 20. Attempts to locate the file in National Archives and Records Administration were, in 2016, unsuccessful. Starr notes that Helen Parsons suspected Kahl, then in Texas and disgruntled at being displaced as Priestess.

4. Hugh B. Urban, *Magia Sexualis*, p. 117.

5. Testimony of Parsons as informant T-1: 'Frank J. Malina with alias Dr Frank J. Malina', file no. 100–8196, 28 November 1942, p. 3, Malina FBI.

6. Detailed in: Arthur Wittenberg, 'Dr Sidney Weinbaum, Espionage–R', 65–4846, 5 April 1949, p. 4, Weinbaum FBI.

7. Jack Parsons as informant T-1 is found in: author redacted, 'Frank J. Malina with alias Dr Frank J. Malina, Internal Security–R',100–8196, 28 November 1942, p. 3. Malina FBI. Three years previously was November 1939.

8. Report by Arthur Wittenberg, 'Dr Sidney Weinbaum, Espionage–R', 65–4846, 5 April 1949, p. 4, Weinbaum FBI.

9. Letter from Jack Parsons to Theodore von Kármán, 30 September 1948, Box 22, Folder 35, TVK.

10. OTO and *Liber Oz* circulation are deemed responsible in a letter from Jack Parsons to Karl Germer, 19 June 1949, DD8, GYC.

11. Letter from Jack Parsons to Theodore von Kármán, 20 September 1948, Box 22, Folder 35, TVK.

12. Details of the memo in which this 13 March interview is recorded have been redacted from Parsons' FBI file. Details of the 11 January interview in: Arthur C. Wittenburg, 'Dr Hsue-Shen Tsien, Espionage–R', file no. 65–4857, Tsien FBI.

13. Letter from Jack Parsons to Marjorie 'Candy' Cameron, 9 February 1950, typed by Jane Wolfe and recopied from that typescript, DD7, GYC.

14. Letter from Jack Parsons to Karl Germer, 19 June 1949, DD8, GYC.

15. Evidence of Candy's anti-Semitism and her desire to live in Spain in: Martin P. Starr, *The Unknown God*, pp. 328 and 326.

16. 'Manifesto of the Witchcraft', DD9, GYC.

17. Letter from Jack Parsons to Marjorie Cameron Parsons, 12 February 1950, DD7, GYC.

18. Boyer's name is redacted in the current FBI file but the name is given in George Pendle, *Strange Angel*, p. 292.
19. 'John Whiteside Parsons – Espionage–IS', LA 65–5131, 3 January 1951, Parsons FBI.
20. Much of Parsons' FBI file is concerned with this case. See for instance: author redacted, 'John Whiteside Parsons – Espionage–IS', LA 65–5131, 22 November 1950, Parsons FBI.
21. George Pendle, *Strange Angel*, p. 294.
22. Report: author redacted, 'John Whiteside Parsons – Espionage–IS', LA 65–5131, 13 November 1950, Parsons FBI.
23. Letter from John Tenney Mason, Industrial Employment Review Board to Jack Parsons, 9 January 1952, copy in Parsons FBI file.
24. Pendle cites a letter from Jane to Grady McMurty for this conversation: George Pendle, *Strange Angel*, p. 296.
25. Jane Wolfe, 'Hollywood', *In the Continuum*, vol. 4, no. 9, 1991, pp. 35–45 (p. 43).
26. Letter from Theodore von Kármán to Ward Jewell, 2 April 1952, Box 22, Folder 35, TVK.
27. Frey's comments reported in Jane Wolfe, 'Hollywood', *In the Continuum*, vol. 4, no. 10, 1991, pp. 34–43 (p. 34).
28. This letter from Jack Parsons to Karl Germer is retrospectively (mis)dated by its recipient 2 November 1953 but otherwise flagged as Parsons' last letter, DD8, GYC.
29. George Pendle, *Strange Angel*, p. 4.
30. 'Weird Blast in Pasadena', *The Mirror*, Wednesday 18 June 1952, p. 3.
31. Ibid.
32. 'Slain Scientist was Priest in Weird Sex Cult', *The Mirror*, Friday 20 June 1952, p. 3.
33. 'Frey feels as I did from the start that Jack … deliberately took the fatal step': Jane Wolfe, 'Hollywood' *In the Continuum*, vol. 4, no. 10, 1991, pp. 34–43, (p. 34).
34. Smith letter quoted in Martin P. Starr, *The Unknown God*, p. 327.
35. Jane Wolfe, 'Hollywood', *In the Continuum*, vol. 4, no. 9, 1991, pp. 35–45 (p. 44).

Chapter 16: Taking the Fifth

1. Letter from Frank Malina to parents, 14 October, 1950, Box 22, Folder 10, FJM.
2. Letter From Frank Malina to parents, 12 June 1950, Box 22, Folder 9, FJM.
3. Letter from Frank Malina to parents, 16 July 1950, Box 22, Folder 10, FJM.

4. This coincidence of surveillance with 'Project X' is suggested in Leonard Caveny, 'Martin Summerfield and His Princeton University Propulsion and Combustion Laboratory', 47th AIAA/ASME/SAE/ASEE Joint Propulsion Conference & Exhibit. 2011, available from arc.aiaa.org, 15 March 2017; see also Anthony Chong, *Flying Wings & Radical Things: Northrop's Secret Aerospace Projects and Concepts, 1939–1994*, Forest Lake, NM, Speciality Press, 2016, p. 28.

5. Report by H. T. Burk, 'Martin J. Summerfield, Espionage–R', file 65–1399, 15 June 1949, p. 2, Summerfield FBI.

6. Ibid., p. 2.

7. Author redacted, 'Frank J. Malina – Internal Security–R', file 100–8196, 13 December 1950, p. 2. Malina FBI.

8. Letter from Frank Malina to parents, 23 July 1950, MFC.

9. Letter from Agnes Schneider, American Consul, Paris to Frank Malina, 14 November 1951, MFC. This letter is following an appeal. Malina was advised of refusal by airgram on 26 May 1951; see author redacted, 'Frank J. Malina – Internal Security–R', WFO 105–668, 31 January 1952, Malina FBI.

10. 'Marjorie's notes on FJM', p. 41, MFC.

11. Memo from SAC Los Angeles to FBI Director, 'Martin J. Summerfield – Espionage–R', date obscured but probably January 1951, Summerfield FBI.

12. Philip J. Reilly 'Martin J. Summerfield – Espionage–R', file no. 65–4852, 30 January 1951, p. 3, Summerfield FBI.

13. Memo from Director SAC to FBI Director, 'Martin J. Summerfield – Espionage–R', Bufile 65–58451, 9 June 1951, p. 2, Summerfield FBI.

14. Philip J. Reilly, 'Martin J. Summerfield – Espionage–R', file no. 65–4852, 30 January 1951, p. 6, Summerfield FBI.

15. Details of 'mail cover' in memo from SAC Newark to FBI Director, 'Martin J. Summerfield – Espionage–R', 11 April 1951, Summerfield FBI.

16. Summerfield lost clearance in October 1951; SAC Baltimore to FBI Director, 'American Rocket Society', 1 July 1952, Summerfield FBI.

17. Iris Chang, *Thread of the Silkworm*, p. 168.

18. This provision for concentration camps was never actually used. See Masumi Izumi, 'Alienable Citizenship: Race, Loyalty and the Law in the Age of American Concentration Camps, 1941–1971', *Asian American Law Journal*, vol 13, 2006, pp. 1–35; Carole Boyce Davies, 'Deportable subjects: US immigration laws and the criminalizing of Communism', *South Atlantic Quarterly*, vol. 100, no. 4, 2001, pp. 949–66.

19. William L. Ryan and Sam Summerlin, *The China Cloud*, p. 109.

20. Letter from Richard N. Lewis to Edward S. Lesnick, 3 December 1989, GLC.

21. Letter from Richard N. Lewis to John A. Perkins, President, University of Delaware, 5 February 1951, GLC.
22. Memo from SAC, Los Angeles to Director FBI, 4 January 1952, 'Frank Malina, Internal Security–R', Malina FBI.
23. Jan Wunderman journal, 28 May 1996, KWC.
24. Ibid.
25. This detail from Iris Chang, *Thread of the Silkworm*, p. 181.
26. Quite a few reports are missing from the Tsien FBI file for the spring of 1951. This account of the INS hearing from: letterhead memorandum: 'Tsien Hsue Shen – Internal Security – Chinese', 1 March 1967, Tsien FBI.
27. Memo from James McInerney to Director FBI, 'Frank Malina – Internal Security–R', 16 April 1952, Malina FBI.
28. Memo from 'Frank J. Malina – Internal Security–R', 19 August 1952, p. 2. Malina FBI.
29. Memo from A. H. Belmont to W. A. Branigan, 'Frank J. Malina – Internal Security–R', 3 October 1952, p. 2, Malina FBI.
30. redacted author, 'Frank Joseph Malina, Internal Security–R; Fraud Against the Government', PX 46–306, 1 June 1953, p. 2, Malina FBI.
31. Ibid., p. 3.
32. Memo from SAC Los Angeles to Director FBI, 'Frank J. Malina – Internal Security–R', 10 October, 1952, Malina FBI.
33. Memo to A. H. Belmont from W. A. Branigan, 'Frank J. Malina – Internal Security–R; Fraud Against the Government', 18 December 1952, p. 2. Malina FBI.
34. Memo, 'Frank J. Malina – Internal Security–R; Fraud Against the Government', 29 Janaury 1953, p. 6, Malina FBI.
35. Letter from Sidney Weinbaum to Linus Pauling, 8 December 1950, copy in ICP.
36. Letter from Sidney Weinbaum to Lina Weinbaum, 24 September 1951, Jacob Bendix Private Collection.
37. Ellen Schrecker, *Many Are the Crimes*, p. ix.
38. Letter from Lina Weinbaum to Frank Malina, 10 February 1952, Box 3, Folder 35, FJM.
39. Letter from Sidney quoted in letter from Lina Weinbaum to Linus Pauling, 28 December 1951, Box 433, Folder 7, LPP.
40. Letter from Sidney Weinbaum to Linus Pauling, 11 April 1952, copy in ICP.
41. Letter from Lina Weinbaum to Frank Malina, 10 February 1952, Box 3, Folder 35, FJM.
42. Ibid.
43. Letter from Sidney Weinbaum to Linus Pauling, 11 November 1952, copy in ICP.

44. CNDI testimony, reported in letter from R. B. Hood to Director FBI, 'Dr Sidney Weinbaum, Espionage–R', 16 May 1949, Weinbaum FBI.

45. Letter from Sidney Weinbaum to Linus Pauling, 11 November 1952, copy in ICP.

46. Joel Braslow, *Mental Ills and Bodily Cures: Psychiatric Treatment in the First Half of the Twentieth Century*, Berkeley, University of California Press, 1997, p. 101. Bizarrely, the psychiatrist originally in charge of implementing ECT at Stockton, Aaron Rosanoff, was the father of Sidney Weinbaum's hostile witness, Richard Rosanoff.

47. Between 1947 and 1954, 232 Stockton patients were lobotomised. Ibid., p. 128.

Chapter 17: The Rosenberg Connection

1. Michael H. Gorn, *The Universal Man: Theodore von Kármán's life in Aeronautics*, Washington, Smithsonian Institution Press, 1992, p. 125.

2. This is evident from numerous entries from Frank's letters to his parents in the early 1950s.

3. Letter from Frank Malina to parents, 23 October 1950, Box 22, Folder 10, FJM.

4. CIC memo, 12 September 1950, 'Possible Communist, California Institute of Technology', von Kármán FBI.

5. This is detailed in a memo from V. P. Keay to A. H. Belmont, 'Theodore von Kármán', 16 April 1951, von Kármán FBI.

6. Von Kármán denied his party membership but writes of his ministerial position in *The Wind and Beyond*, p. 92.

7. Memo from SAC Los Angeles to Director FBI, 'Theodore von Kármán', 30 October 1950, von Kármán FBI.

8. Robert J. Lamphere and Tom Shachtman, *The FBI-KGB War: A Special Agent's Story*, London, W. H. Allen, 1987, p. 178.

9. John Earl Haynes and Harvey Klehr, *Venona*, p. 295. Though they were convicted of conspiracy to commit espionage, the question of atomic espionage was not explicit in the charge.

10. This is detailed in John P. Andrews, memo: 'Theodore von Kármán – Internal Security–R', 27 April 1951, p. 11, von Kármán FBI.

11. Perl's wife did not know according to a KGB report cited in Harvey Klehr, John Earl Haynes, and Alexander Vassiliev, *Spies: The Rise and Fall of the KGB in America*, New Haven, Yale University Press, 2009, pp. 339–40.

12. Steven T. Usdin, *Engineering Communism: how two Americans spied for Stalin and founded the Soviet Silicon Valley*, New Haven, Yale University Press, 2008, p. 53.

13. Ibid., p. 87.

14. John A. Harrington, 10 February 1953, 'Unknown subject; unknown consultant at Aswan Dam, Egypt, 1946–1949', NY 65–15384, von Kármán FBI.
15. Steven T. Usdin, *Engineering Communism*, p. 33.
16. Ibid., p. 73.
17. 'Notes made during trip on US East Coast', MFC, unpaginated.
18. Records of Columbia record Perl as employed as a Technical Assistant from February 1947 to June 1948, see: John A. Harrington, 10 February 1953, 'Unknown subject; unknown consultant at Aswan Dam, Egypt, 1946–1949', NY 65–15384, p.16, von Kármán FBI.
19. Ronald Radosh and Joyce Milton, *The Rosenberg File: a Search for the Truth*, New Haven, Yale University Press, 1997, p. 123. The exact dates detailed in Perl's testimony to Rosenberg Grand Jury, 15 August 1950, p. 4, Record Group 118, Rosenberg Case Files, 7/22/1947–1/13/1985.
20. Redacted author, 'Unknown subject; unknown consultant at Aswan Dam, Egypt, 1946–1949', WFO 65–5545, 22 July 1952, p. 8, von Kármán FBI.
21. John A. Harrington, 'Unknown subject; unknown consultant at Aswan Dam, Egypt, 1946–1949', NY 65–15384, 10 February 1953, p. 16, von Kármán FBI.
22. Ronald Radosh and Joyce Milton, *The Rosenberg File*, p. 299.
23. Perl borrowing von Kármán's car at this time is detailed in John P. Andrews, 'Theodore von Kármán – Internal Security–R', LA 105–863, 27 April 1951, p. 11. von Kármán FBI.
24. Steven T. Usdin, *Engineering Communism*, p. 98.
25. Sam Roberts, *The Brother: The Untold Story of the Rosenberg Case*, New York, Simon and Schuster, 2014, p. 510.
26. The account of this weekend is taken from Steven Usdin's extraordinary interview with Sobell, published in Ronald Radosh and Steven Usdin, 'The Sobell Confession', *The Weekly Standard*, 28 March 2011, accessed online 24 May 2018.
27. Radosh and Milton, *The Rosenberg File*, p. 30; Haynes and Klehr, *Venona*, p. 10.
28. Harvey Klehr, John Earl Haynes, and Alexander Vassiliev, *Spies*, p. 347.
29. Testimony of William Perl, 18 August 1950, Rosenberg Grand Jury, p. 9439, Record Group 118, Rosenberg Case Files, 7/22/1947–1/13/1985, NARA.
30. This dialogue from Iris Chang's interview with Andrew Fejer 12/7/91, Tape 29, side 1, ICP. Malina's friend Andrew Fejer, another Unit 122 regular, was struggling with clearance issues at the University of Toledo, and had made exactly the same request of von Kármán – with the same response.
31. Memo from FBI Director to James McInerney, 'Theodore von Kármán – Internal Security–R', 3 February 1951, von Kármán FBI.

32. To A. H. Belmont from C. E. Heinrich, 'Theodore von Kármán – Internal Security–R', 22 January 1951, p. 7, von Kármán FBI.

33. Letter from Frank Malina to parents, 21 June 1953, MFC.

34. Radosh and Milton, *The Rosenberg File*; John Earl Haynes and Harvey Klehr, *Venona*.

35. Ronald Radosh and Steve Usdin, 'The Sobell Confession'.

36. Ibid.

37. John A. Harrington, 'Unknown subject; unknown consultant at Aswan Dam, Egypt, 1946–1949; Espionage–R', NY 65–15384, 13 January 1953, p. 4, von Kármán FBI.

38. Memo from SAC New York to SAC Newark, 'Harry Vorti Aivazian, Security of Government Employees', 23 January 1957, p. 10, von Kármán FBI.

39. See John A. Harrington, 'Unknown subject; unknown consultant at Aswan Dam, Egypt, 1946–1949; Espionage–R', NY 65–15384, 13 January 1953, von Kármán FBI.

40. John A. Harrington, 'Unknown subject; unknown consultant at Aswan Dam, Egypt, 1946–1949', NY 65–15384, 14 August 1952, p. 16, von Kármán FBI.

41. John A. Harrington, 'Unknown subject; unknown consultant at Aswan Dam, Egypt, 1946–1949; Espionage–R', NY 65–15384, 13 January 1953, p. 2, von Kármán FBI.

42. Letter from Frank Malina to parents, 23 October 1952, Box 22, Folder 9, FJM.

43. Memo from Mr R. J. Lamphere to Mr W. A. Branigan, 'Unknown subject; unknown consultant at Aswan Dam, Egypt, 1946–1949; Espionage–R', 17 February 1953, p. 2, von Kármán FBI; see also Robert J. Lamphere and Tom Shachtman, *The FBI-KGB War*, p. 250.

44. Harvey Klehr, John Earl Haynes, and Alexander Vassiliev, *Spies*, p. 34f.

45. See Allen Weinstein, Alexander Vassiliev, and Bill Wallace, *The Haunted Wood: Soviet espionage in America – the Stalin era*, New York, Random House, 1999; John Earl Haynes and Harvey Klehr, 'Introduction, Alexander Vassiliev's Notebooks: Provenance and Documentation of Soviet Intelligence Activities in the United States', 2009, History and Public Policy Program Digital Archive, Alexander Vassiliev Papers, Manuscript Division, Library of Congress.

46. This detail from Marjorie Malina, 'Notes on FJM' document, MFC.

47. Theodore von Kármán, *The Wind and Beyond* (1967).

48. Jonathan C. McDowell, 'The edge of space: Revisiting the Karman Line', *Acta Astronautica,* vol. 151, 2018, pp. 668–77.

49. Ibid.

50. For an account of these engineers, see Asif A. Siddiqi, *The Red Rockets Glare: space flight and the Soviet imagination, 1857–1957*, New York, Cambridge University Press, 2010, p. 155f.

Chapter 18: Light Engineering

1. Letter from Frank Malina to parents, 15 February 1953, MFC.
2. Theodore von Kármán, *The Wind and Beyond*, p. 318.
3. Frank Malina Oral History interview with R. Cargill Hall, 29 October 1968, p. 5, copy in ICP.
4. Theodore von Kármán, *The Wind and Beyond*, p. 315.
5. Letter from Andy Haley to Frank Malina, 13 June 1950, Box 2, Folder 6, FJM.
6. 'Marjorie's notes on FJM', p. 37, MFC.
7. Letter from Andy Haley to Frank Malina, 11 November 1953, Box 2, Folder 6, FJM.
8. Details of share price and various sales of stock are detailed in correspondence with Andy Haley, see for instance: Andy Haley to Frank Malina, 10 December 1954; cable from Andy Haley to Frank Malina, 29 April 1955, Box 2, Folder 6, FJM.
9. Letter from Frank Malina to Theodore von Kármán, 20 February 1957, Box 19, Folder 27, TVK.
10. 'Marjorie notes on FJM', p. 42, MFC. This detail is notably absent from the memoir, Eric Roll, *Crowded Hours*, London, Faber and Faber, 1985.
11. Letter from Home Office, author and addressee both illegible, dated 25 August 1954, MFC.
12. Stephen Twigge and Alan Macmillan, 'Britain, the United States, and the development of NATO strategy, 1950–1964', *Journal of Strategic Studies*, vol. 19, no. 2, 1996, pp. 260–81 (p. 263).
13. Army Ballistic Missile Agency, *Development of the Corporal: the embryo of the army missile program vol. I*, ABMA unclassified report Redstone Arsenal, Alabama, 1961, p. 263.
14. Fraser MacDonald, 'Geopolitics and "the vision thing": regarding Britain and America's first nuclear missile', *Transactions of the Institute of British Geographers*, vol. 31, no. 1, 2006, pp. 53–71.
15. Advert, 'Corgi Toys: percussion warhead', provenance unknown, purchased on eBay; see also Marcel R. Van Cleemput, *The Great Book of Corgi 1956–1983*, London, New Cavendish Books, 1989, p. 50.
16. This is an aggregate figure from the sales of the Corgi Corporal in various permutations from being a stand-alone toy to being sold as part of a gift set; ibid.

17. Corporal toy ephemera is discussed in Fraser MacDonald, 'Rocketry and the popular geopolitics of space exploration, 1944–1962', in Simon Naylor and James R. Ryan, ed., *New Spaces of Exploration: Geographies of Discovery in the Twentieth Century*, London, IB Tauris, 2010, pp. 196–221.

18. Letter from Andy Haley to Frank Malina, 17 August 1954, Box 2, Folder 6, FJM.

19. Haley wrote to Bill Pickering, Louis Dunn, and Homer J. Stewart among others asking them to support his nomination of Malina. None did. See letter from Haley to Frank Malina, 23 May 1962, Box 2, Folder 9, FJM.

20. Florence Fowler Lyons, 'Rocket Society Bestows Honor on Malina', *The Ledger*, 6 December 1962, p. 4.

21. Letter from Andy Haley to Frank Malina, 21 January 1963, 1962, Box 2, Folder 9, FJM.

22. Ibid.

23. Alan Rogers, 'Passports and politics: the courts and the Cold War', *Historian*, vol. 47, no. 4, 1985, pp. 497–511.

24. *Kent v. Dulles*, 357 US 116 (No. 481), decided 16 June 1958.

25. Letter from Frank Malina to Andy Haley, 9 June 1959, Box 2, Folder 8, FJM.

26. Letter from Frank Malina to Andy Haley, 10 January 1959, Box 2, Folder 8, FJM.

27. Homer J. Stewart interview with John L. Greenberg, October–November 1982, CAL, p. 70.

28. Letter from Donald G. Agger to Frank Malina, 9 April, 1980, Box 5, Folder 1, FJM.

29. Unable to get butalbital in the US, Summerfield had asked Malina to send them; letter from Martin Summerfield to Frank Malina, 31 January 1963, MFC.

30. Letter from Frank Malina to Donald Agger, 1 June 1980, Box 5, Folder 1, FJM.

31. This quote is Chang reporting on the conversation: letter from Iris Chang to Jan Wunderman, 15 April 1993, ICP; original emphasis.

32. Ibid.

33. Jerry Grey, *Enterprise*, New York, William Morrow and Co., 1979, p. 119; see draft letter marked 'NOT SENT' from Frank Malina to Jerry Grey, 21 November 1979, Box 5, Folder 26, FJM.

34. Walter R. Dornberger, 'The German V-2', *Technology and Culture*, vol. 4, no. 4, 1963, pp. 393–409; Wernher von Braun, 'The Redstone, Jupiter, and Juno', *Technology and Culture*, vol. 4, no. 4, 1963, pp. 452–465; Wernher von Braun and Frederick I. Ordway III, *History of Rocketry and Space Travel*, New York, Crowell Publishing, 1966, p. 308.

35. Letter from Frank Malina to Eugeme Emme, 5 December 1976, Box 5, Folder 20, FJM. 'The important part that Goddard played in American rocket development is being built up out of all proportion,' complained Malina; see letter from Frank Malina to Eugene Emme, 24 January 1966, Box 5, Folder 20, FJM.

36. Ibid.

37. Letter from Arthur C. Clarke to Frank Malina, 20 December 1965, Box 5, Folder 17. The next non-fiction book was Arthur C. Clarke, *The Promise of Space*, New York, Harper & Row Publishers, 1968.

38. Wernher von Braun and Frederick I. Ordway III, *History of Rocketry and Space Travel*, p. 104.

39. Letter from Frank Malina to Frederick C. Durant III, 26 January 1968, Box 5, Folder 19, FJM.

40. Letter from Frank Malina to Walter Powell, 21 December 1968, Box 4, Folder 1, FJM.

41. Frank J. Malina, 'Comments on the Occasion of the Commemoration of the First Test of a Rocket Motor by the GALCIT Rocket Research Group on 31 October 1936', 31 October 1968, Box 4, Folder 1 FJM.

42. John F. Bluth, 'Biographical notes on Louis Dunn', 17 July 1998, CL#12-1651, JPL.

43. Letter from H. S. Tsien to Frank Malina, 8 December 1954, FJM.

44. Testimony of Sol Penner in Iris Chang, *Thread of the Silkworm*, p. 191.

45. Tsien quoted in 'Jet Propulsion Scientist Sailing to Red China: Dr Hsue-shen Tsien Ends Long, Honorable Career Here to Help People of Own Nation', *Los Angeles Times*, 18 September 1955, pagination unknown.

46. Letter from H. S. Tsien to Frank Malina, August 1971, Box 3, Folder 29, FJM.

47. 'Li Yuchang: lixuesuo zaonian de ren he shi' [Li Yuchang: people and events in the early history of the Institute of Mechanics], an interview with Li Yuchang by Xiong Weimin, 19, 28 May 2015, Beijing, in Xiong, Weimin. *Duiyu lishi, kexuejia youhuashuo: 20 shiji zhongguo kexuejie de ren yu shi* [Scientists have something to say about history: People and events in the circle of science in 20th century China], Beijing, Dongfang Press, 2017, pp. 225–48, (p. 228), trans. by Zuoyue Wang, 12 July 2018.

48. Iris Chang, *Thread of the Silkworm*, p. 192.

49. See entry for 16 August 1984, Jan Wunderman journal, KWC.

50. J. Edgar Hoover, *Masters of Deceit: The Story of Communism in America and how to Fight It*, New York, Henry Holt and Company, 1968, p. 293.

51. See undated entry, approx. 1993, Jan Wunderman journal, KWC.

52. Letter from Gus Albrecht to Liljan Wunderman, 11 January 1956, MWC.

53. Iris Chang, *Thread of the Silkworm*, p. 179; author interview with Ena Dubnoff, 18 August 2014.

54. James Jackson and Martin Grotjahn, 'The concurrent psychotherapy of a latent schizophrenic and his wife', *Psychiatry*, vol. 22, no. 2, 1959, pp. 153–160; this paper was identified by Ena Dubnoff as referring to her parents.

55. Undated typed flier, posted around Sherman Oaks, signed by 'James Jackson MD, Publisher', Ena Dubnoff private collection.

56. 'Ex-psychiatrist jailed after 2 year law battle' *Van Nuys News*, 7 January 1975, p. 8.

57. Author interview with Ena Dubnoff, 18 August 2014.

58. Letter from Sidney Weinbaum to Linus Pauling, 16 September 1953, Box 433, Folder 8, LPP.

59. Memo from SAC San Francisco to Director, FBI, 'Lina Litinskaya Weinbaum 18 December 1964', Lina Weinbaum FBI.

60. Hotel owner testimony in memo from SAC Los Angeles to Director, FBI, 'Lina Litinskaya Weinbaum', 22 January 1959, Lina Weinbaum FBI.

61. Letter from Lina Weinbaum to Linus Pauling, 19 January 1960, Box 433, Folder 12, LPP.

62. Sidney Weinbaum interview with Mary Terrall, 15, 20 and 25 August 1985, CAL, p. 56.

63. Frank J. Malina, 'Reflections of an artist-engineer on the arts-science interface' *Impact of Science on Society*, XXIV, no. 1, 1974, pp. 19–29, in Jacob Bendix private collection.

64. Roger F. Malina, 'Frank J. Malina: Astronautical Pioneer Dedicated to International Cooperation and the Peaceful Uses of Outer Space', 57th International Astronautical Congress, 2006.

65. Letter from Frank Malina to his parents, 31 October 1953, Box 22, Folder 10, FJM.

66. Frank J. Malina, 'Electric Light as a Medium in the Visual Fine Arts: A Memoir', *Leonardo*, vol. 8, no. 2, 1975, pp. 109–20 (p. 110); for a wider discussion of Malina's art, see W. Patrick McCray, *Wiring Art Work: Artists, Engineers, and Collaborations Between Creative Cultures Since 1960*, Cambridge, The MIT Press, forthcoming.

67. 'He used to think of himself as a professional astronautical engineer with art as a hobby': Frank Popper, 'Frank Malina, Artist and Scientist: Works from 1936 to 1963', undated, accessed 12 February 2015, http://www.olats.org/pionniers/malina/arts/monographUS.php

68. Roger Malina, 'Jack Parsons and Frank Malina: Strange Angel and Sparks of Genius', blogpost, http://malina.diatrope.com/2018/06/15/jack-parsons-and-frank-malina-strange-angel-and-sparks-of-genius/ 15 June 2018.

69. Roger Malina, 'Jack Parsons and Frank Malina'; Fabrice Lapelletrie, 'Life of Frank Joseph Malina', *point-line-universe: a retrospective exhibition of Frank J. Malina*, Museum Kampa, 2007, catalogue, p. 20.

70. Fabrice Lapelletrie, 'Life of Frank Joseph Malina', p. 21.

71. Letter from Jan Wunderman to FJM, 5 April 1972, MFC.

72. Jan Wunderman journal, undated April 1972, KWC.

73. Jan Wunderman journal, undated, 6 May 1981, KWC; Jan was wrong about his age. He was 68.

74. Jan Wunderman, 'My Background and Intent as a Painter of Nonfigurative or Abstract Pictures', *Leonardo,* vol. 14, no. 4, 1981, pp. 305–7.

75. Jan Wunderman journal, 11 June 1999, KWC.

76. Jan Wunderman journal, 13 November 1981, KWC.

77. All dialogue from Jan Wunderman journals, 10 November 1981, KWC; additional contextual detail from Karen Wunderman.

78. The edited collection from this conference with a tribute to Malina is: Martin Pollock, ed., *Common Denominators in Art and Science*, Aberdeen, Aberdeen University Press, 1983.

79. Jan Wunderman journal, 13 November 1981, KWC.

80. Personal correspondence with Karen Wunderman, 29 August 2017.

81. Jan Wunderman journal, 13 November 1981, KWC.

82. Ibid.

83. Berkeley launches detailed in: Jonathan McDowell, 'Rockets for Extrasolar X-rays', unpublished manuscript on author's website, 4 February 2013, p. 15.

Chapter 19: Afterword

1. Jan Wunderman journals, undated, 1992–8, KWC.

2. Letter from Frank and Marjorie Malina to Frank's parents, 15 February 1953, MFC; 'Marjorie's Notes on FJM' document, MFC. p. 2.

3. Jan Wunderman oral history with Benjamin Zibit, 22 April 1996, p. 7, CAL.

4. Letter from Jan Wunderman to Iris Chang, 6 April 1992, ICP.

5. Jan Wunderman journals, 11 June 1999, KWC.

6. Letter from Frank Malina to Liljan Malina, 6 February 1947, MWC.

7. Allen Weinstein, Alexander Vassiliev and Bill Wallace, *The Haunted Wood*, p. 29.

8. Ibid., p. 184.

9. It's important to note, in particular, Clayton Koppes, *JPL and the American Space Program.*

10. Letter from Frank Malina to parents, 6 July 1947, MFC.

11. Ian Hacking, 'Weapons research and the form of scientific knowledge', *Canadian Journal of Philosophy*, vol. 16, 1986, pp. 237–60.

12. Nathalia Holt, *Rise of the Rocket Girls*; Margot Lee Shetterly, *Hidden Figures*, London, HarperCollins, 2017.

13. Letter from Frank Malina to C. Stark Draper, 2 August 1976, Box 5, Folder 18, FJM.

14. Letter from Ray Bradbury to Frank Malina, 31 August 1980, Box 5, Folder 11, FJM.

15. Letter from Frank Malina to Eugene Emme, 3 January 1974, Box 5, Folder 20, FJM.

LIST OF ILLUSTRATIONS

In text images

p. 39 Two sketches from Malina letters, letter from Frank Malina to Liljan Malina, 12 September 1938, letter from Frank Malina to Liljan Malina, 16 September 1938. (Malina Family Collection).

p. 110 Extract of letter from Liljan Malina to Frank Malina, undated but sequence suggests December 1943. (Malina Family Collection)

p. 187 Extract of memo from SAC to FBI Director, 're Sidney J. Weinbaum, Espionage-R', 16 March 1949, Weinbaum FBI.

p. 192 Sidney Weinbaum's JPL Personnel Security Questionaire, loose copy, apparently unattached to a specific memo, in Weinbaum FBI file.

p. 227 'Sex Madness' Cult Slain Pasadena Scientist Revealed, headline in the *Mirror*, 7 June, 1952.

p. 252 Report by John A. Harrington, 'Unknown Subject; Unknown Consultant at Aswan Dam, Egypt, 1946–1949', 10 February 1953, von Kármán FBI file.

Plate section

1. Theodore von Kármán, c. 1938. (Photo: Caltech Archives)
2. The nativity scene: methyl alcohol and gaseous oxygen test, Arroyo Seco, 1936. (Photo: Caltech Archives)
3. Frank Malina and Jack Parsons, Arroyo Seco, 1936. (Photo: Malina Family collection)
4. Frank and Liljan Malina, wedding day portrait. (Photo: Karen Wunderman private collection)
5. Helen and Jack Parsons, with friends. (Photo: Malina Family Collection)

ACKNOWLEDGEMENTS

From the outset of this long project, I've been sustained by the encouragement and insight of so many people. Foremost here are the protagonists' families, whose kindness and openness made the book possible. My heartfelt thanks to Roger Malina for his hospitality and access to his family archives; to Karen Wunderman and Marc Wunderman for their generosity with time and with their mother's papers; to Jacob Bendix for sharing memories of his grandfather, Sidney Weinbaum; and to Wendy Prober Cohen for the Berkus family history and for her perceptive remarks on an early draft. And to the many others with whom I have been in touch: David Altman, Jody Altman, Eve Borsook, Alice Brunner, Wendel Brunner, Joan Churchill, the late Elaine Lustig Cohen, Ena Dubnoff, Steve Dubnoff, Linda Payne Dunn, Reg Gadney, Leane Hines-Hill, Tracy Lewis, Gil Lewis, Lynn Forman Maginnis, Alan Malina, Susan Pile, Harriet Sturtevant Shapiro, Patricia Canright Smith, Barbara Mogel Stewart, Leon Trilling, Larry Wilson and Ilona Wiss. I'm grateful too for the help and interest of other scholars: Ewen Chardronnet, Oliver Dunnett, Harvey Klehr, Fabrice Lapelletrie, Roger Launius, Scott McClellan, Patrick McCray, Simon Morrison, Mike Neufeld, Wyatt D. Phillips, Martin Starr, Steve Usdin, Mike Weaver, Simon Winchester and Benjamin Zibit. Special mention must be made of Audra Wolfe and Zuoyue Wang, whose expertise improved the book no end.

For early inspiration, thanks to Clare Forster, Nicky Forster, David Gilbert, David Glass, Alice Gorman, Keith Hart, Beryl Hartley, Mike Heffernan, the late Anna Jakomulska, Anja Kanngieser, Duncan Lunan, David Matless, Lee Mackinnon, James R. Ryan and Melanie Thomson. Coming into the orbit of Nicola Triscott and Rob La Frenais at Arts Catalyst was transformative; the same is true of Simon

Clews at the University of Melbourne's Writing Centre; and Norman MacLeod and Andy Mackinnon at Taigh Chearsabhagh Museum & Arts Centre. For miscellaneous but essential support over the years, my thanks to Faten Adam, James Annal, Julian Baker, Ailsa Bathgate, Francis Bickmore, Emily Brady, Severin Carrell, Jamie Chambers, Brent Everitt, Gavin Francis, Ben Garlick, Iga Gozdowska, Mary Holmes, Rachel Hunt, Jane Jacobs, Hamish Kallin, Eric Laurier, Hayden Lorimer, Anne MacLeod, Ealasaid Munro, Simon Naylor, Genevieve Patenaude, Jamie Pearce, Meredith Rose, Niamh Shortt, Richard Sobolewski, Dan Swanton, Nancy Wachowich, Abigail Williams, Charlie Withers and Iain Woodhouse.

I've been brilliantly served by the team who first saw promise in this book and worked hard to bring it to publication: Benjamin Adams at PublicAffairs, George Lucas at Inkwell Management and, most particularly, James Macdonald Lockhart at Antony Harwood Ltd and Ed Lake at Profile. Thank you all.

I'd like to acknowledge small grant support from the Royal Society of Edinburgh, the Carnegie Trust for the Universities of Scotland and from Edinburgh University's Institute of Geography. I'm grateful to William Breeze and the OTO for permission to quote from Aleister Crowley's literary estate, and to Scott Hobbs and Susan Pile from the Cameron-Parsons Foundation for Jack Parsons' extracts. For archival assistance, I'm indebted to Julie A. Cooper and Micky Honchell at JPL, Loma Karklins at Caltech Archives, Callie Bowdish and Sylvia Baldwin at UC Santa Barbara Special Collections, Geoff Somnitz and Chris Petersen at Oregon State University Special Collections, Philip Young and Clare Lappin at the Warburg Institute, and Bill Keegan of Heritage Consultants.

Special thanks to Ishbel MacDonald and Mairead MacDonald for their fond eye-rolling about this project. Underneath it all, they are generally in favour of books, a quality they inherited from their four grandparents, whose support has been concrete and enduring.

My greatest debts in writing, in life and in love always exceed acknowledgement; she knows what I mean.

INDEX

Credit: Iga Godzowska

Fraser MacDonald is a lecturer in human geography at the University of Edinburgh, where he teaches historical geography and the history of science. He has a regular byline at the *Guardian* and has also written for *Aeon* magazine, the *Herald*, the *Age*, the *Australian*, and the *LRB Books* blog, among other publications.

PublicAffairs is a publishing house founded in 1997. It is a tribute to the standards, values, and flair of three persons who have served as mentors to countless reporters, writers, editors, and book people of all kinds, including me.

I. F. STONE, proprietor of *I. F. Stone's Weekly*, combined a commitment to the First Amendment with entrepreneurial zeal and reporting skill and became one of the great independent journalists in American history. At the age of eighty, Izzy published *The Trial of Socrates*, which was a national bestseller. He wrote the book after he taught himself ancient Greek.

BENJAMIN C. BRADLEE was for nearly thirty years the charismatic editorial leader of *The Washington Post*. It was Ben who gave the *Post* the range and courage to pursue such historic issues as Watergate. He supported his reporters with a tenacity that made them fearless and it is no accident that so many became authors of influential, best-selling books.

ROBERT L. BERNSTEIN, the chief executive of Random House for more than a quarter century, guided one of the nation's premier publishing houses. Bob was personally responsible for many books of political dissent and argument that challenged tyranny around the globe. He is also the founder and longtime chair of Human Rights Watch, one of the most respected human rights organizations in the world.

•　　•　　•

For fifty years, the banner of Public Affairs Press was carried by its owner Morris B. Schnapper, who published Gandhi, Nasser, Toynbee, Truman, and about 1,500 other authors. In 1983, Schnapper was described by *The Washington Post* as "a redoubtable gadfly." His legacy will endure in the books to come.

Peter Osnos, *Founder*

The Management
of
Oral History
Sound Archives

The Management
of
Oral History
Sound Archives

FREDERICK J. STIELOW

GREENWOOD PRESS
New York • Westport, Connecticut • London

Library of Congress Cataloging-in-Publication Data

Stielow, Frederick J., 1946-
   The management of oral history sound archives.

   Bibliography: p.
   Includes index.
   1. Archives, Audio-visual.   2. Oral history.
I. Title.
CD973.2.S74   1986   907'.2   85-14716
ISBN 0-313-24442-1 (lib. bdg. : alk. paper)

Library of Congress Catalog Card Number: 85-14716
ISBN: 0-313-24442-1

First published in 1986

Greenwood Press, Inc.
88 Post Road West
Westport, Connecticut 06881

Printed in the United States of America

The paper used in this book complies with the
Permanent Paper Standard issued by the National
Information Standards Organization (Z39.48-1984).

10 9 8 7 6 5 4 3 2 1

Dedicated to the memories of Richard M. Dorson and Innocent Terrebonne—the scholar and the folk—both of whom relished a good fight.

I have nothing to say
and I am saying it   and that is
poetry   as I need it
   The space of time   is organized
      We need not fear these   silences,—
we may love them.
                  John Cage, *Silence*

CONTENTS

ILLUSTRATIONS

ABBREVIATIONS

UMCP: University of Maryland at College Park
OHMAR: Oral History Association of the Mid-Atlantic States
AASLH: American Association for State and Local History
OHA: Oral History Association
ISBD: International Standard Bibliographic Description
AACR2: Anglo-American Cataloguing Rules, Second Edition
LCSH: Library of Congress Subject Headings
COM: Computer Output Microfilm
MARC: Machine Readable Cataloging
TAPE: Time Access Pertinent Excerpts
ips: inch per second
RAM: Random Access Memory
ASCII: American Standard Code for Information Exchange
RDBMS: Relational Data Base Management System
FMS: File Management System
DBMS: Data Base Management System
S/N: Signal-to-Noise
db: decibels
MOL: Maximum Output Level
CRT: Cathode Ray Tubes
PFS: Personal Filing System
CDs: Compact Disks
THIC: Tape Head Interface Committee
TMS: Tape Management System
IEC: International Electrotechnical Committee

PREFACE

Millions of people are now using magnetic tape recorders – from teenagers pirating popular songs from the airways and individuals sending personalized oral letters to court and business reporters capturing the text of meetings. In addition, hundreds of professional scholars, archivists, and librarians have been joined by an even greater number of amateur genealogists and local historians, plus an array of senior citizens and students, in projects to gather the elite and folk memories of modern times. Yet few of these people – even historians – are aware of the organizational principles and technological factors needed to preserve and make their efforts available for future generations. Instead, much of that more serious work is in effect being abandoned and rendered useless for posterity.

This book was written partly to help redress this situation. It is aimed at those who are charged with the stewardship and administration of such collections in archives, historical societies, and especially libraries around the country. In a more general sense, the work should be of value to anyone engaged in oral history and folklore collection projects – or even to concerned individuals trying to save their tapes and facilitate their later use.

The following pages attempt to blend a general introduction and indoctrination into the value of sound collections with more advanced concepts in the field. Although somewhat introductory and designed for the most inexpensive implementation, this work strives to convey the most appropriate and bibliographically correct approaches presently available. It provides the reader with the practical information necessary for organizing and managing a sound archive, as well as the theoretical concepts that should underlie such an endeavor. In addition to exploring the range of activities from public service and community outreach to technical processing and preservation, it also offers the sound archivist or librarian the first systematic attempt to introduce data processing applications to the process.

The reader might benefit from an understanding of my own background and experience in regard to these subjects. Although the "trickster" in me cries out to proclaim that this work is mine, all mine – honesty also compels an acknowledgment of others who have contributed to it. Following a stint in an underground data processing installation in Germany, my formal interests in oral history and folklore began in graduate school at Indiana University. There I benefited from contacts with fellow students with an interest in oral history, including Gerald Handfield and John Moe, and professors, including Richard Kirkendall. Oral history in the early 1970s, however, was still too elitist to facilitate my research plans. Fortunately, the program also brought me into contact with noted folklorists, such as Henry Glassie, and above all folklorist/historian Richard Dorson. Dorson's presence allowed me to undertake a holistic history of a small French-American settlement in Louisiana, a study that would have been impossible without oral interviews.

My more particular concerns for sound archives stemmed from the opportunity to organize the holdings of an Archives of Acadian and Creole Folklore/Oral History, which Barry Ancelet began to assemble at the University of Southwestern Louisiana. The frustrations involved and information gathered during that process first led me to see the need for this current study. At the same time, I was fortunate to encounter oral historian Joel Gardner (who awakened me to the great changes in this field) and a number of folklore archivists, such as Sandy Ives, Mark Glazer, Wayland Hand, Janet Langois, and Richard Thill. A National Endowment for the Humanities grant for the archive also brought me in contact with Canadians Sr. Catherine

Jolicoeur and Carole Saulnier, as well as a month's study at the Library of Congress' Archive of Folk Culture with Joseph Hickerson and Gerald Parsons and contact with Gerald Gibson in the Motion Picture and Recorded Sound Division.

The final conceptualization came in the summer of 1983 during my Fulbright lectureship in Italy and was helped by the criticism of Mary Gibson. I began the writing in December of that year and substantially completed it by November of 1984. During the writing, I was able to try out some of the ideas on students in my Introduction to Curatorship and Records Management class and a specialized workshop in sound archives at the University of Maryland at College Park (UMCP). Equally important were contacts, such as Mary Jo Deering and Donald Ritchie, in the Oral History Association of the Mid-Atlantic States (OHMAR). Martha Ross, 1984-1985 president of the Oral History Association and adjunct professor for oral history at UMCP, was also extremely helpful (and, of course, gracious) in her comments on sections of the manuscript, as were Gary Marchionini of UMCP, Susan Swartzburg of Rutgers, and Lee Wisel of Columbia Union College.

Last but not least, commendations must be put forth for my graduate assistants—Bridget Toledo and Herbert Swanson. In addition, Pamela Schroeder merits mention for producing the more attractive illustrations in the text (the unattractive ones are mine) and, above all, Golda Haines—who merits another bottle of Spumanti for finishing this typing. And a note of praise is extended to the editors at Greenwood Press into whose hands I commend this manuscript. Finally, I need to list those institutions I have visited, or whose inhouse manuals and guides I was able to secure for the writing of this text:

Archive de Folklore, Laval University

Archive of Acadian and Creole Folklore/Oral History, University of Southwestern Louisiana

Archive of Folk Culture, Library of Congress

Archive of Traditional Music, Indiana University

Louis Watson Chappell Archive, University of West Virginia

Columbia Oral History Project, Columbia University

Computerized Folklore Archive, University of Detroit
Folklore Archives, Indiana University
Folklore Archive, Wayne State University
Folklore Archives, University of North Carolina
Northeast Archives of Folklore and Oral History, University of Maine
Regional Oral History Office, University of California at Berkeley
Rio Grande Archive, Pan American University

The Management
of
Oral History
Sound Archives

TO SET THE STAGE

I pressed the lever over to its extreme position. The night came
like the turning out of a lamp, and in another moment came
tomorrow. The laboratory grew faint and hazy, then fainter and
fainter. Tomorrow night came black, then day again, night again,
day again, faster and faster still. An eddying murmur filled my
ears, and a strange, dumb confusedness descended on my mind.
                                    —H. G. Wells, *The Time Machine*

By the time H. G. Wells published *The Time Machine* in 1895,
humankind had already learned how to conquer time with far
less disorienting—yet equally powerful—machines. First, the
camera allowed humans to freeze a visual moment and carry it
forward indefinitely. Then, in 1877, Thomas Edison provided
the skill to capture conversations and play them back at will
with his cylinder recorders; thus, humankind gained the revolutionary command to re-experience the past second by second.

As French semiologist Roland Barthes taught, the most important factors in a culture are often overlooked because of their
very familiarity, and such may be the fate of the camera and
recorder. Nevertheless, our relationship with the past has been
irretrievably altered by these technologies. History and events are
made more personal, and time itself is telescoped closer than
through purely written accounts. To begin to concentrate on

sound recordings, listen to Franklin Delano Roosevelt pronounce "A Day of Infamy"; to a Caruso recording; or to a newscaster screaming that a president has been shot in Dallas and wonder at the sounds of a Gettysburg Address or the words of a prophet.

In addition to such personalization, the formal study of humankind itself has also been enhanced by the development of new research techniques that employ audio recorders. Ethnologists and folklorists, for example, early on embraced this machinery to capture fading oral traditions from "primitive" peoples, who were being altered by contacts with an increasingly industrial and literate world. The other methodology of particular interest for this manuscript arose after World War II. Then historians concerned over the loss of written documentation with the newer communications and transportation devices ironically turned to the newly emergent tape recorder for a solution and created a process known as oral history.

The number of oral history and folklore collections began to expand more rapidly following the appearance of the portable cassette recorder in the mid-1960s. Not only professional scholars, but also hundreds of archivists, genealogists, librarians, local historians, students, and senior citizens began to participate in collection efforts. Such projects have even gone beyond adding to the documentary heritage and have often extended in a direct fashion to help build community identities and individual self-worth. All too often, however, the physical products of those endeavors have been left as useless caches in someone's desk: the results of poor planning or the large expenses involved in full processing. The succeeding pages seek to provide a remedy to this dilemma through a form of sound archiving, which, though simple and low cost, is also in keeping with the highest bibliographic and preservation standards.

THE AUDIENCE AND THE MODEL

As already suggested, this volume should be of interest to anyone dealing with sound collections. Within that broadly defined audience, it speaks directly to those involved in oral history or folklore recording projects. The work is more specifically addressed to those de facto or professional archivists and librarians who have been placed in charge of the information from such efforts and are somewhat baffled by the lack of standardization and training on its care and handling.

Rather than a comparison of existing systems, this work presents an idealized system that is drawn from the information sciences and the small universe of sound archives. The model itself is adaptable to the management of most sound collections and videotapes as well. The concentration, however, is on magnetic sound recordings rather than on other media or paper transcripts. Magnetic tape is stressed as the primary source of information and because of its dominance and flexibility in fieldwork, its malleability for copying other media, and its suitability for preservation.

The reader will be exposed to theory and practice through an archival/library approach but one still rooted in the historical development of folklore and oral history. The design is aimed specifically at unique field recordings located in one repository, rather than formally published sessions that are found in numerous repositories with their own set of bibliographic controls. While adaptable to large operations, the envisioned repository is a fairly small-core establishment in which services are highly personalized and use is generally restricted to inhouse.

Like any hypothetical exemplar, this one possesses a number of affectations. For example, as the focus is on the post-recording processing of the tapes, the archive may seem to be set in an exaggerated center of the overall collection effort. The intended image is also of a broadly based repository that receives a variety of donations, but is normally responsible to some higher authority—be that a board of directors or the head of a library. Furthermore, because the material in question is of an unpublished nature, the title of archivist will be given precedence over that of librarian.

The archivists in this scenario function as examples of the new information professionals. More than mere keepers of data, they are seen as conscious intermediaries and stewards of a cultural heritage, who promote communication within a complex chain of informants, interviewers, researchers, the public, tapes, machines, and a bureaucracy. Although primarily concerned with the information processing and preservation aspects of stewardship, this exemplar, too, recognizes that archivists function as scholars in their own right.

A MANAGEMENT PERSPECTIVE

To be successful stewards, sound archivists must also act as managers. This work recognizes the importance of such a role and such key concerns as bureaucratic and community relations, as well as budgetary realities. Yet, this book is not a budgetary document or management science or operations research text, but a more practical manual in line with the planning and control functions of management. Moreover, some of the manager's traditional duties in a business or profit sense are neglected because of the nonprofit makeup of most sound repositories. Similarly, the small size and tendency toward "one-person shops" for such collections has mitigated against a section on staffing.

Because many sound archivists do not come from the information sciences and business, it might be well to prepare them for some of the schematic techniques in the book. The third chapter, for example, presents a systems analysis method for the organization of an interview project; the example given calls for a carefully laid out approach to a problem, which takes the reader through sequential stages from initial definition, to analysis, to design, to implementation, and to evaluation. Not only is that approach a useful guide, but it also is part of a "language" that can further communication with institutional authorities and granting agencies. In addition, the flowcharting (see Figure 1:1 for terminology) employed can work to similar ends in structuring the archivist's approach and aiding in communications.

Forms Design

The value of such admittedly simplistic techniques becomes more evident in a discussion of forms design, which is also part of the necessary preparation for this text. Forms delimit the range of information available and serve to freeze the activities of an archive along specific channels for years to come. Lack of

Figure 1:1
Flowchart Symbols

---

1. The Input/Output Symbol: a parallelogram representing making information available for processing/or recording processed information.

---

2. The Magnetic Tape Symbol: a specialized input/output symbol for magnetic storage media, like cassette and reel to reel tapes.

---

3. The Process Symbol: a rectangle with a great range of application that illustrates the execution of operations, including arithmetic calculations and data storage, transfer, or creation.

---

4. The Direction of Flow Symbol: a simple arrow that indicates the direction of information flow and connects other symbols. Note, normal direction is left to right and top to bottom.

---

5. The Decision Symbol: a diamond that shows where a choice must be made in the process.

---

6. The Connector Symbol: a small circle to provide continuity for continuation to the next page or elsewhere in the chart, which helps to avoid cluttering. A letter or number inside is used for reference.

---

7. The Auxiliary Process Symbol: a small square, which we will use to show additional factors that enter into a process, often unintentionally.

---

foresight can prove quite expensive and damaging, especially for later automation. Unfortunately, inhouse forms are generally created in a rather haphazard fashion in response to specific needs. Most of us have merely adapted preexisting forms without full analysis and have borrowed unabashedly from other institutions.

The idea for the systematic control of forms became directly tied to big business and the federal government following the post-World War II explosion of modern records. In essence, these efforts seek to provide a more economical, rational, and less redundant direction for record keeping. Although the concepts of forms analysis are quite straightforward, scholarly collectors with an antipathy toward business administration have frequently overlooked them. The elements in forms analysis may be seen as determining the following:

1. Are the form and individual items necessary?
2. Is the information duplicated elsewhere?
3. How can the form be simplified or combined with others?
4. What can be done to improve its efficiency and use?
5. How can one check its later efficiency and value?

Equally pertinent for the beginning sound archivist is an acquaintance with the principles of form design. Especially with those forms that come in contact with the public, one should be aware of or seek advice on graphics and printing layouts. Beyond the collection of information, such documents also help to produce an image of an institution. As to the specific theoretical elements in the design of forms:

1. Study the purpose and use of a form from the user's perspective.
2. Maintain a simple and pleasing design that eliminates unnecessary information and manual recording.
3. Design even manual systems for adaptability to data processing entry.
4. Arrange information in a logical sequence.
5. Preprint constant data and allow sufficient space for recording information.

6. Use natural language and avoid abbreviations and coding whenever possible.
7. Standardize, including paper stock and instruction terminology.
8. Consult with users and others about the applicability of a design and be willing to redesign if necessary.

The forms presented in this manual are designed for production on a typewriter, but, of course, they are reduced for this presentation. To return the model to full size, the standard formula calls for 3 × 5 spacing: that is, three horizontal lines (typewriter double-spacing) to the inch and five characters per inch for entry. The reader may also wish to investigate printing forms with a more efficient box design, called ULC for the placement of the headers in the upper left-hand corner. Yet, any suggestions are merely guides. The best records can only be designed after a thorough examination of user's needs, institutional demands, and current exigencies such as staffing, time, and equipment.

Formatting techniques for computerization will be more fully discussed in a following chapter, but a few cautionary comments should be made here. In particular, not all forms should or need to be entered into a data base. Rather, the proper functioning of an installation rests in an integrated system that employs manual and automatic techniques in areas where they are most cost- and use-efficient. Note, too, that legal releases and contracts at present require a formal signature, which cannot be automated. In general, most single-use and peripheral or temporary notes might be best served on paper, as are those more easily completed on site. By contrast, those forms addressed on repeated occasions and information repeated on more than one form are amenable to data processing.

OVERVIEW

In terms of the makeup of the text, Chapters 1 and 2 provide the initial background and intellectual components to sound archiving and the disciplines that gave rise to it. Chapter 1 indicates the underlying model and direction of the work, as well as introduces some of its tools and techniques. The following chapter focuses on the special attributes of oral communication and an as yet unrealized potential for its taped accounts. In addition, Chapter 2 emphasizes in more detail the historical context of the field. For example, it discusses the increasing convergence of interests between folklorists and oral historians. Oral historians have altered their initially elitist thrust to include wider social topics, whereas folklorists have moved in a similar direction and away from merley gathering survivals (like Child ballads and fairy tales). But that chapter also concentrates on laying the theoretical cornerstone for a distinct profession of sound archivy beyond the roots of its parent disciplines.

With Chapter 3, the text begins to center on the practical day-to-day aspects of administering a sound collection. Although that chapter focuses on areas of contact with the public within the archive, it also argues for the importance of planning, notably of the archive entering the scene with its forms and advice even before the interview. In addition, this segment deals with a maze of ethical and legal relationships, especially in the areas of copyright, fair use, and privacy legislation.

Chapters 4 and 5 extend the argument for preparation into the technical processing of collections. Beyond suggestions on the importance of standardized forms and terminology, the major contribution of these chapters is the idea of three stages of description—collection, item, and transcript. Collections are assigned in a sequential fashion in relation to their supposed value and use and the resources of the repository. Thus, all collections worthy of retention receive a collection description; this parallels the form used for standard library book cataloging. The second, or item-level, should be given to the bulk of the holdings and may be compared to an in-depth chapter outline. With sufficient time and money, the most significant materials can be made more accessible through partial or full transcription—the third level.

All three levels are amenable to automation. Collection-level description is suitable for MARC formatting and the online bibliographic exchanges of the library world. But, as Chapter 5 explains, the last two levels are particularly hospitable to microcomputer applications. Inhouse microcomputers offer a number of advantages in terms of cost, retrievability, and security, as well as for research. Archivists, however, will need to be cognizant of specific hardware and software requirements. For example, they should be aware of the potential of data base management systems for the second stage of processing and word processing software for transcriptions.

The final chapter presents information about the equipment and magnetic tape, as well as a practical solution to the tension between playing and still preserving the recording. Although this discussion will include technical detail, it should be understandable even to the neophyte. In essence, archivists are struggling to preserve their materials against the attacks of humankind, machines, and the environment—especially changes in heat and humidity. To counter those perils, one may employ some rather common-sense and simplistic measures such as the production of separate storage and user tapes and the use of air-conditioning.

SELECT BIBLIOGRAPHY

In each chapter of this book, the Select Bibliography provides the reader with complete citations to quotations and the studies used in writing the chapter, as well as with suggestions for further reading.

Clark, Robert L., Jr., ed. *Archives-Library Relations.* New York: Bowker, 1976.

Connors, Tracey, ed. *The Nonprofit Organization Handbook.* New York: McGraw-Hill, 1980.

Drucker, Peter F. *Management.* New York: Harper and Row, 1974.

Harvard Business Review. *Management of Nonprofit Organizations.* Boston: Harvard Business Review, 1981.

Maedke, Wilmer, Mary Robek, and Gerald Brown. *Information and Records Management.* Encino, Calif.: Glencoe, 1981.

Nygren, William. *Business Forms Management.* New York: American Management Association, 1980.

Rath, Frederick, and Merrilyn R. O'Connell. *Administration.* Nashville: American Association for State and Local History (AASLH), 1980.

Thompson, Enid. *Local History Collections.* Nashville: AASLH, 1978.

Trask, David, and Robert Pomeroy, III, eds. *The Craft of Public History.* Westport, CT: Greenwood Press, 1983.

Wells, H. G. *The Time Machine.* New York: Henry Holt, 1895.

TOWARD A THEORY OF
SOUND ARCHIVES

> As soon as it is remembered that philosophizing does not consist in
> addressing fantastic beings in fantastic language, but that those to
> whom the philosopher addresses himself are human beings . . .
> then it will be evident that the ideal of a persistent striving is the
> only view of life that does not carry with it an inevitable disil-
> lusionment.
>
> —Soren Kierkegaard

Those seeking only a practical "how to" manual may turn to the
next chapter, but a major purpose of this work is to identify an
underlying intellectual component for sound archiving. Without
exploring this element, the field will never achieve its potential;
instead, it will remain little more than an assortment of isolated,
passive collections. Yet sound archives do offer barely tapped
and, it may be suggested, still unknown benefits for our
documentary heritage. For example, they can provide access to
unprecedented slices of communal memory, as well as a
redress for information lost through the ephemeral nature of
electronic communications.

Even in the context of the burgeoning "Information Age," there
is still an unconsciousness of the potential of sound archives in
theory and practice. This unawareness stems largely from the
historical origins of sound archives as mere adjuncts to collection

efforts. In part too, it derives from the uncertain status ascribed to such resource centers in the minds of many practitioners. Yet another contributing factor is the very recency of recorded sound, but beneath the surface of this are deeper "McLuhanesque" implications involved with the introduction of any medium.

This chapter does not pretend to set out a full-blown general theory of sound archives, but presents only elements that contribute toward a discussion of such a theory and the idea of multiple, special theories for specific types of holdings. The approach is contextual, with the complexity of the problem pointed out and some of the significant factors toward a solution indicated. Thus, analysis will be limited and normally confined to a pragmatic or instrumental viewpoint. Still, readers should be aware, first, of the broader, ethical issues which are raised for any information agency as part of the superstructure and, second, of the increasingly important dialectic between power and knowledge.

This introductory chapter begins by placing sound archives in historical context and then looks to the effects of the media, as well as possible contributions from other disciplines. It recommends that a distinctive body of knowledge can only arise from a synthesis of subject area expertise and the tools of information management. For archivists, that union also reflects their dual persona as scholars and stewards; its realization can provide a new direction to their researches and further the quadripartite duty of stewardship – the acquisition, control, preservation, and promotion of the information within their holdings.

THE ROOTS OF SOUND ARCHIVES

A theory of sound archives cannot be considered apart from the historical development of the field and the dual parentage of folklore/social anthropology and history/oral history. In general, the academic interests of those disciplines have effectively delimited sound archives during most of their history. In particular, the collections of scholars provided the basis for most archives and their first taxonomies. Moreover, folklorists and historians normally served as the first archivists, and their pre-

delictions determined why, how, and what was to be preserved –
or not preserved.

To gain a historical perspective, one must first examine the
inception of the parent disciplines in the universities during the
Age of Enlightenment and Romanticism and the nineteenth-
century professionalization of knowledge – activities that gave
rise to new fields of study and the eventual replacement of the
classical liberal arts curriculum. With that context in place, one
gives particular attention to developments following the effective birth of sound archiving after the Second World War.

THE FOLKLORE ELEMENT

Folklore and its closely related sister discipline, social
anthropology, initially partook of an elitist blend of romanticism
and scholarship. Arising at a time of great intellectual fervor and
the birth of the modern university, they were also part of the
political context of nationalism and the Industrial Revolution.
Early folklorists drew from the seminal concepts of John
Gottfried von Herder in *Uber den Ursprung der Sprache* (1772);
they searched for deeply rooted "racial" identities through
survivals in the speech patterns and oral and musical traditions
of their country's peasants.

Those who believe in the benign neutrality of information
should take a Nietzschean lesson from the propaganda potential
of folk collections. As early as the 1830s, sufficient accounts of
peasant verbal and musical performances had accumulated to
lead to their organization as valued demonstrations of a long-
standing heritage. In reality, those holdings appeared as
byproducts of fieldwork, but such afterthoughts fit neatly into
the contemporary Western pattern of national libraries and
museums. Moreover, folk collections were particularly suited to
the mentality, limited resources, and search for national identity
that characterized Scandinavia and Finland, home to the bulk of
early folklore archives. Norway was a prime example; this area,
struggling to escape from centuries of foreign domination,
actively promulgated the traditions of its sagas and consciously
built a linguistic identity from folk sources.

Nationalism also provided a push for centralizing efforts in the more technologically advanced areas of Europe, such as England, France, and Germany. But the confluence of scholarly inquisitiveness and the military might and colonial aspirations of those same countries also stimulated anthropologists among the native populations of conquered territories to begin collections. The imperialist governments used the results of these endeavors to help control subject peoples, while similar exploitations at home contained the additional danger of xenophobic reactions. As Richard Dorson explains in *Folklore and Folklife* (1972), even the naive romanticism that motivated the first generation of collectors could not guarantee against possible misuse:

This impulse stirred the Grimms in Germany, Asbjornsen and Moe in Norway, Lonnrot and the Krohns in Finland, Uvk Karadzic in Serbia, Douglas Hyde in Ireland, and other illustrious scholars. As a stimulus to imaginative research and innocent national pride, this quest for a heritage had its virtues, but in extreme form it became entwined with political ideology and virulent nationalism, especially in Nazi Germany and Soviet Russia.

From a contemporary viewpoint, one might be somewhat shocked by such excesses and by the racism and paternalism inherent in the earliest ventures. Others might wonder at the use of romantic gleanings as agents of social control in a campaign for nationalism, a campaign that would strangely invoke village or family traditions to support the creation of a national identity. A similar paradox lies in the efforts of print-oriented scholars – the product of one of the most individualistic eras in history – searching for nonprogressive, communal identities through previously maligned peasant traditions.

Beyond those ironies, folklorists and anthropologists continue to live with more positive contributions of the past. To archivists, that legacy is particularly evident in the continuation of the taxonomies that the first fieldworkers imposed to order personal collections: for example, the numbering system assigned by Francis Child in the 1850s remains a standard identifier for ballads.

The most important classification tools, however, emerged in the twentieth century through the historical-geographical or Finnish school of folklorists. This Finnish method was far more self-consciously academic than that of Romantics, and it demanded the formation of a more "scientific" apparatus. Its initial concentration was on the collection of numerous variants of the same oral tradition in a search to abstract or identify an original "Ur-type." For the archivist, its key contributions were the Aarne-Thompson type-index for folktales and Stith Thompson's *Motif Index of Folk Literature* (1932). Of the latter, Robert Plant Armstrong would comment in "Content Analysis in Folkloristics" (1972):

Any discussion of the analysis of substance of folklore material inevitably invokes initial mention of Stith Thompson's motif index. A motif is a substantive unit (e.g., "the devil falls into the well") which appears frequently in folklore, and is found in very wide cultural distribution. . . . Thompson has presented and discussed them in six large volumes. Here is a system devised for the purpose of cataloging the elements of certain kinds of narrative situations, and it has accrued to itself the great respect to which its breadth of scholarship entitles it. It is, in fact, a kind of trait list, analogous in essence to the lists of cultural traits once in fashion in ethnology.

In addition to such general reference systems and the tools of individual collectors, folklore archivists might also have recourse to the taxonomies created for large repositories like the Frank C. Brown Collection at North Carolina. Such institutions must frequently manage a melange of different genres and even media. For example, let us consider the outline developed at the Swedish Institute for Dialect and Folklore Research:

A. Settlement and dwelling
B. Livelihood and household support
C. Communication and trade
D. The community
E. Human life
F. Nature

G. Folk medicine

H. Time and division of time

I. Principles and rules of popular belief and practice

J. Myths

K. Historical tradition

L. Individual thoughts and memories

M. Popular oral literature

N. Music

O. Athletics, dramatics, playing, dancing

P. Pastimes, card games, betting, casting lots, toys

Q. Architecture

R. Special ethnic units

S. Swedish culture in other countries

T. Traditions about foreign countries and people

U. Additional

As George List describes in his excellent article "Archiving" (1972) in the Dorson volume already cited, specialized repositories for cylinder recordings appeared around the turn of this century. The first was the Austrian Academy of Science's Phonogramm Archive in Vienna, but it was soon followed by similar accumulations in Berlin, London, and Paris, the capital cities of the leading imperialistic powers. In the United States, ethnologists for the Department of Interior were engaged in capturing the languages and traditions of the conquered native populace as early as 1893. Within a generation, anthropological recordings were being grouped at institutes in Washington, D.C., New York, and Chicago.

The first truly dedicated folklore archive in America, however, awaited Robert Gordon's cylinder collection which launched the Archive of Folk-Song (now Folk Culture) at the Library of Congress in 1928. That repository rose to its current preeminent position during the mid-1930s under the directorship of John Lomax. Benefiting from Gordon's introduction of electronic disk recorders, Lomax first concentrated on saving Western cowboy songs, but soon focused on the preservation of Afro-American folk music, for example, his well known discovery of Hudie "Leadbelly" Ledbetter. Lomax also joined in starting the Depression era projects under the Folk-Song and Folklore Department in the Works Progress Administration, later headed by B. A. Botkin.

But the real breakthrough came with the post-World War II appearance of the tape recorder and the coeval emergence of graduate folklore programs in the United States. With the tape recorder and professionally trained folklorists, modern sound archives could be established in the United States. And, by 1958, a sufficient number of archivists were evident to support a separate journal – *The Folklore and Folk Music Archivist* (1958-1968).

Folklorists had proved quite willing to accept sound recording. Ironically, they adopted a new technology to preserve traditions being attacked by other technologies and the forces of industrialization. Remarkably, ethnomusicologists with notational schema that were geometrically more precise than the best narrative transcription led in the adoption. Folklorists had recognized the futility of totally reconstituting an oral performance through any scribal media even before the close of the nineteenth century. Their embracing of audio recorders would be scheduled by the availability of the technology but predicated on a felt need within the field (see Chapter 6 for an overview of the technology). Thus, although the machinery was only slowly perfected, folklore theory was predisposed to the oral text and rapidly altered to include audio recordings. According to Scottish Folklorist Donald A. MacDonald ("Fieldwork," 1972):

Most modern theory and practices are based on the assumption that the days of paper recording texts are now past except in the most exceptional circumstances. The advent of sophisticated machinery has taken much of the drudgery out of collecting – and, what is more important, it has also removed much of the inaccuracy. The function of paperwork should, by and large, be confined to note-taking, transcription, documentation of recordings, setting the scene, and filling in other detail that, for one reason or another, it has been found impossible to incorporate in the sound record. Granted that even the professional phonetician will usually concede the difficulty of producing a totally objectively accurate transcription, even from captive material, the case for the use of machinery in all possible situations scarcely needs stating.

The Oral History Component

In addition to the development of folklore sound archives, the postwar era also witnessed the emergence of a "rediscovered" technique for historical research—the oral interview. Largely conceived by Allan Nevins at Columbia University in 1948, oral history was intended to supplement traditional written sources. Nevins and his fellows were consciously reacting against a perceived decline in the art of letter-writing and the transitory nature of such modern means of communication as the telephone. Drawing on their knowledge of journalism and the nineteenth-century work of Hubert Bancroft, they set out to compensate for the losses by gathering oral memoirs.

According to Nevins, this technique "was born of modern invention and technology." Of its beginnings, he commented in "Oral History" (1966):

They had enthusiasm, these planners. It was partly the enthusiasm of ignorance; the undertaking looked deceptively easy. Anyway, they set to work, at first with pencil and pad, later with wire recorders, later still with early tape recording machines. They found that the task needed a great deal of money, and money was hard to get. It needed system, planning, conscientiousness, the skill that comes with experience, and above all integrity. It was more complicated and laborious than they had dreamed. The results were sometimes poor, but hard effort sometimes made them dazzling.

Nevins' idea was a revolutionary adaptation that ran counter to the prejudices of most academic historians of the time. Instead of only identifying, sifting, and manipulating remnant documents, historians were creating a resource that could directly answer their questions. Despite that important distinction, the oral history of Nevins and the Columbia Oral History Project was still intrinsically tied to the paper-bound, elitist focus of the profession; it could hardly be another way and secure legitimacy.

Again in an apparent paradox, the study of history had proceeded from the same traditions and division of knowledge that generated folklore. Unlike folklore, however, historians soon came to rely almost entirely on written texts. As Paul Thompson describes in *The Voice of the Past: Oral History* (1978), one of the basic theoretical works in oral history, historians occasionally went to extremes in this reliance.

> The notion that the document is not mere paper, but reality, is here converted into a macabre gothic delusion, a romantic nightmare. But it is nevertheless one of the psychological assumptions which underpin the documentary empirical tradition in history generally. . . . It was documentary tradition which emerged during the nineteenth century as the central discipline of a new professional history. Its roots go back to the negative scepticism of the Enlightenment as well as to the archival dreams of the romantics.

Historians were in fact developing a critical method with the best of all possible intentions. Following Leopold von Ranke's naive dictum to reconstruct the past as it really was ("*wie es eigentlich gewesen ist*"), historians strained for objectivity. In particular, those working in the somewhat suspect realm of the recent past soon came to disassociate themselves from potentially polluting personal contacts with biased participants. Such factors, in combination with psychological predilections, training needs, and the desire for professional specialization, produced an almost exclusively documentary method. In a further example of the triumph of the nation-state, academic historians also began to concentrate on "great men" and key events of global or national import.

Out of such a context, it is not surprising that the initial interviews centered on leaders from the Franklin Delano Roosevelt Administration. Because the interviewers were historians trying to correct for losses in the written record, it follows that they tried to recreate the written format of the memoir. Unfortunately, it also follows that the subconscious prejudices for print surfaced through an unquestioned preference for the transcript as *the* primary product. The result of that fixation and the high costs of early tapes was the erasure of most early recordings, a practice that was not reversed until the 1960s.

The Columbia project under Nevins and under the later leadership of Louis Starr, Elizabeth Mason, and Ron Grele set the standards for the field. The Regional Oral History Office at the University of California at Berkeley, for example, was begun in 1954 by the distinguished Willa Baum and has continued to follow the Columbia memoir model. In its promotional literature that office has described its practices and the memoir.

An oral history memoir is a recorded and transcribed series of interviews carefully designed to cover the major stages and events in the life and work of the selected individual to convey the uniqueness of his personality as well as his contributions. . . . An oral history study is a set of interviews of varying length by a number of individuals who have observed the same aspect of human endeavor from varying viewpoints. Memoirists review their transcripts after editing by the interviewer. Transcripts are then retyped, indexed, bound with photographs and illustrative materials, and placed in the Bancroft Library and other suitable locations. The memoirist receives a copy for his own use. The Bancroft Library safeguards and administers the use of the memoir.

Although transcript-bound procedures were successful, they were so costly that at first efforts were limited to large universities and institutions like Berkeley and UCLA or, later, the presidential libraries. But the value of oral history could not be denied. Stirred by the cheaper technology of the cassette recorder, the movement quickly grew during the 1960s and by 1966 gave birth to a distinct Oral History Association out of a conference in Lake Arrowhead, California.

During the same period, the American historical profession itself began to change. In keeping with the heightened social awareness of the era, historians awoke to the value of a "history from the bottom up." American historians and graduate schools also shifted their focus from presidents and battles to a social history of people and places. Following the lead of the French *Annalistes,* historians started to investigate theory and techniques from other fields and to accept the propriety of a holistic approach to research. First to Africanists but then to Americanists, oral history appeared a useful and occasionally a necessary tool.

By the 1970s, the presence of inexpensive and truly portable recorders, as well as the phenomenal commercial success of *Roots* and the *Foxfire* series, had insured that oral history would not be contained among professional practitioners and in a few repositories. Concerned amateurs quickly discovered the technique for genealogical and family history research. They were joined by similar endeavors from community groups and private and public institutions. Scholars from other fields also adapted the technique, moving beyond biographical memoirs to investigations of specific events and places.

Folklorists were among the first nonhistorians to awaken to the potential of guided interviews. It is important to note that American folklorists also entered a dynamic phase in the 1960s and 1970s. Fueled by a populist climate of opinion and a more demanding level of scholarship, folklorists expanded their scope beyond survivals and into the creative life of the community. They experimented with different analytical approaches from psychological theory to the structuralism of Claude Lévi-Strauss and Vladimir Propp. Others, led by Richard Dorson, allied historical tools to those of the folklorist, including those of the oral interview. The landmark in that union and for nonelite studies in oral history was the 1970 publication of Lynwood Montell's *The Saga of Coe Ridge*. Although significant differences still exist, folklore and oral history have continued to blend. The outstanding work in this regard for sound archives was the appearance in 1974 of Edward "Sandy" Ives' *A Manual for Fieldworkers* (later issued as *The Tape Recorded Interview*), a manual based on his experiences at the Northeast Archives of Folklore and Oral History.

By the end of the sixties, the multiplicity of oral history projects also began to attract the attention of the nation's libraries and archives. Although at first unprepared for the attention, these repositories eventually assumed their logical role in the movement. (A fuller discussion of their contributions and theory is presented in Chapters 3 and 4.)

THE MEDIA AND THE EVENT

Although a historical perspective is essential, a theory for sound archives should also take into account the media and the format of the recording session. In reality, a sound archive will involve a wide variety of media, receiving deposits that typically might also include photographs, manuscripts, and material culture artifacts. The heart of this exegesis, however, properly rests with scribal transcripts and taped interviews and performances.

Since Marshall McLuhan first burst on the scene, scholars have actively pursued the differences between print and oral cultures. As McLuhanite Walter Ong summarizes in *The Presence of the Word* (1970):

> The development of writing and print ultimately fostered the break-up of feudal societies and the rise of individualism. Writing and printing created the isolated thinker, the man with the book, and downgraded the network of personal loyalties which oral cultures favor as matrices of communication and as principles of social unity.

Sound archives require a parallel focus on the transcript versus the audio recording, but those forms should also be recognized as part of a print-influenced culture. For example, a transcript can be seen as a transitional device that attempts to provide a bridge between oral and written communication. Oral history literature reveals considerable attention to the development of the transcript and the problems arising from the multiple contributions of informants, interviewers, and transcriber/ editors. Some of the difficulties should be recognized as typically scribal: for example, the transposition of letters, missing lines, and variant texts. Others are caused by the implicit impossibility of totally converting spoken conversation by a written means. Even Ferdinand de Saussure, the leading linguistic theorist of the twentieth century, could state in an essay in defense of graphic representations (see *Course in General Linguistics,* 1959):

> Language and writing are two distinct systems of signs; the second exists for the sole purpose of representing the first. The linguistic object is not both the written and the spoken forms of words; the spoken forms alone constitute the object. But the spoken word is so intimately bound to its written image that the latter manages to usurp the main role. People attach even more importance to the written image of a vocal sign than to the sign itself. A similar mistake would be in thinking that more can be learned about someone by looking at his photograph than by viewing him directly.

The obvious answer to the dilemma of transcripts is simply to recognize that they are a secondary resource – that the tape itself is primary. Such an answer is definitional and logically an unassailable rejoinder to theoretical debates over the problems of transcribing. At the same time, the specious debate over the validity of oral testimony should also be summarily dismissed. That residue of print dependence flies in the face of common logic and our legal system of oaths. Moreover, it somehow supposes that the human behind the written record is more prone to "truth" than the same individual in speaking.

Still, transcripts do offer important benefits and must be incorporated into any archival theory. In practice, transcripts are the accepted standards for most oral history collections. As oral historian Martha Ross notes in her courses and lectures, even partial accounts of key excerpts can greatly speed research and do not require the potentially troublesome mechanical interface of a tape recorder. In addition, written documents allow for random access and ease of review over the serially accessed tapes. Finally, scribal media are more amenable to the tools of literary analysis, from textual criticism to deconstructionism.

If we can accept the primacy of the tape recording, then it is incumbent to recognize that not all recording sessions are equal. Contents are affected by the type of material sought and the culture in which they are collected. In *Oral Tradition: A Study in Historical Methodology* (1965), Jan Vansina has produced a basic text for any sound archivist and a more in-depth analysis of these sentiments. As he notes:

Oral traditions are conditioned by the society in which they flourish. It follows therefore that no oral tradition can transcend the boundaries of the social system in which it exists. It is spatially limited to the area circumscribed by the geographic boundaries of the society in question, and limited in time to that society's generation depth. The historical information that can be obtained from oral traditions is therefore always of limited nature and has a certain bias. This is equally true of many written sources, and written and oral sources are very much alike in this respect. In both cases, the factor which most imparts bias and imposes limitations is the political system.

For Western archivists, the society in question is a literate one in which print has already impinged on oral traditions. Roman Jakobson has written on the liberating aspects of writing, which frees the individual performer from the censorship required for conformity to the communal memory. In this vein, Italian oral historian Allessandro Portelli writes in "The time of My Life" (1980):

The hegemony of writing liberates the oral tradition from the burden of memory. The preservation of texts, the certification of events and contracts, are carried out better elsewhere. It is less necessary to freeze oral speech into forms that can be kept intact across time (hence, the gradual decrease in traditional folklore – slower than many expected, but yet true). . . . What was once the obstacle oral discourse was trying to overcome, now becomes its main asset: the close link between speech and time enhances the values of immediacy, improvisation, spontaneity over the machinery of writing. For the very fact that writing exists, oral narration may now to a certain extent flow and change with time.

Although there is always change over time, alterations within oral traditions have been accelerated by the rise of literacy, as well as by the effects of the mass media. Within most of the United States, one can project that the genres for accurately preserving the communal memory have all but disappeared. (This is not to say that people are forgetting the past, but that the mnemonic keys and formal structures of the folk memory have generally vanished.) But other forms of oral tradition do exist, and their importance cannot be minimized. At the least, as Barbara Allen and Lynwood Montell have shown in *From Memory to History: Using Oral Sources in Local Research* (1984), archivists have a responsibility to recognize the infiltration of apocryphal "travelling legends" (e.g., "The Vanishing Hitchhiker") into purportedly factual accounts. Archivists should be able to differentiate between legend and fact, but, more importantly, they will require techniques to highlight the appearance of oral performances as distinct from interviews, and perhaps distinguish among specific musical and vocal genres.

The interview must itself be seen as a unique format because of its directed nature. Even though performances can be recorded in an interview, the interview itself is still a very different "event" than the simple recording of a performance. In essence, the interviewer is injected more forcefully into the session and alters its content. To Portelli, this effect is even apparent in the collection of oral tradition:

(I am not saying that the historian's interference *ruins* the source, but only that it *changes* it): for instance, he [the storyteller] may feel for once stimulated to give the best of himself, to think things over, to organize them into a coherent whole, to search for the best way to express them in language. This is one way in which the historian may very concretely be said to have *produced* the document. . . . The final result of this meeting is a transferral of the oral source into written form. This assures that the words will be preserved; but also freezes their fluidity (the Italian language has a very meaningful verb to describe what happens to sounds on tape – they are "incisi", i.e., "carved"). Thus, no matter how much we may talk about ourselves as "oral historians", the very technology of our work is to turn oral into written words, to freeze fluid material at an arbitrary point in time. This is perhaps neither "good" nor "bad", and there is probably nothing to be done about it; but we ought at least to be aware.

Although he was speaking primarily of transcripts, Portelli's mention of the tape is instructive. Like the artificial freezing of one text in time, a number of parallels exist between the effect of the tape and the transcript as media. But the transcript is derivative; in McLuhanesque terms, it could be seen as a survival from an older print era. Again, the taped interview may be posited as a transitional device – one deliberately created to preserve a culture's letter-writing and memoirs, forms that are being supplanted by modern communications.

Theoretical considerations must also be paid to other aspects of the tape recording as a medium. This medium is not "the message," but it does have an effect. For example, however unobtrusively positioned, the machine is intrusive to the encounter. In particular, a recorder tends to formalize an interview far more than any simple pen and paper session because the participants are aware of its presence and its future reproducibility. Although increasing familiarity has lessened the "performance" traits among informants, the machine still imposes a distinctive time frame based on its length. Hence, the normal interview session lasts an hour, with a brief break for flipping the typical C60 cassette. The tape recorder is also a "cooler" medium than writing during review sessions, for it demands a higher degree of attention as well as a more conscious effort to begin the review.

Other considerations might be paid to the difference between the visual focus of writing as opposed to the aural one of tape, an effect that might presage a shift in the general consciousness in keeping with McLuhan's concept of a "global village." Implied there, too, is the necessity of including a dynamic or evolutionary construct within any theory. Such flexibility is further demanded by the introduction of newer media for sound archives: for example, the effects of computers, which are described in Chapter 5, or the more performance-oriented medium of videotaping.

SCHOLARSHIP, STEWARDSHIP, AND POTENTIALS

A contextual theory for sound archives may embrace an even wider range of factors. Whereas knowledge and research skills were compartmentalized during the nineteenth century, we are today living in a period of blurring disciplinary lines and multivariate analysis. Such change heralds new possibilities and demands. On one hand, archivists must become aware of and incorporate concepts from a variety of fields beyond their parent disciplines. On the other, archivists must recognize that the potential of their field is not secondary and passive, but rather can enhance a number of fields and even create its own form of scholarship.

Other disciplines, from linguistics and cognitive psychology to the study of popular culture, may contribute to sound archivy. Linguistics and semiology, for example, would place the archive as an agency within a chain of communication and would lead us to understand that communication proceeds as an interactive process of mutually recognizable "signs" (e.g., words, gestures without any intrinsic meaning) that are exchanged through a medium in an approximate form. Understanding is conditioned by the experiences of the participants and is limited by the media and the possibility of outside interference. The epistemological implications here alert us to the complexity and uncertainty of communication, for, as Magoroh Maruyama suggests in Kathleen Woodward's *The Myths of Information* (1980), such a contextual framework presupposes that

1. the universe is basically heterogeneous;
2. the universe consists of interrelations and interactions of events, and everything occurs in a context which may vary from event to event; therefore, the value of a verbal or nonverbal message lies in its relation to its context, its interrelations with other messages, etc.;
3. differences or disagreements within a message or between messages convey useful information because they reflect the richness of the interrelations, just as in binocular vision, the differentials between the two images enable the brain to compute an invisible dimension;
4. "objective" meaning is useless; there is no universal meaning; each piece of information must be interpreted in the context of other pieces of information and in terms of the given situation.

The top section of Figure 2:1 graphically illustrates the individual interview in keeping with a theory of signs. It seeks to demonstrate the exchange of information between individuals with their own framework and also to recognize the "noise" or distractions that can enter the situation. The rest of the figure shows the interview as it is captured within the limitations of the recording medium. As semiological analysis would proclaim, it is vital that archivists understand they are dealing with the taped account and its noise – not with the interview itself. This acknowledgment is a bipartite one of the interview as an event and of the data exchanged.

Other studies may contribute to sound archives, yet these repositories might also hold answers to some of the more vexing problems of the contributors. Much of this ability is ensconced in the almost untouched cumulative potential of an archive's holdings to reveal change over time. For example, oral historians have begun to inquire about the functions of human memory; however, medical science and cognitive psychology are far from definitive in that regard. Although rarely mentioned in the literature, the dominant two-stage model of long- and short-term

Figure 2:1
Communication and Sound Archiving

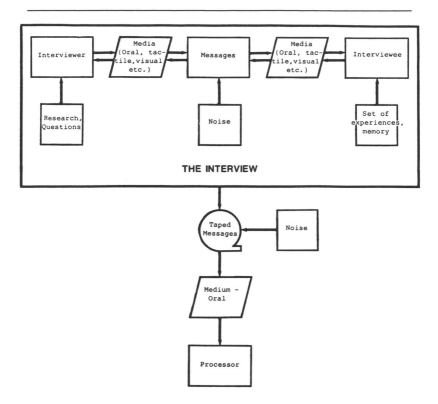

memory dates back to William James at the turn of the century and has a number of very real limitations. Furthermore, only in 1985 did scientists announce that their previous belief in the progressive death of brain cells with age was skewed and inaccurate. According to Walter Reitman in *Models of Human Memory* (1970), memory studies inherently are limited by the simplifications required by experimental methods:

The simplifications involved in producing these memory models and data typically are quite severe, raising questions about their relevance to everyday memory phenomena. . . . Consider some of the differences between everyday memory activities and the memory tasks required of subjects in the laboratory. Little of what we recall under everyday conditions is deliberately rehearsed for that purpose. Under everyday circumstances we take in information not as an isolated operation, but in conjunction with and in the course of other ongoing activities. Typically, that information consists of facts, statements, and patterns, not isolated nonsense syllables and words.

Certainly, oral historians can benefit from scientific inquiries into memory, and those studies are indeed growing in sophistication. But could not our knowledge of memory benefit from an inspection of the holdings in sound archives? Where else are there such extensive, direct incursions into human memory? Much like census returns, sound collections may provide unplanned secondary or tertiary uses beyond their initial implementation. The possibilities are an exciting extension from the current synchronic analysis of individual projects to broad diachronic comparisons through the entire archive as a data base: applications that will be facilitated by computer technology and the ability to distinguish patterns within large accumulations of data.

In general, the research of sound archivists today only mimics that of their primary academic discipline, yet archivists are in a unique position to explore the future and present potentials of their collections. For example, their intimate knowledge of the donations and the subject discipline could be employed in the construction of an "archaeology" of interviewing, which examines the techniques and frameworks employed by practitioners in terms of an intellectual history of the representative fields. If the potential for diachronic analysis develops, the future might hold the intriguing prospect of the archive functioning as a laboratory for ongoing analysis. As Thomas Charlton also predicts about videotape in a recent issue of the *American Archivist* (1984):

Revisionist schools of historians may rise and fall on their relative abilities to interpret both audio and video historical interviews as primary personal documents. . . . This type of analysis may be practiced in the library or archives by tomorrow's scholar. At the very least, videotaped interviews will create opportunities for additional speculation about the true meaning of historical evidence.

Yet the scholarship of sound archivists as archivists needs to be differentiated from that of the interviewer. In particular, it should be integrated with the specific functions of their peculiar archive: perhaps as part of a research design to further collection development and fill in perceived "holes"; for the more technically minded in the retentive properties of the storage media; or in terms of the intellectual control of holdings.

The effective separation of the work of an archivist from that of the interviewer is more evident in the routine activities of preservation and organization. This functional distinction provides the practical and theoretical basis for the following chapters in this book.

The concept of appraisal—of deciding the value of an object for retention—provides a key theoretical distinction for archival practices. Therefore, in terms of preservation, the archivist is defined not only by managing the practical considerations of machine wear and tape architecture, but also by selecting what will be saved for posterity. Based on an uncertain formula of institutional commitments, needs, and finances, plus the archivist's subject area expertise, those decisions may also extend to the speed and degree of description provided for accepted material. As Chapter 4 details, other important considerations are the organizational principles utilized for the physical arrangement of the holding, the type of descriptive apparatus, and the criteria for the exchange of information. Proper stewardship, however, implies more than preservational and organizational concerns; it also suggests an awareness of social responsibilities and the cultural role of the archive.

At a pragmatic level, the archive's role must be determined in regard to a specific mission. For example, the oral history project at the John F. Kennedy Presidential Library can be expected to be historically oriented, with a distinct biographical and political flavor. In a similar vein, the Archives of Acadian and Creole Folklore/Oral History may link the two fields and has a special ethnic thrust defined in its title. In contrast, the Wayne State University Folklore Archives in Detroit, because of its university ties and location, can be expected to add the research concerns and student activities of a folklore department to its geographic focus.

Each archive is unique, and archival theory will have to allow for these specializations in emphasis and topics. But its very uniqueness also implies a special responsibility to those people, groups, or places that are the focal points in the collections. Archives share a general ability to promote and build communal identity among their subjects. In part, that capacity is established through the attention and implied value accorded to the subjects by interviewing. The archive itself further legitimates or affirms the subject's importance by its very presence, for the archive may stand much like a monument to a people. Such an application has worked for the elites in the culture for generations but holds special significance for those outside the political and social mainstream. As Paul Thompson describes in *The Voice of the Past* (1978):

The possibility of using history for such a constructive social and personal purpose comes from the intrinsic nature of the oral approach. It is about individual lives – and any life is of interest. And it depends upon speech, not upon the much more demanding and restricting skill of writing. Moreover, the tape recorder not only allows history to be taken down in spoken words but also presented through them. . . . [It] always brings the past into the present with extraordinary immediacy. The words may be idiosyncratically phrased, but all the more expressive for that. They breathe life into history.

In the simplest terms, a sound archive is an institution created to preserve, make available, and develop a documentary heritage through concentration on deliberately structured interviews and performances made by recording devices. Its theory must be an elastic one, responding to new factors and incorporating concepts from numerous contributors. Personally, the crowning elements in that theory should be humanistic ones. Such elements are not easily written or voiced, yet they may be extracted from those expressed in the following quotations: the first from the oldest of historical accounts and the second cited by MacDonald and copied by folklorist J. H. Delargy in Ireland in the 1920s and only a month before the respondent's death:

Herodatus of Thurii here sets down his inquiries toward the end that the things done by men should not be forgotten with the passage of time and that the great and marvelous exploits, performed by both Greeks and barbarians, should not lose their radiance.

I suppose you will bring out a book of these stories someday. I have told you now all the tales I can remember, and I am glad. . . . I hope that they will shorten the night for those who read them or hear them being read, and let them not forget me in their prayers, nor the old people from whom I myself learned them.

SELECT BIBLIOGRAPHY

With the exception of a group of European oral historians, the theoretical implications of the oral record have been and are being more fully explored by folklore sound archivists than by their counterparts in oral history. In general, my attitudes on this matter have been shaped largely by the folklorists and my earlier tutelage under Richard M. Dorson.

The key works for this chapter include:

Allen, Barbara, and Lynwood Montell. *From Memory to History: Using Oral Sources in Local Research*. Nashville: AASLH, 1984.

Armstrong, Robert Plant. "Content Analysis in Folkloristics." Pp. 173-93. In Pierre Maranda, ed. *Mythology*. Middlesex, England: Penguin, 1972.

Baum, Willa, and David Dunaway. *Oral History: An Interdisciplinary Anthology*. Nashville: AASLH, 1984.

Charlton, Thomas. "Videotaped Oral Histories." *American Archivist*, 47 (1984): 228-36.

Dorson, Richard, ed. *Folklore and Folklife*. Chicago: University of Chicago, 1972.

Gardner, Joel. *Oral History for Louisiana*. Baton Rouge: Louisiana State Archives, 1981.

Grele, Ron. *Envelopes of Sound*. Chicago: Precedent, 1975.

Henige, David. *Oral Historiography*. New York: Longman, 1982.

Jakobson, Roman. *Roman Jakobson, Selected Writings*. The Hague: Mouton, 1966.

Lévi-Strauss, Claude. *Structural Anthropology*. New York: Basic Books, 1963.

List, George. "Archiving." Pp. 445-54. In Dorson, ed. *Folklore and Folklife*.

MacDonald, Donald. "Fieldwork." Pp. 407-30. In Dorson, ed. *Folklore and Folklife*.

McLuhan, Marshall. *Gutenberg Galaxy.* Toronto: University of Toronto, 1962.

Maruyama, Magoroh. "Information and Communication in Poly-Epistemological Systems." In Kathleen Woodward, ed. *The Myths of Information.* Madison, WI: Coda Press, 1980.

Montell, Lynwood. *The Saga of Coe Ridge.* Nashville: University of Tennessee, 1970.

Nettle, Bruno. *Folk and Traditional Music of the Western Continents.* 2d ed. Englewood Cliffs, NJ: Prentice-Hall, 1973.

Nevins, Allen. "Oral History." *Wilson Library Bulletin* 40 (1966):600-01.

Oblinger, Carl. *Interviewing the People of Pennsylvania.* Harrisburg, PA: Historical and Museum Commission, 1978.

Ong, Walter. *The Presence of the Word.* New York: Simon and Schuster, 1970.

Portelli, Allesandro. "The Time of My Life." *Papers of the International Oral History Conference, 1980.* This series contains a number of important contributions to theory.

Propp, Vladimir. *The Morphology of the Folk Tale.* Bloomington: Indiana University, 1968, 1928.

Reitman, Walter. "What Does It Take to Remember." In Donald Norman, ed. *Models of Human Memory.* New York: Academic Press, 1970.

Saussure, Ferdinand de. *Course in General Linguistics.* New York: Philosophical Library, 1959.

Thill, Richard, ed. *Washington Conference on Folklife and Automated Records.* This volume is currently in preparation and brings together some very significant building blocks for the field.

Thompson, Paul. *The Voice of the Past: Oral History.* New York: Oxford University Press, 1978.

Thompson, Stith. *Motif Index of Folk Literature.* Bloomington: Indiana University, 1932; see also, *The Types of the Folktale, The Folktale,* etc.

"Toward New Perspectives in Folklore", Journal of American Folklore, 84 1982.

Vansina, Jan. *Oral Tradition: A Study in Historical Methodology.* Chicago: Aldine, 1965.

RELATIONS WITH THE
PUBLIC AND THE LAW

With the theoretical foundations established, we may now turn to the more practical day-to-day aspects of sound archiving. This chapter and all the following chapters present an integrated system that draws on the organizational techniques of the archival and library communities. In addition, it is argued that the archive should take a more active role in the entire process of collecting, especially through the pre-interview distribution of forms and through advice on technical and organizational questions.

In this chapter, we may conveniently view the sound archive in library terms as a noncirculating, closed-stack special collection. The first section of the chapter focuses on the interrelationships of this entity with its public – a public that includes projects, interviewees, interviewers, researchers, the community, and outside authorities. The complexities involved in servicing that compound public cannot be minimized, especially in regard to an ethical and legal labyrinth of rights and responsibilities. The reader is cautioned to recognize that public services do not stand alone; they are part of a dynamic with technical processing. The problem for the writer is where to break into this process; one answer is to examine points of contact during the creation of an outside interviewing project.

A SYSTEMS APPROACH TO PROGRAM PLANNING

Although a sound archive may be the adjunct of a single collecting effort, most such archives are composite repositories for a number of projects, individual efforts, or continuing deposits from academic programs. Lamentably, the archive is most often only at the end of a conveyor belt, receiving deposits but having little input in their formulation. To reverse that tendency, archivists should begin to publicize the archive as a resource center and inject the repository at an early stage in any classroom or other project. Once the idea for an oral history or folklore project appears, the most important consideration for the organizers or instructor is to sit down to careful planning. Logically, sound archivists with their knowledge and specific expertise can prove a valuable asset to such sessions, as well as guaranteeing an interest in the final disposition and preservation of the product. Another significant contribution could be a system approach as an exploratory tactic, which offers particular promise for later presentations to parent organizations and granting agencies. In addition, a typical five-stage model of systems analysis (Definition, Analysis, Design, Implementation, and Evaluation) implies a very rational mode, which aids in administration and helps avoid many of the typical pitfalls for the new project.

The initial or Definition phase begins with a precise statement of the general theme and mission of the project. Many experts also recommend the early establishment of an advisory board to share the responsibility and to gain necessary historical, legal, and technical expertise. Again, if a depository has not been involved from the start, the project leaders should immediately locate a likely one in an archive, library, or museum. Contacts with the appropriate professional associations and educational institutions may also prove helpful. In essence, this phase concentrates on the collection of information, such as the possibility of similar projects. Typical efforts will also include collecting data on funding, budget, equipment, and references, as well as a list of potential interviewees and a rough outline or chronology of the phenomena under study.

Once the information has been gathered, the need is to ascertain feasibility. Although the central focus of this Analysis phase is to remain flexible, one must look toward achievable goals in terms of time, money, personnel, and institutional realities. Can the mission be reasonably satisfied with the resources at hand? Are the results worth the efforts? If we assume positive answers to these and similar questions, operations can turn to the design of the actual project.

During the Design phase, the archive acts as a reference and guide to suitable equipment, forms, and procedures. The project leaders will construct a general timetable for completion, as well as develop a scenario outlining central events, issues, and personalities. The timetable must reflect the futility of attempting to cover the entire historical spectrum and the practical realities of geographic dispersal and change. An effective design may be quite complex and may need to respond to changing exigencies. For example, as a result of his experiences at the Kennedy Presidential Library as reported in *An Oral History Program Manual* (1974), William Moss considers:

Planning for a major program is likely to shift back and forth from a broad to a narrow focus, depending in part on who is available to be interviewed. Just where the priorities of concentration should go is constantly being reviewed and reconstructed. . . . There are practical limitations to prevent a program from ranging too freely and too distantly from its central focus, but the conceptual limitations are more difficult to determine and more difficult to enforce.

As Moss notes, the selection of informants and the order and priority of their interviews are among the most crucial steps in this process. Archivists, interviewers, and project coordinators also need to realize that these selections help to define a portion of the constituency, a public, for the collection in the future. Yet, it is patently impossible to interview every possible source; moreover, those who conduct community or ethnic studies need to be particularly aware of the dangers of alienating certain sectors through unintentional oversights or faulty protocols.

During the Design phase, interviewers themselves use the project's scenario to help structure their own extensive background searches. From such sources, they then formulate an outline or set of questions for their dialogues.

With the initial planning and preparation completed, the project enters the actual Implementation or interviewing stage. Within limits, this stage should be sufficiently malleable to respond to new leads and information. Since publicity is normally considered as one of the major functions of a project during this stage, any participating archive will want to insure that the place of final deposit is prominently displayed. Equally important, the archive must ascertain that interviewers secure sufficient background information and the necessary legal agreements during or as soon as possible after their interviews. Implementation for the archive, however, is generally later and begins with an appraisal of the deposits for retention and level of processing, which will be discussed in the next chapter.

The Evaluation phase for the interviewers begins with a post-session review and abstracting of the contents of the tape. In keeping with an ordered approach, efforts in this matter are well directed toward completing predistributed forms from the archive. To aid in an overall evaluation of a project, one might profit from the guidelines promulgated by the Oral History Association (OHA) at its 1979 Wingspread Conference under the chair of Enid Douglass and William Moss. A complete set of guidelines is available through the OHA, but its specific criteria for a project are as follows:

Purposes and Objectives

Are the purposes clearly set forth? How realistic are they?

What factors demonstrate a significant need for the project?

What is the research design? How clear and realistic is it?

Are the terms, conditions, and objectives of funding clearly made known to allow the user of the interviews to judge the potential effect of such funding on the scholarly integrity of the project? Is the allocation of funds adequate to allow the project goals to be accomplished?

How do institutional relationships affect the purposes and objectives?

Selection of Interviewers and Interviewees

In what way are the interviewers and interviewees appropriate (or inappropriate) to the purposes and objectives?

What are the significant omissions, and why were they omitted?

Records and Provenance

What are the policies and provisions for maintaining a record of the provenance of interviews? What can be done to improve them?

How are records, policies, and procedures made known to interviewers, interviewees, staff, and users?

How does the system of records enhance the usefulness of the interviews and safeguard the rights of those involved?

Availability of Materials

How accurate and specific is the publicizing of the interviews?

How is information about interviews directed to likely users?

How have the interviews been used?

Finding Aids

What is the overall design for finding aids?

Are the finding aids adequate and appropriate?

How available are the finding aids?

Management, Qualifications, and Training

How effective is the management of the program/project?

What provisions are there for supervision and staff review?

What are the qualifications for staff positions?

What provisions are there for systematic and effective training?

What improvements could be made in the management of the program/project?

SERVICES AND SECURITY

Although the evaluation process for archives is similar to that for projects, there are a number of subtle differences. Most of these distinctions are essentially a matter of timing. The project concentrates on the creation of the information source and the archives on the post-production access and preservation of the data, as well as on dealings with later researchers.

A major facet of archival work involves performing public services inhouse, such as answering reference questions and pointing to and securing proper materials. To such ends, archivists normally act as subject area experts with an intimate knowledge of their holdings and a scholarly grasp of the information's potential. In addition, archivists should have a collection of pertinent reference works at hand for researchers and themselves. Moreover, depositories should maintain handouts describing their services, facilities, finding aids, policies, and raison d'être. These handouts do not need to be extensive, but they can prove very useful in introducing patrons to procedures and in managing the most frequent reference questions.

Both public services and security can be facilitated by formal procedures and registries for the use of the archives. Many depositories demand that researchers sign in daily in a guest book or similar instrument. Archives are closed stacks and frequently require check-in and check-out or circulation slips for each request. Beyond those, the most important record is a researcher registration form. This form identifies researchers and their goals and also informs them of your rules. In addition, it may serve as a contract to bind their observance of such rules (see Figure 3:1.) In general, regulatory documents are also very beneficial in evaluating the archive's public services and allocating resources.

In addition to inhouse service, archivists should be concerned with outreach to their public, as a portion of their duty as stewards. The archive should not remain a silent, passive repository. The archivist and institution can play a larger cultural role in building a community's identity and sense of self-worth. More than a collection of information, the archive also represents a cultural monument and can serve as a museum of verbal impressions. Part of its role, for example, may and should extend to the production of displays and traveling exhibits.

If they are to meet the complete meaning of stewardship, therefore, archivists must actively promote the documentary heritage. As suggested by the previous section, efforts in this vein include

Figure 3:1
Researcher Registration Contract

In exchange for the ability to use the collections of this archive, I agree to conform to the regulations and standards listed below and that the information supplied by me is correct.

1. The researcher agrees to act in an ethical manner and to abide by the pertinent copyright, libel, and privacy statutes, including--

    A. Accepting responsibility for securing copyright releases.
    B. Acquiescing to the right of the archivist to restrict certain materials as part of privileged communication and conservation needs, donor contracts, or by law.
    C. Agreeing to indemnify and hold the archive and staff harmless for any misuse of its materials by the researcher.

2. The researcher agrees to give proper citation and acknowledgement of the archive in all publications which make use of its holdings.

3. The researcher acknowledges the right of the archive to inspect items brought onto its premises and agrees to treat all collections, equipment, and facilities with proper care, including--

    A. No smoking, or drinking, or eating.
    B. No use of pencil erasers.
    C. No introduction of magnetic devices.
    D. No archival materials may be removed without permission.

4. The researcher notes that a fuller explanation of archival policies is readily available for use.

5. Researcher Information

    Name:

    Home Address:

    Local Address:

    Institution:

    Purpose:

    Identification:

                        Signature:

                        Date:

                        Witness:

                                        Reg. Form 1984

cooperation with collecting projects, classes, and individual collectors. Beyond the intricacies of pleasing donors and working within a bureaucracy, archivists may also engage in such efforts through contacts with the news media and the amateur/professional scholarly and humanistic or arts community. For example, it is surprisingly easy to place accounts with the local press and in most markets even to appear on local talk shows on television. Similarly, local arts, genealogical, historical, and other societies will fervently embrace a kindred soul–even someone with the strange title of archivist. Although the dangers of overcommitment always exist, opportunities for promotion should not be ignored and are indeed part of the archivist's job.

In the same fashion, it may prove useful to produce short promotional tapes for radio spots and public presentations or as an introduction to the holdings. Some repositories, notably the Archive of Folk Culture at the Library of Congress, have also produced commercial recordings, but such ventures and even the normal duplication services offered by many archives must be carefully monitored. The danger of infringing on the performer's copyright looms with any reproduction for use outside the archives. Furthermore, any nonprofit installation will want to insure its protected status and concentrate primarily on educational and informational services. Thus, those desiring to deal in duplication or sales must take special precautions, even beyond those described in the following section. Finally, we might briefly mention the value of distributing a general book catalog or more specialized descriptions of the collections to the scholarly audience.

THE LAW AND ETHICS

The presence of later researchers produces other responsibilities beyond those of the archives to their donors and client projects. Significantly, archives as repositories have the luxury of explicit protections under the law that have not been stated for projects. Sound archivists must, unfortunately, understand that there are no final answers to any of the legal questions involved. American jurisprudence is an intricate, occasionally contradictory, maze of federal, state, and local enactments, as well as judicial decrees and executive pronouncements. One's own safety demands an interest and consultation with appropriate agents in an organization. Whatever the case, the literature and case law for sound collections are remarkably thin and inconclusive.

The basic answer to the legal enigma is to move in good faith. Archivists should act in accordance with the sentiments of the academic world and democratic legislatures–sentiments that recognize the need to preserve information and the public's right of access to same. Archivists as professionals should also act as though materials in their possession were subject to protection as privileged communication. If applicable, archivists will want to adhere to the standards of organizations like the Oral History Association and American Library Association, and to those of an inhouse council or a board of trustees–which may favorably impress a court.

The archivist's responsibility is to cloak himself or herself in a mantle of professional authority and show proper intent. Such steps cannot be passive but must be actively proclaimed and visible. One's methods should include a written policy statement and constant attention to detail. In addition, the archivist should as a matter of course document his or her actions and secure the necessary legal instruments from donors, interviewees, interviewers, and researchers.

The 1976 U.S. Copyright Act and Contracts

Copyright, Title 17 of the U.S. Code, is clearly the area of law generating the most concern and paranoia in sound archives. But what does this enactment state or cover with regard to the holdings of sound archives? First, we might remember that before 1971 unpublished recordings were not even captured by copyright statute. The 1976 law seems to deny coverage to any unpublished tapes made before 1972: for, as stated in Section 702 (b), "Only those sound recordings fixed and published on or before February 15, 1972 are eligible for registration." Hence, those items excluded appear to enter public domain and may be legally used without restriction, although ethical considerations may dictate caution.

To be certain of coverage under the code, recordings should have at least a copyright symbol and date affixed: for example, the symbol © and the year are placed on tape labels. In addition, one might wish to place a general notification of copyright on the researcher registration forms. Under Section 407, copies need not be deposited under the Register of Copyrights unless the Registrar demands it or unless one is bringing a lawsuit for damages. The copyright duration is fifty years after the last surviving author's death, except for "work for hire" which extends to seventy-five years from publication or one hundred years from creation—whichever comes first. Note that copyrights subsist in both the informant and interviewer.

If the archive does not own the copyright, it can seek permission from the owner for use—for example, when a project merely lends a collection to the archives. While ethically less acceptable, the Limitations on Exclusive Rights, Section 110 (8), seems to allow the archive use even when it does not or physically cannot locate the holders to obtain permission. The law makes special allowances in recognition of the research, educational, and preservational mission of the archive, but only as long as that does not extend to commercial advantage or financial gain. In addition, Section 112, Ephemeral Recordings, states that it is not an infringement to make more than one copy providing:

a.1 The copy or phonorecord is retained and used solely by the transmitting organization that made it; and

.2 the copy or phonorecord is used solely for transmitting the organization's own transmission within its local service area, or for purposes of archival preservation or security; and

.3 unless preserved exclusively for archival purposes, destroyed within six months.

b. except for one copy for archival preservation or security . . .

c. not an infringement of a program embodying a performance with no direct or indirect charges

d. not an infringement for a government body or other non-profit organization entitled to transmit under 110 (8) to make no more than 10 copies, or to permit the use of any such copy by any governmental or non-profit organization not entitled.

In further acknowledgment of the need to strike a balance between the public's right to knowledge and the author's right to protection and profit, exemptions are granted under Section 107, Fair Use. Although the law does not define the term, it does indicate some of the purposes under the purview of fair use, including "criticism, comment, news reporting, teaching, (including multiple copies for classroom use), scholarship or research." This section also sets out four criteria for judging "fairness."

1. The purpose and character of the employment, in particular if it is a commercial versus nonprofit educational nature.
2. The nature of the copyrighted work.
3. The amount or proportion of the excerpt used in relation to the whole.
4. The effects on the market in regard to the work's potential value.

We might also note that under Section 504 employees or agents of a "non-profit educational institution, library, or archives" are immune from statutory damages for copyright infringement, providing they were acting under the scope of their employ and had reasonable grounds for believing the action fell under fair use.

The paramount issue for archives, however, is ownership of the copyright. As Section 202 indicates, the gift of an unpublished work does not automatically transfer copyright, which is a distinct intellectual property right. In Section 201 (d), we find that possession can be exchanged through a legal conveyance, or by will, or by intestate succession. And according to Section 204 (a), any conveyance must be in writing and signed by the owners of the copyright or authorized agent. Yet Section 203 (3) states that this transferral is revocable during a five-year span some thirty-five years from the date of execution.

Beyond that last hitch in the law, we deal with three major forms of possible ownership conveyance—a release, a deed of gift, and a contract. The first two forms are rather easily overturned; thus, the archivist should concentrate on contracts. Such instruments have five general characteristics. First, they demand legally and mentally competent parties. Second, the duties or exchange must be acceptable under the law (e.g., one cannot make a legal contract to murder or rob). Third, the contract may be written or oral. Fourth, the contractees must have a mutual understanding of the terms. Fifth, a contract calls for a consideration or exchange of value; without this interaction we have only a gift.

Although no instrument is perfect, it does appear that one can construct a simple, yet adequate, conveyance in keeping with contract law. The sample shown in Figure 3:2, for instance, is easily embellished and short enough to be grafted to a full interview worksheet (see Figure 4:9). In essence, it requires identification of the donor and institution as the principals in the transaction, together with a statement to show the exchange of

Figure 3:2
Sample Contract for Copyright Exchange

I,_____, donate the

intellectual property rights and copyright of my interview

on _____ 19__ to _____ in

exchange for its efforts to preserve and make the

contents available for research.

Restrictions:

                              Signed:

                              Date:

                              Witness:

value from both sides and a signature block for the donor. In addition, many archives will want to add a statement attesting to the importance of preserving such oral accounts.

The form should also include a rather inconspicuous area for adding restrictions. The presence of this last field is indicative of the need to allow for special treatment for informants. Such precautions serve to insulate the parties and the archive and may allow more candid interviews. Although some resist the depository's assignment of additional restrictions beyond those of the participants, the archivist will certainly encounter instances in which even the strongest commitment to freedom of information will be taxed. Total restrictions on any use, of course, are untenable, and the archivist would do well to require specific time frames – for example, twenty years from entry and not five years after someone's death. While always troublesome, such limitations must be supported as a necessary right of the archive and as a way of facilitating the collection of an accurate record.

The archivist should remember several other facts about copyright law. First, because copyright is vested in both the interviewer and interviewee, the archivist must secure contracts from both parties. In addition, performance rights (like a musical recording) are more protected and require more care than simple interviews, especially in regard to possible profit-making ventures. Finally, because the law has not been fully interpreted by the courts, the archivist will need to keep abreast of any later judgments and newer interpretations.

Libel, Access, and Freedom of Information

In addition to copyright, archivists must keep abreast of and plan for other laws that vacillate between the public's right to access and the individual's right to privacy and property. Because of the peculiarities of locale and institution, each archive will have to examine its unique legal environment. But we can offer some general insights into such matters as libel and privacy in contrast to freedom of expression and information.

The law of libel specifically requires the "publication" of defamatory material about a person that would subject him or her to damage – as in causing ridicule, contempt, hatred, or ostracism. Normally, the injured party must be alive and identifiable and generally not a well-known or public figure. The defense can rest on the truthfulness of the statement, the fairness of the criticism, the public's right to know, and the recency of publication (generally charges must be brought within a year of publication). Fortunately, depository staffs are often protected from prosecution for the presence of libelous or slanderous materials in their charge. Moreover, the absence of malice and the research orientation of the original gathering process vitiate most of the legal dangers. Any archive, however, would be wise to maintain a degree of decorum and not publicize the presence of defamatory statements.

Like libel laws, privacy laws usually do not extend to public figures and require a living person and publication or publicity. The right is a personal one. It allows an individual to seek redress for an invasion of privacy, the disclosure of which has offended ordinary sensibilities. The legal defenses are much the same as those for libel. Privacy laws, however, may also extend repositories some protections: for example, Florida has included the sanctity of library registration and circulation records under its enactment. In addition to safeguarding the confidentiality of such records from all but court subpoena, the rights to privacy can also be cited in the researcher registration form (see Figure 3:1) to help protect against possible lawsuits. Although of doubtful legal standing, some archives have also gone to the extreme of demanding prepublication control on any citations from their holdings as part of the registration agreement.

Other archives must take special precautions because of their institutional or political affiliations. Those archives related to the federal government must comply with the Freedom of Information Act, which might make it impossible for many government employees to place restrictions on their interviews. Thus, federal and many business repositories must make extra security efforts in regard to sensitive oral documents.

Of more general concern are guarantees to First Amendment rights to free expression. Such rights provide the essential underpinnings for collection efforts but have come under specific attack in regard to obscenity. Although the Supreme Court in *Miller v. California*, 1973, did include a vague and potentially dangerous community standards test for obscenity, archival holdings are probably safe under the exemption for materials with redeeming social values held since the Roth decision in 1957. Still, it is wise to take precautions and remember that many collections may contain statements that could offend certain elements in the community. The rule of thumb is to maintain a high level of decorum and professionalism in the management of such materials.

For those who are unaware of the contents of many folklore interviews, a mere listing of the first chapter in Wayland Hand's organization of the Frank C. Brown Collection might prove instructive of potential dangers:

BIRTH: Where Children Come From 3 - Fertility, Sterility 4 -Conception, Contraception 5 - Pregnancy, Confinement 5 - Miscarriage, Abortion 7 - Labor 7 - Afterbirth 12 - Afterpains, Childbed Ailments, Tabus 14 - Nursing, Weaning 16 - Prenatal Influences 17 - Birthmarks 18 - Deformities 22 - Birthday 24 - Number of Children 25 - Sex 27 - Twins 28 - Naming 29 - Physical Attributes, Growth, etc. 30 - Dispostion 34 -Character, Talents 35 - Fortune 40 - Miscellaneous 44 - Health 64. AILMENTS AND REMEDIES: Bed Wetting 47 - Bowlegs 48 - Choking 48 - Colic 48 - Convulsions 50 - Croup 50 - Deafness 51 - Eyes, Cross-Eyes 51 - Fits 52 - Hernia 52 - Hives 52 - Indian Fire 53 - Indigestion 53 - Lice 53 - Liver 53 - Mouth 54 - Naval 54 - Phthisic 55 - Rash 55 - Rheumatism 56 - Rickets 56 - Slobbering 56 - Speech 57 - Stammering, Stuttering 57 -Stretches 58 - Teething, Teeth 58 - Thrush (Thrash) 64 - Whooping Cough 67 - Worms 68. . . .

The Necessity of Ethical Guidelines

The best defense to legal obstacles will lie in the offensive proclamation of the archive's ethical and legal policies – for example, adopting or adapting the Goals and Guidelines of the OHA, which stand almost alone in the literature on sound archives.

These proposals suggest formal policies to insure that interviewers and interviewees are apprised of procedures and their own rights and interests, as well as of reciprocal legal and ethical responsibilities. In keeping with the previous discussion, another important feature is to assure that proper legal documentation and releases are secured. In terms of oneself and staff, a program also needs to impress the professional's responsibility in such matters. Are all aware of the need to preserve confidentiality where appropriate and to act with a proper sense of decorum in regard to the information secured? Is there a sense of mission and recognition of the goals of the institution and the need to gather and quickly process the material, and make it available?

While the OHA guidelines are an excellent starting point, archivists must address their own peculiar network of responsibilities and produce a document for their own peculiar circumstances. Legal points are important in this analysis, but ethical judgments should take precedence. Theoretically, the law is only a negative force by which the state seeks to enforce adherence to commonly agreed on moral policies. Archivists will want a more active public statement that stresses the mission of the archive, its collection policy, and their own professional qualifications – notably, their skills in appraising and preserving, as well as in maintaining the necessary confidentiality. In addition to the considerations suggested by the OHA for the interviewer and interviewee, any exposition from the archive must also speak to the balance of rights between researchers and the public to information and the individual to privacy. With such elements, a statement of ethics shows legal intent and helps to establish the archivist's authority in a wide range of applications.

Finally, it is imperative that the socially responsible archivist recognize the broader implications of knowledge in society. This recognition suggests conscious awareness of their duty to insure a documentary heritage, as well as knowledge of its positive and negative attributes.

SELECT BIBLIOGRAPHY

As indicated by this chapter, the Oral History Association leads the way in regard to ethical and legal questions; these concerns are evident at its national convention and in the pages of the *Oral History Review*. The proper techniques for reference service and in response to user demands, however, have scarcely been touched. Those interested in such issues may address the sparse literature from the library field or reference work in the humanities.

Asheim, Lester. *The Humanities and the Library*. Chicago: American Library Association, 1975.

Eustis, Truman, III. "Get It in Writing." *Oral History Review* 4 (1076): 6-18.

Filippelli, R. L. "Oral History and the Archives." *American Archivist* 39 (1976): 479-83.

Hand, Wayland, ed. *Frank C. Brown Collection of North Carolina Folklore*. Vols. 6 & 7. Durham, NC: University of North Carolina, 1961, 1964.

Jackson, Eugene, ed. *Special Librarianship*. Metuchen, NJ: Scarecrow, 1980.

Katz, William. *Introduction to Reference Work*. 2 vols. New York: McGraw-Hill, 1974.

Key, Betty McKeever. "Oral History in the Library." *Catholic Library World* 49 (1978).

Moss, William. *An Oral History Program Manual*. New York: Praeger, 1974.

Neuenschwander, John. *Oral History and the Law*. Denton, TX: Oral History Association, 1985.

Oral History Association. *Evaluation Guidelines*. N.p.: Oral History Association, 1980.

Romney, Joseph. "Legal Considerations in Oral History." *Oral History Review* 1 (1973): 66-76.

Romney, Joseph. "Oral History: Law and Libraries." *Drexel Library Quarterly* 15 (1979): 39-49.

Shores, Louis. "The Dimensions of Oral History." *Library Journal* 92 (1967): 979-83.

Stewart, John. "Oral History and Archivists." *American Archivist* 36 (1973): 361-65.

West Publishing Company, comp. *United States Code Annotated: Title 17, Copyright*. St. Paul, Minn.: West, 1977.

THE PROCESSING AND ORGANIZATION OF COLLECTIONS

If the heart and soul of a sound archive are its public services and cultural mission, then its body and blood are the organization and technical processing of its collections. Unfortunately, the seemingly mundane steps in the preparation of holdings are generally slighted in the literature in favor of the more glamorous aspects of interviewing. Yet, assuming the interview was worth conducting, should not the results, the information, be made available?

This chapter and the two following offer an integrated system of processing as a model and standard for sound archives, a field that has evolved rather haphazardly and often without regard to preexisting conventions. This model should retain some degree of flexibility and creativity in relation to an archive's particular holdings, but it should also be noted that the archivist does not need to "reinvent the wheel" in terms of bibliographic standards.

We will begin with a consideration of the "nuts and bolts" – the basic files and guides in a bureaucratic system – as they apply to a sound archive. The chapter then proceeds to a discussion of the technical aspects of appraisal and the acquisition of collections. The bulk of the chapter is concerned with a tripartite schema for the description and retrieval of data found in the tape-recorded interview.

PROCESSING AND ITS CONTROL

Obviously, good record keeping and attention to detail are keystones for any information agency. Files should be consistently arranged and maintained, and normal practice calls for a general overview or processing guide to inhouse procedures. This guide may stand alone or as a subdivision within a larger installation manual, which could also contain public access policies, ethical and legal positions, a disaster plan, and the like. Such records and guides work on a number of levels; for instance, they actually help to give a sense of legitimacy and value to a bureaucratic mentality, a mentality often represented in a board of trustees, a library director, or an academic dean. A guide also proves a useful introduction and check for employees, volunteers, and successors, and its availability even aids in legal defenses. Moreover, the production of a guide is extremely helpful for organizing thoughts and pinpointing problem areas – not to mention later jogs to one's own memory.

Where does one begin with a processing guide? One starting point is with background manuals like this one. Again, the idea is not to reinvent the wheel, but to benefit from the experience and mistakes of others. In addition, most archivists will need to examine existing collections: how many are there, how were they processed, what general themes or topics are addressed, and what is the physical condition of the tapes? If this information is not readily available, then a quick inventory of the holdings is in order (Figure 4:1). Because this step is performed onsite and is of temporary use, the information is generally better recorded in an open-ended form that would not require automated data entry.

Although the arrangement of a processing guide is quite arbitrary, one normally begins with a statement of mission. The rest of the work will focus on the life cycle of the collections from acquisition and conservation processing, to possible discarding – or weeding in library parlance. A guide should be viewed as a living book that not only chronicles current endeavors, but also records alternatives and evolutions within a system. In this vein, it is essential to include a sample run of all forms and a record of when they were adopted.

Figure 4:1
Sample Inventory Form

INSTALLATION NAME
Inventoried by:
Date(s):

COLLECTION Donor/subject	CONTROL NUMBER	DATES	FORMAT(S) (reel, cassette...)	CONDITION (poor, fair, good)	PROCESSED (yes, no, partial)	NOTES

INV 1984

Filing procedures are also arbitrary, but need to be standardized, with the practices recorded for future reference. Although numerous alternatives exist, in general most manual filing systems employ a combination of subject access that is subdivided alphabetically and then chronologically. The typical archive might have separate files or categories for financial records, legal documents, general correspondence, internal correspondence, and supplies and equipment. A strong recommendation is the establishment of uniform alphabetization rules for all records from office files to the card catalog.

More particularly, dealings with specific collections should be isolated in one file, which mirrors the physical distribution or classification of the tapes. In the examples, the disposition of files and tapes is ordered by the provenance of the individual collector or project as the producer of the material. As will be more fully detailed in the section entitled Collection-Level Description in this chapter, all collectors are assigned an identifying number upon the accession of their first donation into the archives, and each of their sessions is married to that number by two additional subsections of the codes. Project files, then, are ordered numerically by the collector's number and subdivided by function – such as donor and informant subfiles, processing records, and information requests. The subfiles may be further differentiated depending on size or frequency of use and are normally arrayed in reverse chronological order with the most recent item first.

An essential element in a well-run sound archive is a log to guide, record, and control the processing of individual collections. The log in Figure 4:2 is designed for the general system advocated in this work and should be adapted for the particular practices within the archive. In an operation with high volume, such a log could be formatted for computer entry, but in most installations and for the purpose of this book a paper medium is adequate. It should be noted that the model presents internal checks, such as dating and initialing, to provide the administrator and processors with a running oversight throughout the procedures.

In conjunction with the log, the flowchart (Figure 4:3) illustrates an overall approach to the technical processing of collections. If we assume that proper files and forms are ready to greet

Figure 4:2
Audio Processing Log

Tape Number:
Collector:
Date of Acquisition:
Number/Speed: Cassettes _____; Reels _____/_____; Discs _____/_____; Cylinders _____.
Condition: Good_____; Fair _____; Poor _____.
Accompanying Materials:

Origins: Gift _____; Loan _____; Purchase _____; Class Project _____; Other _____.
Others Aiding in Acquisition:

Release forms: Interviewer _____; Informants _____.
Level of Processing: Collection _____; Item _____; Transcription _____; Reject _____.
Date:_____ Archivist's Approval _____.

LOG

Item	Date	Initials	Extent or Notes
1. Audio Processing Log:			
2. Project Folder:			
3. Mastered:			
4. User copy:			
5. MARC Record:			
6. Cards Filed			
7. Worksheet:			(_____ By Collector; _____ Archive)
Filed:			
8. Session Screen:			
9. Contents Screen:			
10.* Auxiliary Cards:			
Filed:			
11. Transcript:			
12. Donor Acknowledged			
13. Donor Copy Returned:			(Not requested _____)
14. Later Duplicates:			
15. Weed:			Reasons:
Archivist's Approval:			

*For manual systems in place of 8 and 9.

Figure 4:3
Collection Processing Flowchart

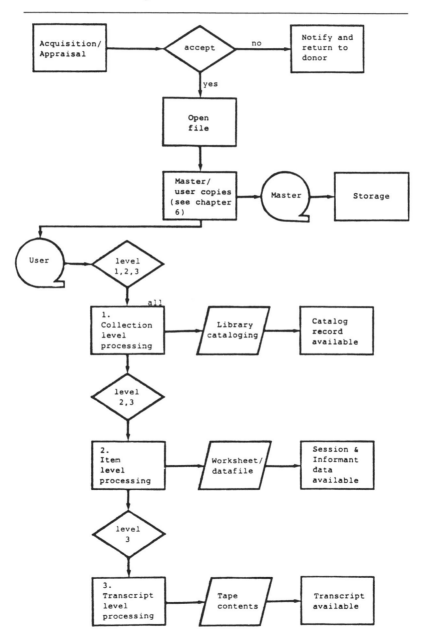

incoming donations, then the initial stage is one of deciding whether or not to retain the donation. Such a deceptively simple question, however, can be quite involved.

ACQUISITIONS AND APPRAISAL

As indicated in earlier chapters, acquisitions and appraisal are complex procedures that often involve previous agreements. Judging the acceptability of a collection, however, should be based primarily on its value in relation to the mission of the institution. This assessment also needs to take into account the archive's own resources, legal questions, and the physical condition of the donation. If the archive cannot afford to process, cannot legally offer access, or cannot hear the contents – why bring the donation into the archive? There may be reasons, but they do need to be articulated and should be stated in a formal collection policy.

Generally, collections are recorded immediately upon entry into the archives in an Acquisitions Register. This device may be either a proper, distinguished-appearing book on view in the public area, or a more informal, working document in the processing area. In either case, a register provides an initial level of control and a chronological listing of additions. The form of such a document is fairly flexible. The archive might merely adapt a commercially produced guest register with blocks for name of donor, date, and, if possible, extent of item. In order to reduce the variety of forms in the archive, another suggestion is to convert the inventory form (see Figure 4:1) for such a purpose.

It is very important to record as much information as possible and to gain legal releases during acquisition. The more delays in securing such materials the more difficult the task. Thus, another recommendation is the immediate opening of a processing log and, if not already begun by the donor, a worksheet (Figures 4:8, 4:9).

After a donation enters the archive's domain, decisions have to be made about retention and the level of description. Again, with many collections, decisions on retention are fait accompli – made well in advance. Prior commitments, institutional realities, and common courtesy, as well as the lack of networking among sound archives, often make the question of appraisal for retention rather moot. Many archives also seem to keep whatever comes in the door. Yet, there are considerations that should lead anyone to reject deposits: in particular the probable continued absence of legal releases, restrictions that would make the donation all but useless, and tapes in very poor or unplayable condition (see Figure 4:4). Again, whatever the case it is important to have a written statement of the acquisition/appraisal or collection development policy on hand as a first line of reference and, possibly, defense.

Figure 4:4
The Process of Appraisal for Retention

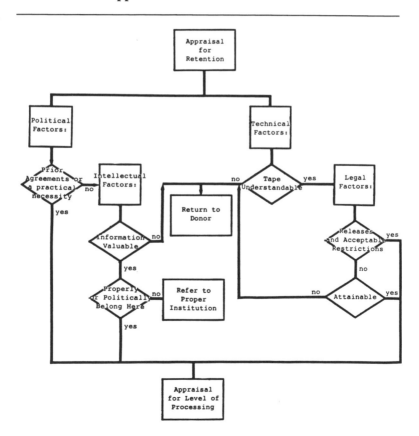

Assuming that the material is accepted, the archive will then assign a collection number, establish a file in the collections area, and send the tapes forward for the production of a master and user copies. As more completely explained in Chapter 6, this reproduction phase is quite important for the purposes of conservation. Only the user copy will be employed during the rest of the descriptive process.

The next stage involves appraising accepted materials in regard to three levels of description. Here, one must not only evaluate their supposed value and use but also acknowledge financial, staffing, and technological exigencies within the archive. If holdings are to be of any use to future researchers, they must as a matter of course receive at least a collection-level description. The bulk should also acquire an inventory or outline to the contents of the tape, a level of analysis that can be greatly facilitated through the preacquisition distribution of standardized work forms. Finally, and with sufficient resources, the most valuable tapes should be supplemented by transcripts of their contents.

COLLECTION-LEVEL DESCRIPTION
(AND UNIT RECORDS)

Over the years, most sound archives have relied on catalog cards as a major form of access into their holdings. Yet few archives have paid any consistent attention to the rules and conventions for cataloging. The cataloging of unpublished sound recordings has actually been treated as a bastard child – caught between the worlds of archives, folklore, history, and libraries. The result is a confusing plethora of systems and a lack of uniformity that often occurs even within individual institutions.

This section explores techniques for reversing these tendencies through a manual system, which is grounded in archival theory and library methodology. In particular, it examines the utility of unit records, which reproduce the same basic cataloging information in the same format on each card. Such reproduction gained currency after the Library of Congress began issuing full printed sets of unit records in 1901 and remains standard in most American libraries. Unit records are placed in that late nineteenth-century innovation–the card catalog. Generally, sound archives will contain a dictionary catalog that groups all author, title, and subject cards together, as well as separate or auxiliary drawers for a shelf list, interviewee file, authority headings, and, perhaps, separate genre items. If the archive is part of a larger library or similar institution with its own catalog, one suggestion is to infiltrate a duplicate set of the records from the dictionary catalog into that of the parent body. Another idea is to note the format of these cards may also be transferred onto paper to produce a book directory of holdings.

Although many archivists might not wish to become overly involved with the intricacies of library cataloging, it is important to realize that international bibliographic standards do exist. These standards are part of a long refinement, dating at least to Antonio Panizzi's 1841 *Rules for the Compiling of the Catalogue* for the British Museum, but coming rapidly to the fore only during the last twenty-five years. Our current canons reflect adherence to the *Statement of Principles* issued by fifty-three participating nations in the 1961 International Conference on Cataloguing Principles in Paris. The present "bible" for these International Standard Bibliographic Description (ISBD) practices is the *Anglo-American Cataloguing Rules*, second edition (*AACR2*). Sound archivists would be well advised to mirror the organization of catalog cards suggested for nonpublished sound recordings in *AACR2* (see Figure 4:5).

Before launching a more detailed–yet still cursory–query, one should realize that *AACR2* practices treat oral history and folklore tapes as though they were books. The taped interview may then be somewhat equated to an edited memoir. Of particular importance for this system is the recognition of the name of the interviewee within the title as the main entry for the work. But a secondary degree of responsibility also rests in the interviewer, who orchestrates the session and is somewhat analogous to an editor.

Classification Code

At present, no generally accepted taxonomy exists for any form of sound recordings. Oral history and folklore tapes, for

Figure 4:5
Sample Catalog Cards (ISBD)

A. Main Entry

Folk Terrebonne, Henri.	Main Entry /
702- Performance by Henri Terrebonne / recorded by	Title
82- Helen Issen.	Author (interviewer)
003 1 sound cassette (60 min.) -- (Cajun Folklife	Physical Description --
Project)	Series (project/class)
In English and French.	Language Note
Recorded in Dupre Library, Lafayette, LA.,	Publication, Distribution,
Jan. 21, 1982.	etc. Note (date/place)
Transcription available.	Other Formats Note
A noted Cajun raconteur in a storytelling	Summary Note
session with legends and jokes about life in rural	
southern Louisiana.	
1. Folklore--Louisiana..2. Louisiana--Folklore.	Tracings (may be on rear)
3. Cajuns--Folklore. I. Issen, Helen. II. Series.	

B. Added Entry

```
                Florence Bit biography project

012-    Jones, Amy.
84-         Interview with Amy Jones / by John Smith
002         1 sound cassette (55 min.) — (Florence Bit
        biography project)

            Recorded in Augusta, Me., Jan. 18, 1984.
            The interviewee describes her relationship as
        personal secretary to Florence Bit, noted ornitho-
        logist, from 1948 to 1956.  She also describes
        life and conditions in Narragansett, R.I. in that
        era.

            1. Bit, Florence. 2. Ornithologists. 3. Nar-
        ragansett, R.I.  I. Smith, John. II. Series.
```

example, have been grouped by type of material, size, or accession number and occasionally by such extant systems as Library of Congress and Dewey Decimal Classification. Yet, because of their nature as unpublished and unique intellectual productions, these collections more properly and effectively fall under the purview of archival theory. Rather than library subject classification, archival procedures order by provenance, or agency of origin, which in this system means the interviewer or perhaps project.

There are many reasons for this type of arrangement. On one hand, any system that attempts to physically locate media through some grand ordering of knowledge (i.e., a subject classification) is doomed to failure in today's increasingly intertwined intellectual universe. Such systems are too inflexible and tend to freeze thought into artificial categories that predominate at the time of their creation. On the other hand, provenance eliminates often time-consuming, subjective judgments on what is "the" proper slot for the work. Instead, provenance reflects an inherent order of production and by itself acts to provide information. The labors of an interviewer or project are pulled together on the shelves like a series; they are not scattered throughout the repository with the possible loss of the interrelationships.

The coding advocated draws from provenance to allot a specific designation to each collector, as well as from museum methodology to allocate a dated accession number to each session. While quite simple in form and use, this schema is very efficacious for coordinating a wide range of activities within the installation. In effect, it is a simple three-position numeric code. Those archives that are part of a larger repository may also distinguish themselves by an introductory alphabetic segment, preferably mnemonic in character. Sample distinctions could include TAPE, FOLK, ORAL, and the like.

The numeric section of the code begins with an effectively limitless segment, which reflects the number given to individual collectors or projects. In general, numbers are assigned in ascending order from the first donation (e.g., 001 = the first collector; 002 = the second . . .). Collaborative efforts with two or more interviewers should receive their own numbers, as should group projects. In addition, one may choose to reserve certain blocks for the group or class projects (all the numbers above 1,000, all those from 100 to 199, and so on). The next two segments borrow from museum acquisitions methods to provide a calendar of a collector's work. The second position in the overall schema is a two-digit slot reflecting the year (83 = 1983, 79 = 1979). Finally, three digits indicate the order of entry that year (005 = the fifth session received, 010 = the tenth session). This last section reflects complete sessions and not hours, for although most interviews are finished in a single sixty-minute tape, some can run considerably longer. On putting the segments together, one would find: 010-83-001 = the tenth collector, in the year 1983, and his or her first deposit; 126-79-011 = collector 126, 1979, the eleventh donation. When filed together by this code, one set of cards then represents a shelf list to the entire holdings.

Elements of the Unit Card and AACR2

Although the classification of sound recordings has not been regularized, the other elements of the unit card are governed by formal ISBD rules. In fact, the degree of expertise required for original cataloging can be quite intimidating to the uninitiated. The mysteries of the cataloger, however, need not be fully plumbed to construct a proper card system. A reading of these sections should provide a sufficient introduction to allow one to at least mimic and standardize in accordance with international canons. Those with the need for more in-depth analysis should begin with *AACR2*, especially Chapter Six – "Sound Recordings." Beyond the general treatment for all sound recordings, the specific pronouncements in Chapter Six on unpublished materials are sparse and can be mentioned in their entirety:

6.11 NONPROCESSED SOUND RECORDINGS

6.11A. Follow the rules for sound recordings (6.1 - 6.10) as far as possible in describing nonprocessed recordings.

6.11B. If the recording has no title proper, formulate a title as instructed in 1.1B7. [Addressed to high school students, discussed good writing.]

6.11C. Do not give any information in the publication, etc., area. Give the date of recording in a note.

6.11D. Make notes on participants in such a recording and the available details of the event recorded, as well as other notes prescribed in 6.7.

The card itself may be viewed as a combination of headings and a series of short paragraphs. The specific ISBD rules for punctuation within the body are quite extensive and occasionally picayunish. In general, one may simply mirror the diacritics and layout in the model cards. For example, typewritten cards are single-spaced throughout with the exception of the double-spaced first note. Typing on the main entry (the interviewee's last name, comma, first name) begins on the fourth line from the top and nine spaces from the left-hand edge of the card.

The first paragraph, or "Body of the Entry," is fairly sparse for nonpublished sound recordings and consists of only a title (the interviewee) and a statement of responsibility (the interviewer). In order to simplify the process of cataloging, it is useful to systematize the entries where possible. In particular, titles are originated inhouse and should be standardized into a formula, such as "Interview with . . ." or "Performance by . . ." The title is followed by a (/) and then the name of the interviewer, which may be preceded with an identifying phrase—"interviewed by," or "recorded by," or simply "by." Since the material is not published and as stated in 6.11C., no further information is presented in this first paragraph.

The second paragraph begins with a physical description of the recording. The rule is to describe the original format, although one might also hedge by limiting this entry to a description of user cassettes and indicate the original in a note. The user cassettes should be C60s (sixty minutes in length) with a normal speed of 1-⅞ inches per second (ips). Assuming the interview is part of a larger project or class assignment, this is described in the second—or series—sentence, which is placed within parentheses.

The note area is key for potential users. It can provide a fairly in-depth description of the collection and actually extend to numberless supplemental cards. One, however, should try to limit oneself to a simplistic explanation that can be confined to one card: hence, not defeating the strengths of this descriptive format. Each of the notes is seen as a separate paragraph. The order for these notes is set out in 6.7B of *AACR2*, but again one should be aware that, with the exception of notes on the date and place of the interview and a brief summary, one's efforts should be restricted. The order of notes in *AACR2* is as shown in the following list.

1. Nature of Artistic Form and Medium of Performance (unless it is apparent from the rest of the description).
2. Language (also unless apparent in the rest of the description).
3. Source of Title Proper (if other than the chief source of information).
4. Variations in Title (also may include a romanization, if from another script, e.g., Cyrillic or Chinese).
5. Parallel Titles and Other Title Information.
6. Statements of Responsibility (those persons or bodies tied to the work that are not given in the statement of responsibility, and also the names of other performers and the medium of performance).
7. Edition and History (of interest, if the session is reproduced from one held in another depository).
8. Publication, Distribution, etc. (This note is essential for recording the date and place of the interview.)
9. Physical Description (if used, for a description of the original format).
10. Accompanying Material.
11. Series (and to more fully explain the project).
12. Dissertations.
13. Audience (mention of the intended audience or their intellectual level).
14. Other Formats Available (used especially to indicate the presence of a transcript).
15. Summary (normally an essential, yet brief, descriptor).
16. Contents (a list of the individual titles and any statement of responsibility not previously included: e.g., a list of songs, or even an outline of the entire interview).
17. Copy Being Described and Library's Holdings (any peculiarities if the item has been copied from another repository; or, if a portion of an incomplete set).

Obviously, the application of the full range of notes could be very time consuming and extend through far too many cards. Similarly, a degree of restraint should be applied to the number of access points for a single collection within the card catalog. In library parlance, indications of the variety of "Added entries" beyond the main entry are referred to as tracings. Tracings are the final element of interest to us in the unit record and are placed at the bottom or on the rear of the card. Each of these tracings is then reflected by separate cards in the file that are topped by tracing heading above the main entry line.

Placed in alphabetical order and introduced by an arabic number, subject entries are the first tracings encountered. The choice of subject headings is easily the most taxing element in any collection-level description; in this regard, the arbitrary nature of language and the need to limit to a reasonable number of topics reveal some of the basic shortcomings of card-catalog efforts. Much has been written about these difficulties and the need for a limiting thesaurus — a master list of acceptable terms and spellings for the entry. In general, the major recourse for such terms should be the two-volume *Library of Congress Subject Headings* (*LCSH*), the standard governing tool for ingress into card catalogs. Historical topics are usually indicated by place and time period, as well as by such thematic divisions as diplomatic or social. The appendices contain excerpts from the Library of Congress' *Subject Cataloging Manual* section of Folklore (H1627 revised 6/13/83), but as it describes the guidelines:

Folklore is broadly defined as those expressive items of culture that are learned orally by imitation or observation, including such things as traditional beliefs, narratives (tales, legends, proverbs, etc.), folk medicine, and other aspects of the expressive performance and communication involved in oral tradition. This is defined in contrast to the concepts designated by the more inclusive subdivision Social life and customs which may include folklore, manners, customs, ceremonies, popular traditions, etc.

This memorandum places stress on the fact that a complex of headings, rather than a single heading is to be assigned to each work of folklore. A number of different topics have been identified as having retrieval value for such works, and each should be brought separately, if possible. They are: (a) the ethnic, national or occupational group which originated the folklore in question, and/or the locality; (b) the special theme of the folklore; (c) the folkloric emphasis or genre.

Unfortunately, *LCSH* nomenclature is a national guide and often provides insufficient differentiation for specialized collections with their own highly developed terminologies. Thus, while *LSCH* should be adhered to as much as possible, one may have to develop one's own supplemental thesaurus, especially for folklore collections. For manual systems, this thesaurus file should be tightly regulated and referred to whenever cataloging. The idea is to distinguish the most recognizable and used terms rather than an undifferentiated plethora of terms that only clutter and actually make retrieval more difficult. With the aid of the computer and at its highest level, such techniques can produce a dynamic tool that also furthers knowledge about the contents of the collections.

In addition to topics from a thesaurus, subject entries also generally extend to the locus of the interview and can include an entry for the biographical subject of an interview. As with the subject thesaurus, a degree of regimentation such as an authority file for place and personal names is useful. Practices in regard to titles, spelling, and arrangement should be very obvious, but those with the need for specificity may refer to Chapters 17 and 18 of *AACR2*.

The next set of tracings begin with a roman numeral. Normally, there is an "author added entry" for the interviewer after roman numeral I. Number II may be used for the title, but, since the main entry already provides such data in a more serviceable fashion, this may be reserved for information from the series statement; thus, II. Series will be used to indicate the name of the project or class to which the interview belongs.

Auxiliary Files and Nonunit Records

The economics of online catalogs, computer output microfilm (COM), and machine readable cataloging (MARC) records are fostering a return to nonunit records, which provide only an abbreviated description and pointers to the complete, main entry. Those without access to printed card sets and typing cards inhouse are also quite aware of the utilities of nonunit cards. As suggested in Figure 4:6, this abbreviated style need only contain the heading, classification number, main entry, and, perhaps, the first paragraph of the full card.

Many sound archives also employ an even more abbreviated card, which consists of a heading and underneath references to relevant collections. Although this type of record does save space, it is less desirable than individual cards for each collection.

Beyond substituting for all but the main entry in the dictionary catalog, nonunit cards are frequently featured in separate auxiliary card files. Such files collate distinct access points into one locus, rather than scattered throughout a larger catalog: for example, a shelf list arranged by classification codes. Other instances generally include separate files for interviewees and collectors or projects. In addition, many folklore archives retain files for locations, language, and ethnicity or race, as well as for item-level descriptions from genres like song titles, jokes, and folktales.

The variety of auxiliary indices is limitless, but their very proliferation is indicative of problems in collection-level description and manual systems. Such records are hard pressed to include a full range of access points and still retain their compact nature. In addition and even with abbreviated entries, card catalogs can grow to be quite bulky, and their entries are inherently redundant. While still a flexible and useful medium, card catalogs are destined to be supplemented and eventually replaced by the more malleable technologies of data processing.

MARC Format and Online Catalogs

Following its full initiation by the Library of Congress in 1968, the MARC format emerged by 1974 as the international framework for bibliographic description. Although not initiated at the time of this writing, by the close of 1984 MARC officials were scheduled to accept a standard manuscript (AMS) format supposedly for all materials in archives. This development is

Figure 4:6
Sample Auxiliary Cards

A. Added Entry

```
                          Added entry. (from tracings)

         TAPE   Jaynes, Amy
         156       Interview with Amy Jaynes / by Samuel
         79-    Smith
         001
```

(may have tracings)

B.

```
              Bird Watchers                    See Reference

         SEE: Ornithologists
```

C. Nonstandard item list

```
                              Entry                  Card file

                                                     Song title

                              "Joli Blond"

         Locations            003-67-009
         (may also have       012-75-001
         time sequence added: 012-75-012
         e.g., 003-67-009:28) 156-83-001
```

part of an explosion of current activities in archival cataloging, which includes an emerging consensus among folklore and other archivists for MARC, and new MARC-based archives formats from the Research Libraries Network and OCLC (Online Catalog Library Center). MARC is simply emerging as the descriptive standard at this level. For those with access, such records not only provide the basis for card, COM, and online systems, but should also serve as the medium for relating and exchanging information about specific collections to those outside the archive.

MARC records consist of an initial set of fixed field descriptors and a number of variable field tag lines. Figure 4:7 reproduces a typical MARC screen from an OCLC (the major online bibliographic network) work form. The fixed fields are generally of

Figure 4:7
MARC (OCLC) Screen*

```
OCLC: NEW              Rec stat: n  Entrd:              Used:
▷ Type: j  Bib lvl: c  Lang: eng  Source: d  Accomp mat:
Repr:       Enc lvl: I  Ctry: xx  Dat tp: u  MEBE: Ø
            Mod rec:    Comp:      Format: n  Prts: n
Desc: r  Int lvl:     LTxt:      Dates:

▷  1 Ø1Ø
▷  2 Ø4Ø       ≠c
▷  3 ØØ7       s ≠b l ≠c m ≠d n ≠e j ≠f l ≠g a ≠h n ≠i n ≠k c
▷  4 Ø33 Ø     19820121 ≠b 4014 ≠c L2
▷  5 Ø41 Ø     engfre
▷  6 Ø43       n-us-la
▷  7 Ø99       Folk 7Ø2-82-ØØ3
▷  8 Ø49
▷  9 245 ØØ    Performance by Henri Terrebonne / ≠c recorded by Helen Issen.
▷1Ø 26Ø        ≠c
▷11 3ØØ        1 sound cassette (60 min.)
▷12 44Ø  Ø     Cajun Folklife Project
▷13 5ØØ        In English and French.
▷14 518        Recorded in Dupre Library, Lafayette, La., Jan. 21, 1982.
▷15 52Ø        A noted Cajun raconteur in a storytelling session with legends
and jokes about life in rural southern Louisiana.
▷16 65Ø  Ø     Folklore ≠z Louisiana
▷17 65Ø  Ø     Louisiana ≠z Folklore
▷18 65Ø  Ø     Cajuns ≠z Folklore
▷19 7ØØ 1Ø     Issen, Helen
```

*See appendices for information on the codes.

less importance, and one should attempt to standardize entries as much as possible. For fuller information, one will soon be able to consult the new MARC AMS and OCLC's AMC (Archives and Manuscript Catalog) formats. In lieu of those formats, this work will employ OCLC's *Sound Recordings Format* as a sample. (A simplified introduction to the pertinent OCLC mnemonics and tag lines for sound collections is supplied in the appendices.)

Unfortunately, although the descriptive and information exchange capacities of MARC formatting have been well developed at the collection level, MARC itself is of only limited value to archives. Such limitations extend to present state of online catalogs; moreover, this new technology calls for an economy of size well beyond the magnitude of most sound archives. In addition, even the best of online systems and the best intentions of MARC's archival format to emulate a hierarchy do not adequately address the needs of archives for item-level retrieval. Because of such factors, the small number of collections (vis-á-vis the millions of book volumes), the absence of comparable automated functions for cooperative cataloging or interlibrary loan, and the inertia from preexisting methods, many sound archives can be expected to continue to rely on the traditional manual catalog.

Still, those who can enter their holdings in MARC form on one of the utilities are urged to do so. They will be providing users and other archivists with valuable background and locations. The final hope along these lines will be a working network of the sound holdings in this country and elsewhere, which is structured from the MARC record.

ITEM-LEVEL PROCESSING

Collection-level cataloging is obviously a necessary and integral step, but it is still a qualified one. The card itself can hold only a limited amount of information and still retain its compactness, and the typical catalog could quickly be overwhelmed by every useful citation in a collection. To expand on an earlier analogy, how large would a card system become if it furnished access to every item in every index of every book in a library? A card catalog should merely furnish brief and limited pointers.

Significantly, too, cards are not configured to reveal where to listen on the cassette. Much like a movie goer, sound researchers are locked into a serially accessed medium. With only collection-level finding aids, they must sit through an entire performance waiting for the right moments. One method for overcoming this dilemma as well as increasing the amount of information about individual sessions involves a second level of description. The key document in these efforts is a worksheet with a joint session description and tape outline (see Figures 4:8, 4:9). The completion of such a form should be a standard and initial procedure for the bulk of an archive's holdings.

Not all of the work, however, has to be isolated in the archives. In order to increase accuracy and reduce the workload, the archivist should facilitate the predistribution of forms to potential interviewers. Interviewers are easily the most appropriate and prepared to synthesize their own sessions on the worksheet. Because the recommended methodology for oral historians includes an immediate post-session review and the methodology for folklorists the compilation of field notes, they should record their information on forms with standardized categories. Moreover, a working relationship with interviewers could also simplify securing legal releases through a similar dispensing of forms. The overall effect is magnified only for those dealing with large class or community projects.

Because of the frequency of use and potential for comparisons with the rest of an archive's holdings, the form should be structured with the idea of data entry. (The next chapter will address the application of data base management systems to such records.) Following the principles of forms management, this work advocates a multi-purpose form, with entries for data about the session and the contents of the tapes. The amount and type of information desired will vary from institution to institution, but certain key categories can be agreed upon.

The session information section should be used initially to gather much of the data that will appear on the catalog card, as well as any supplemental information beyond the confines of the card format. Thus, at the least this section should reflect the name of the interviewer and interviewee and the date and place of interview. After this minimum, the range is endless but might include information on the interviewee's birth, race, ethnicity,

Figure 4:8
Sample Folklore Worksheet

INSTITUTION NAME

TAPE:

Collector: Date:

__Project; __Class: Title

Site City: State: Locus:

Informant__of__; __Performing Group, Title--

Last Name: First:

Maiden/Other: Sex:

Ethnic/Race: Religion:

Occupation(s):

Residence(s) :

Other Details:

__Additional Materials:

Summary:

Time	Genre	Topic/Title

__of__pages   FL WKSHT 1984

Figure 4:9
Oral History Combined Form

I.   Collection Number: _____

I, _____, donate the intellectual contents of my taped interview
described below to the _____ program in exchange
for its efforts to preserve and make that information available in the future.

Restrictions: (signed)_____
                                  (address)_____

(witness)_____   (date) _____

II. Collector:

                                III.   ____Cassettes

    ____Project/Class: ____Reels (speed____)
                                           Recorder brand:

    Informant:

    Place--city: ____Dolby (or:

    Other details: ____Other materials:

IV. Subject/brief description:

Time	Topics

1 of ____, (append additional sheets if necessary)   OH 1984

education, sex, religion, dialect, work, and residence. Although too many data elements might hamper processing, the information one chooses to collect will frame and delimit the range of analysis for the future; decisions need to be very carefully considered.

The second part of the worksheet is central to this system of sound archiving. In essence, it is used to provide a combination of an in-depth outline and item index to the interview. This section begins with a space for a general title/description and below that a series of columns, including one for time or sequence. At its simplest, the time column can reflect the readings from the foot meter on one's machine (see appendices for conversion chart). Note that foot meters can vary widely from one machine to the next, but the basis for all cassette records is 1-⅞ ips. (1.875 inches per second). With some machines, one might substitute a pulse for retrieval and hence record the number of the pulse for indexing. This book recommends an audible time signal (e.g., one minute, one minute ten, one minute twenty . . .) on one channel of the user cassette for pickup on a stereo cassette receiver (see Chapter 6 for details).

Dale Treleven is credited with first proposing this technique, which derived from a method for indexing television news programs. Known as TAPE (Time Access Pertinent Excerpts), the technique was originally proposed as a less expensive and less time-consuming alternative to transcription.

Because folklore archives are generally more highly developed than oral history archives, they require at least one column to indicate genre or broad categories. The column itself suggests a degree of thesaurus control; one should have a master list to limit the type of entries. In general, this list should be developed in regard to the makeup of the archive's collections, and for processing speed no more than ten categories are recommended. Although not totally inclusive and often maddingly overlapping, some frequently encountered genres include:

Autobiography or memorates (of events as experienced by the individual)

beliefs and superstitions

biography

community life

folktales

jokes

legends (told as true but not personally experienced)

recipes (or domestic life)

rhymes and riddles

songs and music

work habits

If concentrating on one genre, one might choose to use this column to differentiate within the category. For example, songs and music could be subdivided in a number of ways by form of performance or in regard to type – such as ballads, blues, bluegrass, gospel, jazz, minstral, ragtime, and work songs. In addition, one might add keys for place and name to call attention to specific locations or individuals.

While ignored in the first simplified examples, professional folklore and ethnomusical archives frequently also demand a column for the specialized classification systems of folklorists. Again some form of thesaurus control is recommended to specify the systems and references used by the archive. Because of space limitations, the archivist might wish to consider some coding device, preferably mnemonic, to indicate the references on the form: for instance, Stith-Thompson motifs (Thn), Child ballad number (Chd), Lehman/Archer Tylor riddle method (L/T), Wayland Hand's superstition references (Hnd).

Though rather simple in its appearance, the contents column can prove difficult. In coordination with the time sequence and genre, the idea is to give a brief schematic of the interview and perhaps an indication of specific items of interests, like place or individual names. One would hope that the interviewer has used the archive's form and provided it with a detailed outline, perhaps locking references to specific items. If that is the case, the archive's job is merely to review, edit for consistency with the rest of its holdings, add in any additional references, and prepare a final typed copy or enter the information into a data base. If not, one may see if the interviewer has followed an outline of questions (a normally recommended procedure) and attempt to use that document as a guide. At the worst one may have to sit through the tape for two or more sessions: the first to take general notes, and the second or additional times to record the format. Whatever the scenario, a final review of the tape and worksheet is quite valuable.

As Figure 4:9 shows, the worksheet can also be easily expanded to contain other information. The illustration, for example, has a section for technical information as well as a legal release.

A completed, typed worksheet is then placed in an appropriate folder within the project file. For manual systems, a duplicate copy of the worksheet should be made for daily use and the original preserved from contact. The worksheets themselves will serve as major finding aids for the contents of the tape and for the makeup of individual sessions. In addition, the forms are the basis for the production of auxiliary files – for example, indices for songs, jokes, and fairytales.

TRANSCRIPTION-LEVEL PROCESSING

A completed worksheet may also provide an outline to the contents of transcripts by simply replacing the time sequence with page numbers; moreover, the form itself can be used for the transcription. Again, throughout much of the early development of sound archives, and still in many oral history collections, the transcript has generally been regarded as the product and predominates in most accounts. Yet, it is intellectually better to view transcription as only a final level or refinement in the process.

Because the transcription of tapes has been so well covered in the literature, this section will be short. Interested readers are directed to some of the standard works on this topic: for example, Willa Baum, *Transcribing and Editing Oral History* (1977), Cullom Davis, Kathryn Black, and Kay MacLean, *Oral History: From Tape to Type* (1977), and, in particular for this current study, Edward Ives, *The Tape Recorded Interview* (1980), David Lance, *An Archive Approach to Oral History* (1984), and William Moss, *Oral History Program Manual* (1974).

In general, the goals of good transcription are to produce a document that is (1) faithful to the tape and (2) understandable by the reader. Yet, a perfectly accurate transcript of every sound and nuance within the interview is impossible. At the most obvious, hence most overlooked, level, we must recognize that we filter out most extraneous background noises as part of the process of communication. The reader should also be aware that transcribing is a multi-stage process. Efforts begin by listening to and typing the information in rough form. The initial transcription then is given to an editor, who checks for accuracy and transcription of typographic errors, and also verifies and corrects spellings. The editor may also wish to add explanatory material in brackets to aid the reader. When the editor is finished, the manuscript is often taken to the interviewee for review. Finally, it undergoes another editorial examination, and a final typing and proofreading.

With the system in this book, one would then convert the outline and index from the sound level of processing to reflect the page numbers of the transcript. The presence of a transcript should be noted on the catalog cards and other access tools. The final transcript, typed on high-quality, acid-free paper, can be bound or placed in archival quality folders within archival boxes and arranged in a separate location in keeping with its classification number.

Beyond the problems of cost and accuracy, archivists will also need to develop clear policies in regard to such issues as the reproduction of dialects or inclusion of false starts and tag questions ("uh's," "ya know,"). The archivist should also recognize that transcription at its best is an art form, one that requires time and practice and a knowledge of the subject. The art form is heightened when working with individuals from cultures with strong oral traditions. For example, in *Passing the Time in Ballymenone* (1982), folklorist Henry Glassie describes his efforts in transcribing accounts from a small village in Northern Ireland:

My transcriptions are not designed to make stories look like prose or poetry, but to make them look like they sound. To that end the most important device is leaving white space on the page to signal silence. The only odd mark I found it necessary to add is a diamond (◇), indicating a smile in the voice, a chuckle in the throat, a laugh in the tale. Stories enter the mind through the ear, not the eye. Transcribing them, I thought more of music than verse, and I worked them over and over, without changing a word, until I felt no disharmony between eye and ear as I simultaneously read my transcription and listened to the tape.

The following short tale, used to show the wit of the locals in besting outsiders, will serve as a quick example of his technique and also as a close to this chapter.

"Well there was some time
this Tommy Martin
he had been in the army.
"And he
was a wee bit curious at one time:
the nerves give up,
and he was in the clinic at Omagh for a *while*.
"And like many other laborin men that used to get their
pay on Saturday night, he got into a row, and he got
jailed for a while.
"But when at one time the work became scarce and he
seen that he'd have to sign on the dole.
"So he went in anyway, into the dole office. The dole
man started questioning him, and he says, You'r an
ex-soldier, he says.
"I'm an ex-soldier, says Tommy.
an ex-convict,
and an ex-lunatic." ◇

SELECT BIBLIOGRAPHY

American National Standards Institute. *American National Standard for Bibliographic Information Interchanges on Magnetic Tape* (ANSI). New York: ANSI, 1976.

Baum, Willa K. *Transcribing and Editing Oral History*. Nashville: AASLH, 1977.

Davis, Cullom, Kathryn Black, and Kay MacLean. *Oral History: From Tape to Tape*. Chicago: American Library Association (ALA), 1977.

Deering, Mary Jo, and Barbara Pomeroy. *Transcribing Without Tears*. Washington, DC: George Washington University, 1976.

Frost, Caroline. *Cataloging Nonbook Materials.* Littleton, CO: Libraries Unlimited, 1983.

Glassie, Henry. *Passing Time in Ballymenone.* Philadelphia: University of Pennsylvania, 1982.

Gorman, Michael, and Paul Winkler, eds. *Anglo-American Cataloguing Rules.* 2d ed. Chicago: ALA, 1978.

Ives, Edward. *The Tape-Recorded Interview.* Knoxville: University of Tennessee, 1980.

Lance, David. *An Archives Approach to Oral History.* London: Imperial War Museum, 1978.

Lance, David, ed. *Sound Archives.* London: International Association of Sound Archives, 1984.

Mason, Elizabeth, and Louis Starr. *The Oral History Collection of Columbia University.* New York: Columbia University, 1979.

Moss, William. *Oral History Program Manual.* New York: Praeger, 1974.

OCLC. Sound Recording Format. Columbus, OH: OCLC, 1980.

Sound Archives Section. *Sound Archives.* Ottawa: Public Archives of Canada, 1979.

Statement of Principles Adopted at the International Conference on Cataloguing Principles, Paris, October 1961. London: International Federation of Libraries Association, 1971.

Wynar, Bohdan. *Introduction to Cataloguing and Classification.* Littleton, CO: Libraries Unlimited, 1976.

MICROCOMPUTER
APPLICATIONS

This chapter supplements the material on the technical processing of collections and should be read in conjunction with the
preceding chapter. To adequately describe sound holdings is so
complex that the question is generally not *whether* to automate
but *when*. Until recently, however, the costs of an inhouse
system have been prohibitive. The advent of the microchip has
alleviated this situation and has placed computers within the
reach of almost every institution. Before proceeding, we should
note that current microcomputers are not a total panacea. Larger
installations might quickly surpass the capacity of today's microcomputers for some of their operations, and even smaller depositories will have to make adjustments for present limitations.

The following pages concentrate on the use of microcomputers
for the transcribing, processing, and retrieving of information
from audio recordings. In general, the theory and system offered
here are also adaptable to a larger mini- or mainframe computer.
"Micros" are selected primarily for their low cost, but also
because of ease of use, flexibility, and security advantages, as
well as superior administrative control. In an additional benefit
for the archivist, the magnetic storage media for the computer
are essentially the same as audio tape; hence, we will be able to
overlap our conservation considerations with Chapter 6.

On a theoretical level, Marshall McLuhan's observations on

the onset of a new communications medium are more than valid for the microcomputer revolution. This fledgling is certainly "cool" in McLuhanesque terms, for it intimately involves the users in its processes; one cannot be a passive observer in the current state of affairs. As McLuhan predicted, the novel form also borrows heavily from the conventions of preexisting media in order to be more comfortable and comprehensive. But the resulting inertia also means that the capabilities of the new medium will not be fully explored for some time.

INTRODUCING THE SYSTEM

With or without McLuhan, one approaches an introduction to data processing with trepidation. Is it possible to extract a simple explanation from a jungle of nonstandardized terminology and undecipherable manuals? Despite such obfuscations, the actual use of the machines has become quite easy and should become even easier in the near future. The key factor to remember is that a workable automated system must rest on a well-designed manual approach. It should also be recognized that computer analysis is linear and implicitly reductionist; hence, it is far less complicated than one's own thinking. Archivists will need to simplify their conceptualizations into a very schematic and precise mode along the lines of a systems analyst, but also much like speaking to an extremely stubborn and literal-minded two year old.

As an administrator, the archivist might also prepare for some of the typical transitions involved in the introduction of an automated system. (These patterns will eventually vanish with growing familiarity.) Problems will appear with the onset of any new system — for instance, in blending into a preexisting organizational structure and its components, but above all today in regard to personnel. It should also be understood that the incorporation of data processing will progress through predictable stages. The first and most traumatic is one of initiation, which requires planning for a "laying on of hands" and a training period; the second is one of contagion and experimentation, which necessitates oversight while the machine is treated as a grand and almost narcotic toy; finally, experts predict a period of integration and maturity, when the system takes its proper place as a useful tool. Although doubtful that anyone can fully prepare for the onset of a communication's revolution, one can understand that automation calls for a real commitment in time and training and for phased acceptance. Furthermore, one must resist the temptation of overly ambitious commitments at the start.

In addition to the preceding considerations, the beginning planning focuses not on the equipment, but on what one needs and would like to see produced by the system. Outputs should define what hardware (the equipment) and software (the magnetically encoded commands) to seek. Because automation at the collection level has already been addressed through the online networks, the concentration is now on the second (the worksheet) and third (the transcript) levels of processing.

Assuming the readers have at least a tangential knowledge of data processing, we can begin to address hardware requirements. Unfortunately, the volatility of the field allows only the most general recommendations. Less than three years ago, for example, a "state-of-the-art" layout would have included an eight bit (the number of magnetic switches to make a character), forty-eight K (or thousands of characters available for operation within the computer) machine. Today composing on a 16 bit, 128 K micro feels somewhat passé. Archivists will be faced with making selections from an uncertain market among products with very little standardization and great problems in compatibility. Furthermore, any choices will soon be technologically bypassed and will surely be less expensive soon after they have been purchased. Yet the benefits are so substantial that one should indeed dive into the maelstrom.

The basic equipment needed for a sound archive is actually quite modest: the microcomputer (or black box); a keyboard for input, a video monitor for seeing entries; an offline storage device, like a floppy disk drive; and a printer to record in hard copy. This study assumes that specific choices will depend on an archive's analysis of the current market, its finances, future expansion, and, perhaps, the existence or planning for equipment elsewhere in the institution. The constantly reoccurring caution is compatibility, which must be viewed in regard to the relation of each of the elements within the microcomputer system.

The best recommendation for an overall purchase is to remember that one is selecting an entire system and will need to base the evaluation on all the various software and hardware elements available. Yet, the most obvious decision is which computer to buy. The microcomputer market itself is in the process of a "shakeout", with several of the weaker producers already falling by the wayside. Although nothing is certain, two major surviving camps seem to have appeared. The first of these is Apple Corporation – or more precisely for our purposes, Apple with a CP/M disk operating system (an initial layer of software that connects – interfaces – and allows for communication among the computer, its peripherals, and other software: hence, a translator of sorts). The second is IBM with the faster 16 or 8/16 bit MS-DOS operating system.

IBM is forging into the lead. The bulk of new software development is aimed at IBM or IBM-compatible machines, especially in the business-oriented packages that are most adaptable to processing. In addition, the new OCLC terminals are IBM machines and are equipped with dual disk drives to act as an independent microcomputer, when not online. If an IBM-compatible microcomputer is chosen, it should be recognized that one is often only receiving an MS-DOS congenial machine. These run the bulk of textual programs but normally differ in graphics from IBM. Another consideration is that one is not limited to IBM or Apple or their look-alikes, only by the degree to which the machinery can help fulfill output requirements. Many archivists and historians, for example, use Kay Pro microcomputers with their low cost and large storage capacity, and many other brands can satisfy our relatively simple demands.

Although actual selection will require an extensive investigation, several other general factors can be mentioned. In particular, one should consider the amount of processing or random access memory (RAM). The object is to obtain the highest number for the money; a reasonable expectation would be for 256 or 512 K (i.e., 256,000 or 512,000 characters). For those who are thinking about the inhouse networking or linking together of several micros, efforts should be concentrated on the 16 or newer 32 bit architectures. The older 8 bit configurations, although acceptable for the normal range of processing, are limited in terms of networking. In addition, they are slower, have a limited RAM of 64 K, and are less adaptable to password security systems.

A keyboard for entering or inputting data and communications is a normal peripheral accompanying the sale of a microcomputer. Such trappings are often overlooked but should be important considerations in any purchase. Is the keyboard sturdily constructed? Do the keys "feel" right (avoid the flat membrane keyboards)? Are there sufficient function keys and a separated number and/or cursor pad? Fortunately, the American Standard Code for Information Exchange (ASCII) character set has become the almost universal standard for microcomputers. Thus, the user does not have to be concerned with problems surmounting the conversion of numbers of letters into binary code, unless perhaps converting from or to a mainframe. The ASCII keyboard is also quite familiar, as it parallels the normal "qwerty" typewriter board.

For the monitor, one can begin to speak of pixels or picture elements, expressed by the number of horizontal and vertical lines on the screen – for example, the IBM PC's 640 × 320 pixel high resolution screen. The higher the numbers the better the picture, but users will want to sit down at a screen and judge it for themselves. Although many of the desktop models come equipped with a 12-inch monitor, others might have smaller, unacceptable sizes or require a separate purchase. Decisions also include choosing between amber and green hues, or even a black-and-white screen or color, which is a more expensive option, especially for the recommended RGB type. At this time, most monitors are cathode ray tubes (CRTs), but the market is now opening for even higher resolution plasma models. Finally, one should also take physical discomfort and eye strain into consideration in the selection process.

As with each element in a microcomputer system, the compatibility of the printer to the rest should be assured. Printers should be chosen in part on the basis of their reliability, but the major decision will be between the speed and lower cost of a dot-matrix versus the higher quality lettering of the more expensive daisy wheel variety and the newly emerging jet ink printers. In the interest of economy, a good quality dot-matrix is more than sufficient, although one should avoid the cheapest thermal paper type of printers. Whatever the selection, it is desirable to have sample runs from the available machines and to purchase one with both friction and a movable track or sprocket feeder – which will allow the running of catalog card stock and regular paper.

The amount of off-line storage is another significant criterion. Until now the most reasonable mass storage devices for the microcomputers have been floppy disks. Originally developed by IBM, "floppies" are available in several sizes from 8 inches to 3 inches, with the most common at 5-¼ inches. At the minimum one should have two double-density floppy disk drives, which offer 160 K each, or preferably double-sided, double-density drives at 320 K. Even those are somewhat limited, and one will want to investigate the purchase of increasingly affordable hard disks in the 5 to 20 MG (megabyte or million) character range – like that with the IBM XT. In any case, we can expect the amount of external and internal or RAM memory to continue to incrementally expand in the coming years.

Although not essential, an archive might also look into securing a modem (modulator-demodulator) for communication with other computerized systems. These devices are used to convert the signals from the machine into wave lengths that can be carried over telephone lines and then reconstituted for another computer.

The first element for the microcomputer, however, is its software. Many experts suggest that software selection is more important than the choice of hardware and, indeed, that one should select the software first. It should also be mentioned that users have the option of programming the system themselves. However, the expertise needed and the time involved for programming, as well as the power of packaged programs, make this option impractical for most small or mid-sized operations. In addition, off-the-shelf software offers the luxury of "factory" support and often of informal user networks that one can contact for help and later embellishments.

With the exception of Ted Durr's first MARCHON, no readily available commercial software is even vaguely designed for sound archives. In the jargon, few "turnkey" programs exist that can be plugged in and with little addition be considered as descriptive for a sound archive. Still, as is discussed in the following sections of this chapter, one can easily adapt existing work processing and data base management systems (DBMSs) or, perhaps, spreadsheets to one's needs and output.

Because the market is so unstable and computer costs decline by an estimated 14 to 16 percent each year, one is somewhat leery of giving price estimates. Still in 1984 dollars, an adequate system—consisting of a 16 bit, 256 K microcomputer; a 12 inch amber monitor; a dot-matrix printer; one floppy disk and a hard disk drive; plus word processing and DBMS software—can be realistically purchased for $5,000. (With some shopping that figure can be reduced to as low as $2,000.)

TRANSCRIBING AND WORD PROCESSING

Most users enter the world of microcomputers through word processing, using the system as an expensive typewriter. Because one of the basic outputs for sound archives is transcription, such an implementation dovetails quite nicely with our purposes. In fact, it is now impossible to recommend the inception of any transcription efforts without word processing; word processing simply offers too many advantages. Through its capacity to store and reuse the initial keyboarding, automated efforts can reduce transcription labor (the major cost) by 40 to 75 percent. The entry speed for final composition is elevated beyond any manual and many electric typewriters. Furthermore, because essentially the same data are used throughout the reviewing and editing phases, the final text should theoretically be cleaner of errors and the omnipresent typo.

The selection of the single best word processor for transcription is as problematic as the choice of the best hardware. In terms of cost and flexibility, users are well advised to ignore the more expensive stand alone word processing in favor of a microcomputer. Current word processing software for microcomputers is amazingly powerful and easy to use, even by comparison to the text-editors used on mainframes less than a decade ago. One obvious factor is whether the package will work with the computer and printer. Another is the ease of use and learning; in the beginning, one will probably want a menu-driven program with helpful prompts on the screen. As with many of today's data processing packages, one can expect to become fairly sophisticated in short order–perhaps even outgrow the capacity of the initial selections–but certainly engaging in practices that were not originally envisioned.

Before choosing, the archivist should be alert to some of the normal "preter-typewriter" capabilities and characteristics of word processing. Video displays should allow for a full 60-to-80-character column width and a "memory-mapped" system, which immediately enters the data onto the screen and allows continuous typing without carriage returns. One will want to insure that the program permits such normal chores as automatic scrolling (moving rapidly forward and backward through the text–much like physically "paging" a book). The typical software has the ability to move blocks of information and contains methods for the simple deletion and correction of text. Because the length of a one-hour interview approaches forty pages of double-spaced typescript, the software must also be able to handle large selections of material. The initial word processor provided with Apple's otherwise excellent MacIntosh, for example, could not manage an individual interview on one file. Other programs offer tradeoffs between the highly desirable automatic creation of a backup file and the concomitant loss of storage space for that procedure. Finally, the program should facilitate easy copying and printing, as well as the transfer of other files–like those from a DBMS–into the text.

Because many of the existing packages meet the minimum criteria, the decision may rest on other factors, such as the reputation of the program and its popularity or ubiquity elsewhere in an institution or the field. Other major considerations are the additional features and utilities offered with the word processor. Moreover, these enhancements further establish the superiority of word processing over manual transcription.

Many of those who employ automated entry, however, concentrate so heavily on the "typewriter-function" as to overlook some of the byproducts of digital encoding. This tendency is consistent with Marshall McLuhan's observations that the potential of a new medium is partially defused by the initial need to closely mimic its predecessor. One of the most powerful and attractive features in a word processor, for instance, is the ability to perform "string" searches and replacements – to find a given word or phrase in the file and automatically replace it. With regard to an oral history or folklore program, this capacity should suggest an unprecedented potential to retrieve this information from a text. In particular, we can automatically scan through files by the name of an individual or place and flag that data for inspection or word count. In addition, this feature permits the ready correction of misspelled terms and assures an easy coordination with the authority files.

The preceding features are further enhanced by the presence of a dictionary or spelling-checker utility. This option maintains a dictionary of standard words and will compare that file to one's entries. By using the Spellstar option of WordStar (the most popular word processing program), for example, one checks against an initial list of 20,000 words. The program will tell the number of words in the source document, the number of different words used, the number of dictionary words encountered, the number of terms outside the dictionary, and the number of misspellings. Beyond these aggregate statistics, the program identifies each of the misspelled words and those not in its dictionary. Moreover, one can add to the main dictionary or create a supplemental file of terms. The ramifications here are immense, including the prospect of automated thesaurus control.

Even more exciting is the prospect of automatic voice encoding from the tape recordings directly into the computer. Although the success rate of futurologists gives pause, experts do predict that affordable voice-actuated systems will be on the market by the 1990s. These will be able to produce a rough draft of transcript but will probably require human intervention on homonyms (e.g., to, too, two) and at other more complicated levels of analysis.

PROCESSING AND DATA BASE
MANAGEMENT SYSTEMS

Until the advent of the black box of automatic transcriptions, the outlining or second level of description should occupy the prime attention of the cost-conscious archivist. This information can be accommodated by some of the better word processors but is better addressed by a data base organization. The advantages of this type of automated approach become especially evident during later researching and reference work. In particular, it allows the archivist more rapid access to the information within the collections and eases comparisons among individual holdings.

DBMS Overview

While obvious, the data base concept and DBMS are still so novel as to require a brief overview before proceeding to direct applications for sound archives. The idea behind the concept is to centralize data in one location and then allow multiple users to address, copy, or reconfigure the information to their own purposes. Such an elegant and simple construct reduces redundancy and the errors inherent to recopying. DBMSs are software to facilitate this storage. In particular, they allow the user to make a form on the screen and, thus, in essence to duplicate a manual form such as the worksheet. But unlike the manual example, the data from the screen format can be manipulated and automatically called forth or transferred.

DBMSs are roughly divided by the manner in which they organize the data for storage (see Figure 5:1). The most complex of these architectures is a network system in which individual nodes of data can be interlinked in tinkertoy fashion to other nodes. One next level down are hierarchies in which data are individually linked as in an organization chart or inverted tree structure with only one path of ownership. Hierarchies and networks, or CODASYL programs, are particularly powerful and well suited to large data bases and the structure of most archival collections. But they are also quite complex, requiring a knowledge of their programming language and, in most cases, probably demanding an outside expert for installation.

If some of the DBMS architectures are beyond the neophyte's skills, others can still be easily applied. The simplest of these is a sequential storage in a file management system (FMS), which is normally constructed out of a simple menu. As illustrated in Figure 5:2, a menu-driven system will begin by offering a number of choices. To create a form, one would press the

Figure 5:1
DBMS Architectures

1. Linear-File Management System: Oral History Session Screen

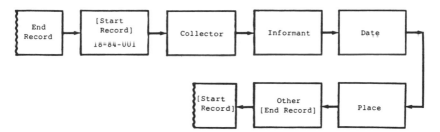

2. Matrix or Relational DBMS: Folklore Contents Screen

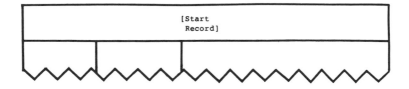

[Start Record] 15-81-002		
Time	Genre	Contents
0:00	Joke	People understanding a DBMS
0:50	Song	"Ode to 8086"
4:10	Legend	Charles Babbage and the Holy Grail
8:30	Memorate	Start of the microchip
11:20	Belief	The big computer in the sky
59:10	Riddle	What is a micro
	[End Record]	

[Start Record]		

Figure 5:1 (continued)

3. Hierarchy: Folklore Contents Screen

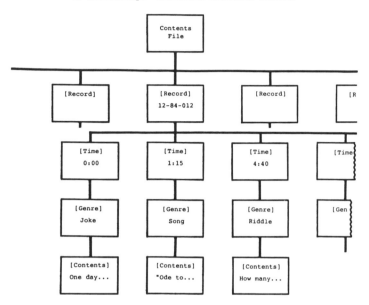

4. Network: Session Screen Relations

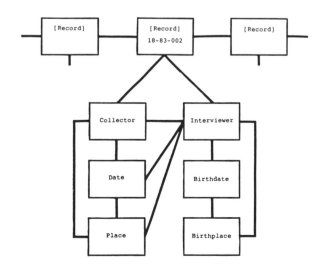

Figure 5:2
Menu-Driven DBMS – Personal Filing System

---

1. DESIGN FILE	4. SEARCH/UPDATE
Create File	
Change Design	
2. Add	5. PRINT
	Print Forms
	Define Print Spec
3. COPY	
Copy Design Only	
Copy Selected Forms	6. REMOVE
Copy Whole Diskette	

number indicated for that command and then proceed to delimit the fields – much like blocking out and placing the headings on a typewritten form. After the screen format is finalized, one is ready to enter the information in the assigned fields or later to perform any of the other tasks available through the menu.

An advanced FMS can handle all of the basic processing chores for a sound archive, but it is limited. For example, it cannot deal with several data bases at the same time. Thus, most archives should probably concentrate on the more powerful relational data base management system (RDBMS). RDBMS formats are created by a list of brief commands, indicating the length and type of the field. Introduced only in 1970 by Dr. E. F. Codd of IBM, the relational approach has quickly risen to prominence, for as David Kruglinski states in *DataBase Management Systems* (1983):

The main reason for the success of the RDBMS is its simplicity. Easier to learn than other data bases, RDBMS allows you to build complex systems one step at a time. The name comes from a mathematical concept called a *relation*, which is simply a table. This table is stored in the computer as a file, or data base, where the records are horizontal rows, and the fields are the vertical columns.

Before addressing specific applications, the text will quickly examine some features that one might examine in choosing a DBMS. Again, the first step is to analyze one's framework – in this case, the size and format of the collections and the worksheet. No current software is perfect, and the archivist must balance costs in regard to a number of variables. How many records will it manage? What is the maximum length of those records? How many fields can be defined within one record and how long can those fields be? What are the system requirements in terms of hardware, operating system, and storage? And what is the speed in regard to access time – the period between a request for data and its appearance – and also for processing large numbers of records?

If the package is acceptable to this point, one may either buy or proceed to a more in-depth analysis. What type of DBMS is it – network, hierarchy, relational, or file management – and is it menu, command, or program driven? Because most of their records are textual, archivists will want to see how the software deals with alphabetic information: in particular, the availability of string and key word searches and of Boolean operators, as well as the possibility of asking simple "yes or no" and "true or false" questions. One extremely valuable talent for more sophisticated ventures is a program's potential for making secondary "user friendly" menus or screens to aid researchers and staff in later endeavors. Certainly, the quality of the documentation supplied or available elsewhere must not be overlooked. For example, the original manual with dBASE 2 – the most popular RDBMS – seemed to require an Egyptologist to decipher, but more than thirty supplemental monographs and a built-in tutorial are now on the market for the perplexed. If the documentation is incomprehensible, one should at least reflect on the purchase.

Borrowing from Kruglinski, some of the more technical features to look for include a data dictionary, which describes the format constructed for the files in the data base. In addition, an uncomplicated query function is essential, so that nonexperts can easily use or enter material. The ability to restructure the data base is another key element, as is the capacity for adroitly correcting errors. One will also want a program that protects against total system failure and that facilitates the creation of reports and the incorporation of data from other programs. On the highly desirable side, one will seek software that can work with more than one file at a time. And although nonnetworked microcomputers assist in security control, password or other privacy features are also increasingly recognized as important. It may also be advisable to select a program that can be used to operate on more than one terminal at a time, although that may presuppose a different operating system (e.g., UNIX).

DBMS Applications

Let us move from these technical considerations to direct applications for the second level of processing: in effect, to the automation of the worksheet. This discussion is intended only to initiate the noncomputer "jocks" into the primary levels of description and retrieval through a DBMS, which will be treated for the simplest file management techniques. Because of the wide variety of programs and their command structures, this exposition merely describes general features. Thus, the implementer is left to construct individual setups based on the unique makeup of each institution and the DBMS selected. Such setups do effectively frame future efforts, but it should be noted that a good DBMS will permit a later reconfiguration and addition of data fields.

As projects such as the Archive of Traditional Music at Indiana University have already discovered, the information from a typical worksheet can often be better manipulated if divided into two screen formats. Such a division is also a practical way of surmounting the size and organizational limitations in some of the DBMSs. The first, or Session Screen, parallels the top section of the worksheet (see Figures 4:8 and 4:9), including information on the participants and location as well as the capsulated description. The Contents Screen concentrates on the outline of the interview that is captured on the lower part of the form. Figures 5:3 and 5:4 demonstrate these formats, whereas Figure 5:1 illustrates the underlying differences in arrangement behind this facade for file management, relational, hierarchy, and network DBMSs.

In a quick hint for those tempted to adapt an unadorned FMS, one might have to demonstrate some creativity to replicate the matrix on the contents section of the worksheet. The Personal Filing System (PFS), for example, has an additions page that allows for the introduction of unstructured material. This page cannot be formatted for columnar entry, but one can enter the data as though it were in columns. That pseudostructure will allow for searching through the key textual material, and one still has the possibility of coordinating for time and genre. Another possibility for columns is to resort to a spreadsheet, like Lotus 1, 2, 3.

Other DBMSs will more easily mimic the original manual form and facilitate automatic entry into the fields. But the screen format for a DBMS is not as flexible as a paper form; for example, one cannot squeeze more information in by writing smaller, nor can marginalia be added anywhere on the form. Many of the early packages are also very limited in dealing with textual material, a tendency that is now being somewhat reversed in the new generation of dBASE 3 and Knowledgeman. Whatever the package, it will demand careful planning and layouts which should take into account the logical paths that a computer system might follow for retrieving data. In particular, the screens should be headed by a distinct number or code, such as the tape number, to ease identification and speed searches. Although screen formats should closely parallel their manual counterparts to simplify data entry, the data may have to be locked into more specific categories and perhaps be slightly rearranged: for example, distinct blocks for street address, city, state, and beginning with the last names (searching initially under a first name is generally useless).

In addition, decisions must be made in regard to the maximum length in characters for each field; perhaps standardizing at fifteen characters for last names, first names, cities, and twenty or more for a street address, and using a two-digit abbreviation

Figure 5:3
Session Screen (Oral History)

```
Tape:

Collector (Last):              C.First:

Project:  Y/N                  Title:

Informant (Last):              I.First:

City:                          State:

Subject:
```

Figure 5:4
Contents Screen (Folklore)

```
Tape:

Time      Genre              Topic
```

for state. Of course, if any abbreviations are used, some control must be exercised, and a distinct pattern may be established for entering dates. (With a number of the DBMSs, this pattern is already built in and some can automatically convert almost any form to the typical numeric fashion of mo/dy/yr, e.g., 05/20/48 for May 20, 1948.)

One final caution is to remember how literal-minded the machines are. For although we can distinguish easily between different dates and places by position, the program will normally assume that anything designated as a date or place is in fact an equivalent. Items that are similar but of a different relational value must then be distinguished on the screen. The date of interview could be "In-date" and a birthdate "Bir-date." Note, too, that searching generally proceeds from left to right through a word, and some packages will prefer to search on only the first three or four letters in a field designator. Thus, terms like birthdate and birthplace or interviewer and interviewee should be avoided in the same screen or record set.

Until this point, the efforts have merely replicated those for manual forms. Additional benefits begin to appear through addressing a spelling checker or separate thesaurus software control and the identification of new terms. Obvious advances also emerge from the indexing capacity of the DBMSs, when searching through unit or multiple records. Such applications allow for the production of permutated or catchword indices, listing all of the important words in the context in which they appeared. When placed in order, the archivist has an immense file of access points into the holdings that realistically exceeds the capacity of any manual catalog. Moreover, such an index can be delimited around certain terms and further defined by the use of Boolean operators (and, or, not) to focus on very specific components and relationships. The end results of such a search may then be arranged for printing as a finding aid or stored for later use as a special search package. The online format surpasses the capacity of paper- or card-bound searches and is more in keeping with the strengths of the computer medium.

The automation of the worksheet can also substitute for collection-level description. Although the adoption of the MARC format is strongly recommended for the initial phase in description, that recommendation is based on MARC's standing as the international bibliographic standard. Beyond that and its serviceability for exchanging cataloging information and the like on published materials, MARC is a somewhat demanding and slightly old-fashioned technique. For example, it calls on the archivist to refer for approved pre-designated access terms, whereas the nature of archival collections favors a post-coordinate construction of terms drawn from the *sui generis* nature of each archive. Thus, the archivist focused within an institution and unconcerned with exchanging information can avoid traditional cataloging; instead the worksheet alone may be automated for online retrieval. At the same time, such a direction certainly would remove the archive from ties with the bibliographic world of the library.

Microcomputer Catalogs

In a similar vein, it may be well to view the prospects of dedicated library cataloging software for microcomputers. At present, however, it is almost impossible to recommend any of the packages. A survey of the numerous purveyors at recent American Library Association meetings, for example, reveals that the costs for these systems and their support is much higher than for comparatively powerful generic DBMSs. More importantly, most of the present products are designed for such specific library chores as circulation systems, book orders, serial control, and shared cataloging – which are not of use in a sound archive. The cataloging modules are also often quite limited in regard to the number of fields, size of the record, and number of records they can easily manage. For instance, Follett Library Book Company with Book Trak, one of the best Apple based systems, can handle no more than 250 records of less than 600 characters each on one diskette.

This situation is destined to change because of ongoing technological and software improvements. Optical disk technology is certainly destined to remove many of the current storage restraints, which hamper microcomputer applications. Storage capacity will likely increase by a factor of ten over present hard disks. Moreover, the intensive marketing of these devices has already begun, as witnessed by the presence of more than a dozen companies in this line at recent conferences such as the Association of Information and Image Management and the Association of Records Managers and Administrators. Although numerous questions still remain with regard to costs and conservation and especially to securing a disk that can be directly encoded inhouse, optical data disk devices (OD3) will likely soon play a major role in all information agencies. (Note that the University of Maryland is currently embarking on pilot projects in order to ascertain the feasibility of using this technology tied to a microcomputer interface and an interactive encyclopedia with touch sensitive screens for displays, instruction, and information retrieval.)

Generic software also has potential for library cataloging on the microcomputer. Some of the current DBMSs could be adapted to replicate MARC recording, hence, conforming to standards and making it possible for nonusers of the online systems to more easily communicate with their fellows. More importantly, the presence of the newer "integrated" software (e.g., Symphony and Knowledgeman), which facilitate the manipulation of the same data through different word processing, spreadsheets, and DBMS software, offers the future possibility of a unified approach to the description of archival collections. Thus, instead of the present "bifurcated" norm of a library catalog approach and archival finding aids, the archivist may be able to unite all in one computerized packet.

Yet even the best of the integrated and library cataloging packages are still too limited for such purposes at the microcomputer level. Breakthroughs, however, are certainly on the horizon: for example, the potential offered by MARCON II of Ted Durr and his AIRS Company. That software appeared in mid-1985 as one of the first to truly manage text with a high level of information science techniques on the microcomputer. Significantly, MARCON II is projected for a MARC conversion module and offers the ease of a menu-driven approach with the superior multiple file capability of a relational data base system. If the pattern of development in computer and information science holds true to form, the reader may expect to find other such textually oriented products on the market by the time this work is out of press.

MANAGEMENT, RESEARCH, AND OTHER APPLICATIONS

In addition to automating the worksheet, reference work, and transcribing, the microcomputer can leave its mark on other aspects within the sound archive. Although the field is still in its infancy in regard to automation, the range of possibilities is immense and is limited only by the imagination. For example, we can look quickly into some of the implications for administration and research.

In terms of administration, one must first take into account the size of the operation and frequency of use. For as size and uses increase, so too does the applicability of data processing. The danger is in overloading the system or becoming so entranced as to apply it in situations where other forms are more suitable. Thus, one could automate the circulation controls for a small sound archive with currently available turnkey software; however, this is hardly justifiable if one is only receiving a handful of patrons a week and not loaning materials. Or the processing log could be placed online for administrative control, but unless it is a large operation with portable machines and bar coding to move along with the collection, the efforts are perhaps better confined to a paper medium.

The preceding applications could prove beneficial for some institutions and produce valuable byproducts for administrative control. The automatic gathering of use statistics can show which collections are being used as guides for conservation purposes and aids in later acquisitions, as well as indicate which need more in-depth description or transcriptions. These and similar indications help support another McLuhan dictum—"data accumulation produces pattern recognition," that is, the accumulation of a large body of information opens the possibility of discerning patterns or relations, which were not easily seen. In simpler terms, the computer can help one to see the forest because it can organize the trees.

The archivist as an administrator will also become aware of some of the other advantages of a microcomputer system. DBMSs, for example, are very useful for budgeting and any accounting purposes. The word processor or DBMS can also maintain mailing lists and help target specific individuals for solicitation or communication and like purposes.

The addition of a modem with the appropriate software permits informal networking between similar institutions. It is hoped that in the near future a telephone bulletin board will be established to allow archivists from around the country to exchange information and make inquiries from their fellows. Such a facility may further obviate the need for MARC entries, and brings to mind McLuhan's concept of a media-shrunk "global village." Moreover, a modem provides a way to enter communication with a mainframe and the intriguing prospects of access to the online information data banks, like Dialog with its broad bibliographic and source files, into a wide variety of subjects.

In addition to duties as an administrator and information manager, the typical archivist is also a scholar. Drawing on the theoretical implications of McLuhan's pattern recognition, some sound archivists are beginning to investigate the use of their encoded data for research. The first efforts in computer-assisted studies for the humanities are normally dated to 1949 and Father Robert Busa's concordance for the eight million words in the works of Thomas Aquinas. In the same year, Bertrand Bronson proposed an automated approach for musical description in regard to the scale, mode, meter, and melodic contours of Child ballads.

Because of their interests in texts, music, and classification, folklore archivists have far outstripped their oral history counterparts in the research applications of data processing. Rather than elaborate on the various lemmatization, cluster analysis, and other manifestations in the field, the reader is directed to the pages of the journal *Computers and the Humanities* or the work being discussed in the ongoing conference series Folklife and Automated Records. This series is sponsored by the Skaggs Foundation and is under the direction of Richard Thill, with the proceedings from the first session scheduled for publication in the near future.

DATA INTEGRITY

The Folklife and Automated Records conference also addressed some of the basic techniques for data processing in the management of collections, but both the conference and much of the literature avoid questions of data integrity. In the almost Neanderthaloid era of the keypunch machine, one could physically examine the cards for damage and even double-check on accuracy through a re-keyboarding on a verifier. Today that level of review has vanished, and the data seem somehow more ephemeral in their magnetic persona. Certain definitional procedures do exist to guarantee that alphabetic, numeric, and chronological data are entered in appropriate fields. Furthermore, parity or other check bits can be added to see if information has been lost. Yet none of those measures insures the validity of the document. Instead, the overwhelmingly textual content of the files causes the archivist to fall back on the tried-and-true visual review of the editor – but with the added luxury of the spelling checker.

Safeguarding of the materials also extends to the physical care and maintenance of the magnetic storage media. Here we must recognize that the digitally encoded material on the disks is more fragile than the analog variety on sound tape recordings. Digital storage is more densely packed with the dropout of a single byte potentially damaging, whereas a concomitant loss on audio tape would go unnoticed.

Precautions begin when entering the data on the microcomputer, for information should be periodically saved to disk or off-line storage during input. The danger is that a sudden surge or loss of power (as in a storm or blackout) will wipe out all of the data from the RAM or temporary internal memory. The basic rule of thumb is redundancy with duplicate copies of all valuable programs and data disks.

The floppy disks themselves should be carefully handled to preclude finger contact on their surfaces. They need to be stored in an upright position within individual envelopes and in nonacidic containers. Unlike some of the cheaper audio tapes, which are occasionally salvaged from old computer tapes, the quality of the disks is generally fairly good. Like tape cassettes, however, completed floppies can be protected against accidental erasures. For the floppies this step consists of placing a strip over the small cutout in the upper portion of their holders. The resultant "write-protected" disks can then only be accessed for their information, but can still serve a variety of purposes after processing, including potential as an inhouse card catalog.

Even beyond the level of care assigned to sound tapes, one should be alert to the perils of static electricity and power surges. Hence, surge protectors and antistatic rugs are advised. In addition, stray magnets, even in the form of temporarily magnetized bobby pins, may wreak havoc. Yet, as will be discussed in the following chapter, magnetic storage media are remarkably tough. For instance, such storage is impervious to depredations from ultraviolet light and more resistant to water damage than paper products. Yet encoded information also presupposes a mechanical interface; one cannot read a tape without the proper hardware and software the way one can read a partially damaged book. Another similar problem is the present dependence on particular types of hardware and software. Such a particularistic reliance in the context of rapidly evolving products implies numerous dangers for the future. In essence, the archivist must be wary, employing paper backups and the precautions already mentioned, as well as those defined in the next chapter.

Rather than a narrative summary, the following list provides a schematic to some of the possible uses of the microcomputer for sound archives. Although most of these uses have been touched on, some—like the potential of executive decision software for appraisal or the use of word processing for editorial projects— remain to be explored.

General Administrative Uses
  Inhouse management
  Promotional ventures
Academic Scholarly Uses
  By archivist
  By researchers
Collection Management Applications
  Appraisal standards (decision-support software)
  Acquisitions and solicitation
  Processing
    Worksheet level (DBMSs, spreadsheets)
    Transcription level (word processing, thesaurus control)

Retrieval
  Augment the production of manual finding aids
  Online catalog
  Monitoring for conservation (tape management software)
  Documentary editing and publication
  Communication with other repositories
  Future prospects: e.g., automated transcription and the unforeseen.

SELECT BIBLIOGRAPHY

The literature on microcomputers alone is too volatile to accurately capture or make recommendations beyond the studies listed below, which most influenced this work. In addition to the maze of current articles in *Byte, Modern Computing,* and other journals, the specific manuals should be consulted for their hardware and software. For computer applications in scholarship, see the journal *Computers and the Humanities.*

Bronson, Bertrand. "Mechanical Help in the Study of Folk Songs." *Journal of American Folklore* 62 (1949):182-87.

Busa, Roberto. *La Terminologia Tomistica dell'Interidrit.* Milano: Fratelli Bocca, 1949.

Byers, Robert A. *Everyman's Database Primer.* Reston, VA: Ashton-Tate, 1982.

Frost, R. A., ed. *Database Management Systems.* New York: McGraw-Hill, 1984.

Geda, Caroline, Erik Austin, and Francis Blouin, eds. *Archivists and Machine-Readable Records.* Chicago: Society of American Archivists, 1980.

Hockey, Susan. *A Guide to Computer Applications in the Humanities.* Baltimore: Johns Hopkins University Press, 1980.

Kesner, Richard, ed. *Information Management, Machine-Readable Records, and Administration.* Chicago: ALA, 1983. A convenient bibliography on automation accounts for archives.

Kruglinski, David. *DataBase Management Systems.* Berkeley, Calif.: Osborne, 1983.

Lincoln, Harry, ed. *The Computer and Music.* Ithaca, NY: Cornell University Press, 1970.

Lucas, Henry. *The Analysis, Design, and Implementation of Information Systems.* New York: McGraw-Hill, 1981.

Lytle, Richard. "Intellectual Access to Archives." *American Archivist* 43 (1980): 64-75.

McCrank, Lawrence, ed. *Automation of the Archives,* White Plains, NY: Knowledge Industries, 1982.

McLuhan, Marshall. *Understanding Media.* New York: McGraw-Hill, 1964.

McWilliams, Peter. *The Word Processing Book.* New York: Ballantine, 1982.

Soergel, Dagobert. *Indexing Language and Thesauri.* Los Angeles: Melville, 1978.

Thill, Richard. *A Basic System for Creation and Processing of a Folklore Archive.* Bloomington: Indiana University, 1978.

Thill, Richard, ed. *Washington Conference on Folklore and Automated Archives.* In publication review and anticipated in early 1986.

CONSERVATION
MANAGEMENT

To be of any use a tape must be played, yet each play helps to destroy its contents. This tension provides the basic dilemma in sound archives, for how can we listen and still preserve our collections? In fact, sound archiving today is full of tradeoffs between ideal goals and practical realities. Even with compromises, however, we can reach a very adequate state through the acceptance of common-sense practices, the production of master and user copies, and the employment of modern air-conditioning and filtering devices.

The ideal physical environment for tapes, for example, is a relatively dust-free room with a constant temperature of 70° F (+ or − 10°) and a constant relative humidity of 45 percent (+ or − 5%). As reported in *A Care and Handling Manual for Magnetic Tape Recording* by the joint government and industry Tape Head Interface Committee (THIC), other general recommendations include:

1. Operating areas should be kept at a slight positive pressure (0.5 lb/in²).

2. Air-conditioning systems should be equipped with electrostatic air filters.

3. All dust-collecting surfaces should be damp wiped (not dusted) periodically.

4. An operating area with a raised floor should have air-conditioning fed from the ceiling and exhausted from the floor. This will remove floor dust instead of blowing it up around equipment.

5. Floors should not be waxed.

6. Vacuuming, if done, should be with the collecting canister and exhaust outside of the area.

7. Floors should be damp mopped, not swept.

8. Buffing machines may be used to clean the operating area but should be restricted to use only during nonoperating hours, if possible. Steel wool pads should never be used with buffing machines.

Such procedures are particularly aimed at eliminating the dangers of particle contamination. Although it is not feasible to follow other suggestions for even a class 100 000 "clean room" (the three general classes are 100 000, 10 000, and 100, with the last and cleanest suggesting no more than 100 particles of 20 microinches per cubic foot), still the sound archivist should seek to reduce contaminants. In addition, experts also suggest minimizing direct contact with human fingers on tapes (use lint-free gloves) because fingerprint oils serve as excellent traps for airborne dust and other contaminants. Similarly, eating and smoking in storage or playing areas is to be forbidden.

The THIC Manual as well as common sense also prohibit introducing stray magnets and recommend following manufacturers' instructions on equipment maintenance – especially cleaning tape transport surfaces (guides, rollers, and record/reproduce heads) at frequent intervals with alcohol or Freon TF solvents. In terms of theory, the reader should note that conservation in these terms extends beyond a focus on the storage medium to include the mechanical interface. Equipment and tapes should be purchased and maintained with the idea of prolonging the life-span of the recorded information.

The most important technical consideration at this time is to avoid any rapid cycling or change in temperature and humidity. Even a single degree or percentage alteration in either factor can change the length in a standard 1200 reel by as much as a foot.

TECHNOLOGICAL FRAMEWORK

July 18, 1877, is given as the official date for the invention of sound recording. Then, Thomas Edison, borrowing from the theoretical work of Charles Cros and Hermann von Helmholz, used an acoustically modulated stylus to cut a "hill and dale" groove on a tin foil cylinder. Even in the first steps, preservation emerged as a problem, for foil allowed very few playbacks. Edison temporarily lost interest in recording until the late 1880s, when Emil Berliner began the successful mass production of phonodisks. This challenge, together with demands from the marketplace, helped spur a number of refinements in both cylinder and disk technology.

The next major historical breakthrough would wait until 1925 and the research of Joseph P. Maxwell and Henry C. Harrison at Bell Laboratories; they were able to adapt the vacuum-tube amplifier to electrically, not acoustically, record on disks. By the early 1930s, the technological flexibility offered by electrical recording opened new horizons for the chronicling of ethnographic and musical performances. These "instantaneous" records used an acetate or nitrate lacquer cover over a thin, normally aluminum disk. While producing high-quality sound duplication, the malleable nature of the acetate or nitrate covers did present significant problems for preservation. In addition, though portable by the standards of the era, the equipment was still rather bulky. In *The Adventures of the Ballad Hunter* (1947), folklorist John Lomax described the setup for his landmark expedition for the Library of Congress in the early 1930s:

As a crown to our discomfort, we also carried a 350 pound recording machine—a cumbersome pile of wire and steel built into the rear of the Ford, two batteries weighing seventy-five pounds each, a microphone, a complicated machine of delicate adjustments, coils of wire, numerous gadgets, besides scores of blank aluminum and celluloid disks, and finally, a multitude of extra parts, of the purpose and place for which neither Alan [his son and also a noted folklorist] nor I had the faintest glimmer of an idea.

Despite subsequent reductions in size, the needed portability and flexibility for field recordings did not come from disk but from magnetic technology. Initially demonstrated in 1898 by Danish inventor Vladimir Poulsen, early work centered on wire recording. The locus for most of the research in the field was in interwar and World War II Germany under the leadership of Kurt Stille. After the war, Americans were able to expropriate much of this Magnetophone knowledge and apply it to both analog and digital recording. A prime contributor would be the Ampex Corporation, which from its birth in 1948 would stimulate the development of commercial recordings and then the now familiar reel-to-reel recorders. By the mid-1950s, these recorders were sufficiently refined to provide the necessary technological underpinnings for field recordings.

In spite of the advancements and longings of many archivists for a reel-to-reel universe, the major technical innovation and popularizer for field recordings emerged later from the tape cassette. Introduced in the mid-1960s by the Dutch Philips Company, the cassette format through its truly portable composition and relatively inexpensive cost revolutionized interviewing and opened the veritable floodgates to amateur and professional interviewers. Once again the innovation was not without difficulties, including special preservation considerations with limited frequency response, annoying machine background noise and tape hiss, and declining audibility with each new generation of re-recording.

Jerry McWilliams has described the underlying process for magnetic sound recording in *The Preservation and Restoration of Sound Recordings* (1975), a work that represents one of the best introductions to preservation for all sound media:

Magnetic recording, whether on wire or tape, operates on the principle that sound waves, converted to a changing electromagnetic field by microphone amplifiers, and an inductive coil (or head), will magnetize iron oxide particles on a moving tape (or steel wire or band) in a manner analogous to the original sound. These magnetized particles (called domains), when moved past a receptor head, will create electrical signals that can be converted back to the original sound. Although the principle is simple, several important signal-processing stages are required to ensure linear distortion-free results. These include the application of a high frequency current (bias) and several stages of equalization, one in recording and another in playback.

Beyond the historical developments, the archivist or librarian should also be cognizant of ongoing improvements in sound recording. Dolby and dbx noise reduction equipment, for example, are eliminating much of the noise or tape hiss associated with cassette recording; they employ special circuitry to code the signal during recording and then in playback. Other machinery includes "time condensers," like Varispeed, which can cut listening time in half through computer-aided manipulation of pauses and yet still retain much of the feel of the original.

Certainly, videotaping, which has been employed at the UCLA Oral History Program since 1973, is gaining favor. Videotaping offers immense possibilities and new horizons: at its most atypical but intriguing level, we have, for instance, John Schuchman's interviews with the deaf at Gallaudet College. For those interested, Brad Jolley's *Videotaping Local History* (1982) provides an excellent preface to the more general techniques. In terms of conservation, however, videotaping multiplies the difficulties. Although advancements are expected, at present no economically justifiable equipment, tape, or method exists for the archival storage and preservation of videotape. One possible solution for video recordings may lay in improvements in videodisk technology. Whether in analog or digital form, the presence of more than 100,000 computer-addressable storage locations on a single, stable storage medium holds great promise for the future. Difficulties center on handling high equipment and mastering charges and on developing a method for the direct encoding of a live performance.

Of related, though more immediate, significance to the sound archivist is the emergence of digital recording. In fact, many of the improvements in magnetic sound recording proceeded from parallel developments in the magnetic storage of data following the Second World War. Ford Kalil of NASA's Goddard Space Center describes the principles in the highly recommended *Magnetic Tape Recording for the Eighties* (1982):

To record digitally, the analog input signal (audio or otherwise) is sampled at a rate exceeding twice its highest frequency and the samples are translated into binary code numbers representing the amplitudes at the instant of sampling. Numbers of 11 to 16 digits may be used, giving accuracies (or signal to noise ratios) of about 2^{11} (66dB) to 2^{16} (96dB). The code numbers are recorded in sequence on tape. When played back, the code numbers become amplitudes; irregularities in timing (tape flutters) are corrected by a buffer memory; and the sample rate is reproduced in perfect crystal clock sequence, thus reconstituting a signal the perfection of which is limited only by the binary code.

The presence of commercial digital recordings and the entrance of industry giants like 3M, Sony, Philips, Technics, Japan Victor, and Soundstream, indicate that this new technology will soon replace phonorecords and even enter archives. Indeed, the Library of Congress is currently experimenting with four-inch compact disks (CDs) as part of their larger videodisk project. Such technology offers a greater capacity in audio signals with the ability to remove mechanical flutter and background noises, as well as the unprecedented prospect of exact copies for generation after generation. But equipment costs and the need for an interim mastering stage suggest that tape recording will continue as the method for field recording for the foreseeable future. One might also add that, as a result of their capital outlays in older equipment, sound archivists are naturally hesitant to invest in new technologies.

Other emerging technologies, like voice-actuated or still unpredicted advances, are sure to appear on the scene. Sound archivists must be aware of their implications and be flexible enough to incorporate pertinent developments when appropriate to their installations. But, if the past can ever be prologue, we should also be ready to anticipate new problems.

EQUIPMENT AND FACILITY CONSIDERATIONS

As with microcomputers, the current volatility of the sound equipment market restrains one from making specific product recommendations; suggestions would simply be outdated by the time they were read. Thus, mention of particular brands is based only on personal experience and is not meant as an endorsement. Anyone purchasing new apparatus must as a matter of course dive into the maze of manufacturers and dealers through experimentation and the help of *Consumers Reports* or other selection aids.

The equipment recommended here may be divided into three classes:

1. Ideal—a full sound and video studio with mixers; digital encoders with tape management system (TMS) software; a laminar-flow, 10 000 class clean room; Nagra recorders for fieldwork; tape cleaner/winder, evaluator/testor, and voice compressor; and so on.
2. Practical—at least one semiprofessional reel-to-reel recorder, and three cassette players (one of which should be a stereo player, preferably with meters to check the recording level and frequency modulation to avoid surges), plus a sufficient number of stereo cassette machines for patron booths.
3. Basic—three cassette recorders (one of which should be stereo).

To focus on the last two levels, one might begin with the fortunate realization that problems of compatibility are not nearly as severe as with microcomputers. Thus, we can center on integrated methodology that can later expand from one level to the next.

Reel-to-reel tape decks represent the major difference between the practical and basic levels of equipment. Reel-to-reels were pioneered in the United States by Ampex and are necessary to achieve archival quality storage. Of the three current grades, sound archivists normally concern themselves with professional or semiprofessional quality and ignore consumer-grade equipment. Yet improvements, the lower cost, and the greater availability of Sony, TEAC, and other consumer-grade decks are such that many a beginning archive should investigate their purchase. In addition to the standard controls (play, record, pause, fast forward, reverse), one will want meters to gauge the recording levels, bias controls, and multi-speed capacity (at least 3-¾ and 7-½ ips), with the ability to make slight velocity corrections. Another important consideration is head configuration; one will have to choose between the greater economics of ¼ track format or the superior performance of ½ track or even full track.

Reel-to-reel decks are also in need should one receive deposits from field recordings made by miniature reel-to-reel recorders, such as those manufactured by Uher, Stellavox, and the highly touted Nagra. These machines are especially valuable for the higher quality demands in recording musical performances.

Assuming that most incoming collections are recorded on cassette players, one would do well to recommend (or loan) decent quality equipment with Dolby, dbx, or other noise reduction; end of tape alarm; and stand alone, preferably lavaliere microphones. Condenser or built-in microphones, for example, capture too much machine noise and should never be used. According to oral historian Joel Gardner, experiments at the Oral History Association Workshop in San Antonio have established very little difference among the recorders themselves, but significant variation in regard to microphone quality. The best microphone one can afford should be purchased. Similarly, demonstrations by Blair Hubbard, a sound engineer with the National Park Service, provide convincing evidence of the superiority of more expensive full-track cassette recordings (as well as the general superiority of reel-to-reels over the best of the cassette recorders). If one can afford full-track capacity, one should double-check that the monaural recorders are actually full and not half track. The archivist's own inhouse machine should include an output jack, standard controls, a recording level indicator, and an input-level volume control, as well as an automatic device to mute volume surges.

Stereo cassette players are used for patron listening, as well as for processing. They should have a control to switch from left to right or on both channels and the same trappings already indicated for monaural machines, although patron machines may have the record button blocked. An extremely useful feature, found on Marantz and some other machines, allows one to fast forward on the time channel (see the final section in this chapter) and hear the times as audible beeps. That is, if one recorded the time every ten seconds, then every six beeps would indicate one minute. Those with a transcribing program will also need a machine with foot pedal controls.

Ancillary equipment includes a microphone, preferably with its own off/on switch, a demagnetizer, and a range of wire connectors, adaptors, and plugs. Although not essential, the strongly recommended list also includes a degausser (bulk eraser), professional splicer, headphones, and a tape cleaner/

winder. In some instances, one may also require an amplifier to boost output and perhaps speakers for large presentations. In addition, it would be better to substitute more expensive tape decks for processing instead of simple cassette players. Although this book does not cover the engineering of recordings to produce a higher quality of sound, it should be noted that some sound archives – notably the Archive de Folklore of Laval University in Canada – also employ sound boards or system equalizers for such ventures.

Again, in terms of equipment maintenance and conservation, THIC and manufacturers' recommendations should be followed. Even with optimum environmental control, however, the abrasions through normal play rub off small portions of the magnetic tape throughout the tape transport system. Thus, periodic cleaning of the magnetic heads, capstan, tape quide, and so on, with Freon TF or isopropyl or methyl alcohol (the roller requires a separate rubber cleaner) is suggested. The solvents should be dispensed from bulk and not aerosol containers in order to minimize metallic particle contamination. If sufficient space exists, lint-free wipes, or cotton swabs should be used. First, any tape should be removed and then solvent should be applied to the cleaning material (never directly to the transport components) and the surfaces – especially around flanged edges – scrubbed. Cleaning continues with new wipes and swabs until they show no evidence of dirt, and then a final pass is made across the heads in the direction of the tape motion. Because the tape heads and other metal parts in contact with the tape are subject to magnetization, a demagnetizer will also be needed at frequent intervals. Less frequent remedies include maintaining proper tape alignment and tension, replacing worn heads, and adjusting the bias and equalization (steps that might require a trained technician).

The need for proper regulation of humidity and temperature has already been stated but cannot be overstressed. Playing conditions demand an ambient temperature of 70° F ($\pm$10) and a relative humidity of 45% ($\pm$5%). These conditions may be continuously monitored on a hygrothermograph or periodically with a far less expensive sling psychrometer. Fortunately, modern air-conditioning units have risen to the occasion, although they might have to be supplemented with a humidifier and/or dehumidifier depending on climatic conditions. It would also be helpful if the air-conditioner were filtered and could maintain a slight positive pressure in playing areas, thereby lessening the intrusion of dust particles, which are also contravened by vacuuming, damp mopping, and damp wiping.

Of greater significance is the need to maintain a constant environment without fluctuations in heat and dampness (Figures 6:1, 6:2). Here one might even desire backup air-conditioning units but at the very least alert maintenance personnel and bureaucrats responsible to the importance of this factor. Such considerations should also play a role in the selection and planning of physical facilities. Thus, attic locations are not advised because of the difficulties in regulating their environment and the prospects of rapid swings in temperature with a power outage. Basements are somewhat more preferable with the plus of more stable temperatures, but they pose a higher risk of flooding and burst pipes. The recommended choice, then, is somewhere in the intermediate floors (only a recurrent dream to most archivists in their basement and attic nooks). Whatever the location, one should be sure to include sufficient electrical power and hookups, especially in the processing area.

Security and Disaster Planning

Facilities should also be designed with the physical security of the equipment and tapes in mind. Archivists need to be aware of potential hazards in planning the layout of their space. Although the danger of accidental erasure from strong magnetic fields is negligible, one should ensure that the materials are not stored under water or steam pipes and away from other conduits—like elevator shafts—that might increase the perils of smoke, fire, or water damage. Smoke detectors and other warning or monitoring devices are also important, as is the installation of automatic fire control systems—especially CO_2 or halon gas—but failing that water sprinklers. In addition, archivists should have a written disaster policy and a cache of emergency gear. The latter paraphernalia can be a remarkably simple array of sponges, mops, plastic sheets, temporary labels, and a wet/dry industrial

Figure 6:1
The Effects of Temperature on Magnetic Tape

A. Playing Conditions

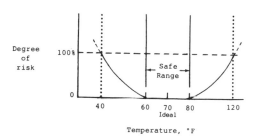

Temperature, °F

B. Storage Conditions

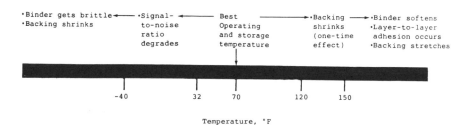

Temperature, °F

Source: THIC, *A Care and Handling Manual for Magnetic Tape Operation.*

Figure 6:2
The Effects of Humidity on Magnetic Tape

A. Playing Conditions

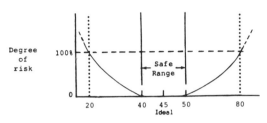

Relative humidity, percent

B. Storage Conditions

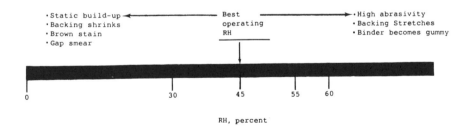

RH, percent

Source: THIC, *A Care and Handling Manual for Magnetic Tape Operation.*

vacuum cleaner. The disaster plan should include the names and telephone numbers of key personnel to contact as well as those of possible experts or contacts at area universities, laboratories, and the like. Obviously, one will want to outline the proper steps to take in a variety of predicaments.

For further information, Sidney B. Geller's *Care and Handling of Computer Magnetic Storage Media* (1983) for the National Bureau of Standards provides some excellent insights on disaster preparedness. Luckily, magnetic tapes are a very tough medium, being quite resistant to attacks, and offer ease of duplication. Although more in-depth guidelines are desirable, general measures in case of a catastrophe begin with a shifting from any area of immediate danger. Rapid salvaging should include a grading or separation of the most important and most severely damaged from the rest, but making sure to retain as much identification of the contents as possible. In flood conditions, one begins with a general drying of the storage area and an immediate natural air-drying of the tapes. Fire damage calls for a careful cleaning away of debris. Normally, one would try to stage all tapes back to proper heat and humidity conditions. Tapes then should be carefully run over for several slow passages over tape winders or cleaners, perhaps first placing the tapes on new reels. Without such equipment, tapes should be run without engaging the heads on the recorders. The final phase is to recopy the contents onto new tapes.

The possibility of damage can also be lessened by proper preplanning for storage. The location of storage sites has already been mentioned, but steps should be taken to insure that the tapes are stored well off the ground and below the top of open shelving. The use of storage cabinets and plastic bags and the returning of tapes to their original containers are also suggested to reduce the introduction of damaging dust particles and the likelihood of damage in a catastrophe. Finally, one of the simplest and most effective defenses is the placement of duplicate copies at another site.

Security also extends to protection against theft. Please note that the very presentation of an archive's concerns and capacity to safeguard its materials is at least as significant a deterrent as its actual implementation. Psychological and physical measures should maintain decorum, but steps such as sign-in registers and call slips for requests are already quite acceptable. In addition, it should be recognized that tape recorders represent a rather easily hidden and marketable temptation, which is heightened as the nature of the research does not make it feasible to extend monitoring to a central control board. Thus, it is doubly important to insure that all patrons register and that the machines are registered with an indelible marking and firmly fastened to their stands.

In terms of layout, the processing area should be securable and separate from that of the researchers. For the purpose of observation, glass dividers between the areas are preferable when a full-time reference staff is not feasible in the listening room. Similarly, one would want to make certain that the sight-lines from the control desk allow for viewing into all sound booths and researcher working areas. Although total security is impossible, Figure 6:3 illustrates a simplified two-room floor plan that incorporates many of the measures already mentioned.

MAGNETIC TAPE

In addition to equipment and facilities, archivists should be aware of the terminology and makeup of magnetic tape. Most archivists are at least vaguely familiar with a tripartite composition—base, binder, and magnetic coating—but many are not cognizant of the vast improvements in this field in recent years. For example, although earlier cellulose acetate backings should be replaced, the current mylar or polyester standard for the base is relatively trouble free. Contemporary binding compounds contain more than adhesive elements, they often hold antistatic agents, fungicides, lubricants, and plasticizers, as well as dispersants and wetting agents. The magnetic oxide particles are the essential element in tape architecture. Ideally, they should be of uniform acicular (needle) shape from 5 to 40 μ inches in length. The THIC Manual describes the manufacturing process:

First the oxide particles are mixed with the plastic binder compound; then this mixture is applied wet to the flexible backing material. (At this point in production, the coating mixture is viscous and the encapsulated

Figure 6:3
Sample Floorplan: Two-Room Archive

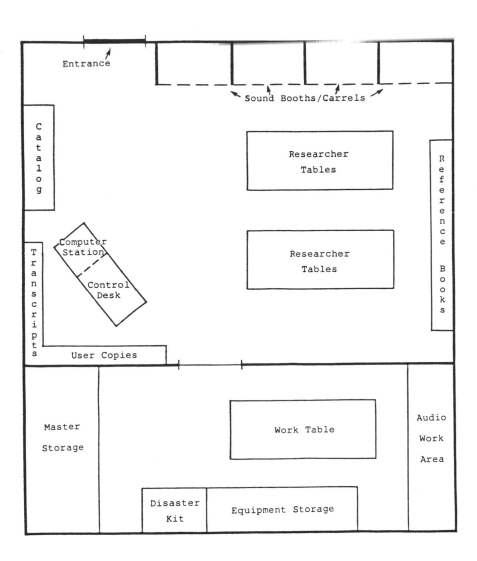

oxide particles can still be moved.) Next, the tape is passed through a magnetic field to physically orient to oxide particles uniformly in a specified location. . . . The coating mixture, with its properly oriented oxide particles is then dried onto the backing material. The coating is subsequently polished to a relatively smooth finish.

Developments have especially proceeded from the computer field and include techniques for more uniformly sized and orderly arranged oxides. A fourth component has been added to some of the higher quality tapes with the emergence of back coating: a finely textured carbon layer is added to minimize static buildup, lower friction levels, limit the shedding of base materials, and reduce distortions from trapped particles. Since 1965 and in addition to perfecting the standard ferric oxide (Fe_2O_3) tapes, scientists have begun to introduce alternative magnetic particles: for example, chromium dioxide, cobalt-doped oxides, particulate iron, and barium feride coatings. The concomitant problems in standardization, bias, and equalization with such innovations have extended to the audio cassette field. (See Figure 6:4 for views of a typical cassette.) To bring some order, the International Electrotechnical Committee (IEC) has devised a four-tiered classification code:

Type I tapes are the traditional ferric oxide tapes, which dominate the market. These require a "normal" bias and 120 microsecond (μs) equalization.

Type II use chromium dioxide or cobalt-enhanced ferric oxide. Developed by Du Pont in the late sixties, these tapes require a higher bias and 70 μs of equalization.

Type III, or ferrichrome tapes, use a dual layer ferric oxide and chromium dioxide for magnetic storage. These tapes are relatively rare and not of general interest.

Type IV, or metal tapes, use pure metal particles and require a special bias setting with a 70 μs equalization.

The rest of this chapter concentrates on Type I ferric oxide tapes as the only currently acceptable archival quality medium. The phrasing used to describe any of these tapes, however, can be somewhat bewildering. We have already encountered such terms as bias—the high frequency signal added at the record

Figure 6:4
Cassette Views

TOP

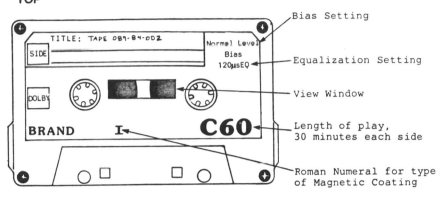

Bias Setting

Equalization Setting

View Window

Length of play,
30 minutes each side

Roman Numeral for type
of Magnetic Coating

OPEN

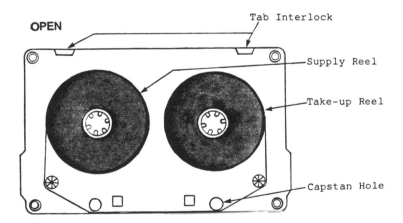

Tab Interlock

Supply Reel

Take-up Reel

Capstan Hole

FRONTAL

Pressure Pad

Tape

head to reduce distortion—and equalization, which adjusts the frequencies to bring out bass and treble tones. The signal-to-noise ratio (S/N), or dynamic range, is another important piece of jargon; it describes in decibels (db) the sonic range between the maximum output level (MOL) before distortion and the bias or hiss noise level. The signal-to-noise ratio is a direct function of the gap between the head and tape and is derived from the formula

$$\text{loss} = \frac{54.6d}{\lambda} \text{ db}$$

where d is the length of separation in inches and λ is the wavelength of signal in inches. For these purposes a minimally acceptable S/N is 60db; therefore, the peak recording levels will be at least 60db higher than the background recording noise.

Other frequently encountered terms include "abrasivity," or degree to which the tape erodes the heads, and phrases to describe the ability of the tape to retain information and the strength of the field to record data. "Retentivity" after saturation (B_R, and for iron oxide tapes generally 1,000 to 1,400 gauss), for example, is often specified, and high levels are associated with high signal output. Experts suggest that "remanence" (ΦR), which records the magnetic flux in relation to a specific tape width in Maxwells (Mx), is a better measure. Another often cited property is "coercivity" (H_C), which indicates the power required to erase a tape in amperes per meter or oersteds (Oe); this ratio is usually twice the encoding field strength, and for standard iron oxide it is typically between 300 and 350 Oe, with higher coercivity linked to better high frequency response.

One surprising factor to arise from the measurement of coercivity is a recognition of the rapid dispersal of erasing strength over very small distances. We can now put to rest many of the apocryphal horror tales of accidental erasures. Although care should be taken to preclude the intrusion of magnets and to design repositories away from items that produce magnetic fields—such as large motors, transformers, and power lines—actual danger is almost nonexistent. Direct contact with any magnetics, of course, will produce instantaneous loss, but with 3 inches of separation even quite powerful source fields produce no adverse consequences. In fact, in an experiment a powerful junkyard magnet, capable of lifting 800 pounds of metal, had no effect on stored signals at a distance of 5 feet and only a 5 percent degradation as close as 1 foot 4 inches.

Preservation Masters, User Copies, and Storage Jargon

In addition to accidental erasures and the factors already indicated, the sound archivist needs to be aware of specific difficulties that may arise in the use or deposit of magnetic tape. Fortunately, the production of reel-to-reel masters and user cassettes alleviates many potential problems for the archivist. Fortunately, too, the jargon describing possible difficulties is more readily understandable than that of tape architecture.

Figure 6:5 illustrates a method for the simultaneous issuance of a preservation master on a reel-to-reel deck and a user copy on a stereo cassette recorder. While treated as an isolated phenomenon, this technique should be seen as part of a larger integrated system of archiving. The specific design draws directly from Dale Trelevan's "TAPE" adaptation at the Wisconsin Historical Society.

In the illustration a cassette player serves as the originating agency, although one could easily substitute a reel-to-reel deck or other output device in the same position. From the source tape a "Y" connector leads into a reel-to-reel recorder. (Alternatively, it might be possible to jack directly through the reel-to-reel, rather than the "Y.") The other channel on the stereo recorder is jacked to a separate cassette recorder with prerecorded time signals. This "time-master" requires one to sit down beforehand with a watch, microphone, and tapeplayer recording the passage of time every 10 seconds: for example, start, 10 seconds, 20 seconds . . . 1 minute, 1 minute 10, 1 minute 20 . . . , or some variation thereof. The beginning of the source tape is then coordinated with the start of the time-master, and all three recorders are begun simultaneously.

Figure 6:6 demonstrates a more simplified approach, which is aimed only at the production of user cassettes.

Figure 6:5
Production of Preservation Reel Masters and User Cassettes

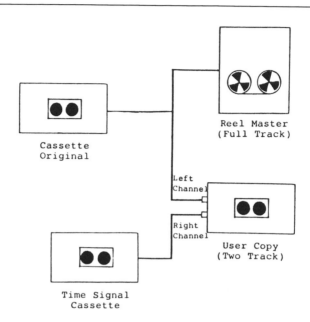

Figure 6:6
Production of User Cassettes

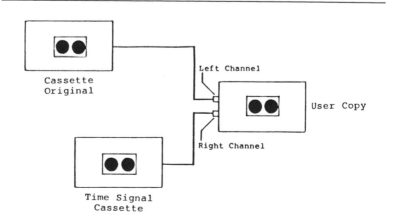

For preparation for storage, archivists will also want to keep contact with fingers and their oils to a minimum. In addition, many choose to place masters within their containers and to seal them in polyethylene bags to keep out dust and reduce the evaporation of binder lubricants. The packages are then stored in collection-number sequence in a vertical position on shelves or suspended within closed cabinets. Although some are now counseling much lower figures, practical considerations dictate that temperature and humidity should generally approximate those of the playing area and within the 70°F ±10, 45% ±5 relative humidity standard. Many, too, counsel the yearly "exercise" of tapes to lessen the possibility of print through from one layer to the next. (Based on the results of a small sample study, Geller recommends a slightly lower storage than playing conditions at 65°F and 40% RH, which could extend the exercise interval to 3.5 years.)

Whatever the scenario, master and originals should be stored away from use if at all possible. While retained, cassette originals have a number of characteristics that preclude relying on them for permanent storage. For example, cassettes employ tape splices with adhesives that can "run" and stick to adjacent layers, as well as collect contaminants. At their best, cassette tapes are no thicker than 0.5 mil and are somewhat prone to tearing and print through. Such tendencies are especially evident in the thinner and longer tapes beyond the C60 level. Finally, although significant improvements have been made, cassettes are simply inferior to reel tapes in terms of fidelity and sound reproduction.

The universally recommended standard for archival storage is 1.5 mil, ferric oxide reel-to-reel tape without leader or splices. Although such a magnetic medium has not been around long enough to totally prove its permanence, enhanced aging tests have convinced experts of its reliability. According to Sidney Geller, only ferric oxide is acceptable, "since its ability to maintain its magnetization for long periods of time has been established. No other presently available particles have yet proven to be of equal long-term stability at this time for similar temperature-humidity ranges." In general, the experts also suggest a minimum speed of 7-½ ips for music and 3-¾ ips for voice recording. At the least, preservation masters should never be registered at a speed lower than that of the original. Certainly, too, it is important to label the reels and containers of the masters and user copies immediately at least with the classification number. All markings should be made with soft tip marking pens. Pencils should not be used and are actually banned from the archive, because of the danger of particle contamination from erasers, wood shavings, and graphite pieces.

In addition to particle contamination and the previously stressed significance of temperature and humidity, archivists must pay close attention to the problems associated with improper winding tensions. Anything other than a smooth, evenly wound tape is evidence of uneven or weak tension. The results are cinching or a folding of tape layers and such related symptoms as spoking or buckling, stepping or scattered wind, and pack shifting; these enhance damage to edges and increase the likelihood of dropouts. The most frequent causes are frequent starts and stops, plus the fluctuations inherent in fast forward and reverse functions. The remedy is a constant run at normal playing or recording speed (2 to 3 ounces of pressure), which results in storage in the recommended played or "tails out" position.

User copies are obviously subject to the same vagaries as preservation masters, as well as the additional ones engendered by active use and frequent human handling. The last is a major factor that favors cassettes for the active mode. In particular, the cassette as user copy offers ease of operation and protective housing superior to a reel format. Cassettes eliminate the threading problems and concomitant dangers of fingerprint contamination that are inherent to reel-to-reel. Cassettes are also generally less expensive and operate at only one speed, as well as having removable tabs to prevent accidental erasures and over-recordings.

Sound archivists will want to take into account many of the precautions for the storage of masters in regard to their user cassettes. A good practice, for example, is to determine that all cassettes are fully rewound on one side at normal playing speed following each address and returned to their containers. In addition, archivists should standardize their purchases as much as possible. Decent quality Type I tapes with screw fasteners for ease of repair are probably most desirable. The length should be C60, which indicates thirty minutes on each side. (This is an approximation, and some of the less expensive brands have been reported for shortchanging by as much as several minutes of tape.) C60 is ideal because it corresponds to the normal length of oral history interviews and at 0.5 mil to the most break-resistant thickness available in the format.

The last element is a periodic inspection of all the technical components of the archive, from user cassettes and preservation masters to the recorders.

In conclusion, the results of the various measures advocated in this chapter are a coordinated response to the dilemma of use versus preservation and the related problems in active and passive phases. Archivists must recognize that the technical facets of sound recording are actually a complex dynamic—a dynamic that includes recorders, tapes, human beings, and the environment. Moreover, archivists should be aware of the specific peculiarities and remedies (and the language used to describe same) of those components as they apply to sound collections. Yet, in a final caution, it should always be remembered that our conservation efforts are part of a larger effort or mission to preserve information and further communication.

SELECT BIBLIOGRAPHY

This chapter is intended as a brief introduction, not as an in-depth technical treatise. Those seeking more detailed information can begin with the works listed below but will also have to turn to more specialized articles in journals such as *IEEE Transactions on Magnetics, Journal of the Audio Engineering Society,* and *Association for Recorded Sound Collections Journal.*

Borwick, John, ed. *Sound Recording Practice.* London: Oxford, 1976.
Cuddihy, Edward. "Aging of Magnetic Recording Tape." *IEEE Transactions on Magnetics* 16 (1980): 236-39.
Cunha, George, Daniel Martin, and Dorothy Cunha. *Conservation of Library Materials.* Metuchen, NJ: Scarecrow, 1972.
Cunha, George, Daniel Martin, and Dorothy Cunha. *Library and Archive Conservation.* Metuchen, NJ: Scarecrow, 1983.

Eargle, John. *Sound Recording.* New York: Van Nostrand Reinhold, 1980.

Geller, Sidney B. *Care and Handling of Computer Magnetic Storage Media.* Washington, DC: Bureau of Standards, 1983.

Jolley, Brad. *Videotaping Local History.* Nashville: AASLH, 1982.

Kalil, Ford, ed. *Magnetic Tape Recording for the Eighties.* Washington, DC: Bureau of Standards, 1982.

Lomax, John. *The Adventures of a Ballad Hunter.* New York: Macmillan, 1947.

Lowman, C. E. *Magnetic Recording.* New York: McGraw Hill, 1972.

McWilliams, Jerry. *The Preservation and Restoration of Sound Recordings.* Nashville: AASLH, 1979.

Pickett, A. G., and M. M. Lemcoe. *Preservation and Storage of Sound Recordings.* Washington, DC: Library of Congress, 1959.

Read, Oliver, and Walter Welch. *From Tinfoil to Stereo.* 2d ed. Indianapolis: H. W. Sams, 1976.

Snel, D. A. *Magnetic Sound Recordings.* Eindhoven, Netherlands: N. V. Philips, 1963.

Swartzburg, Susan, ed. *Conservation in the Library.* Westport, CT: Greenwood, 1983.

THIC Manual in Ford Kalil, ed. *Magnetic Tape Recording for the Eighties,* Washington, DC: Bureau of Standards, 1982.

3M Company. *Retentivity.* St. Paul: 3M, n.d.

Trelevan, Dale. *The State Historical Society of Wisconsin TAPE System.* Madison: Wisconsin State Historical Society, 1979.

Tremaine, Howard. *Audio Cyclopedia.* 2d ed. Indianapolis: H. W. Sams, 1969.

APPENDIX 1

SAMPLE WORKSHEETS

ORAL HISTORY WORKSHEET

Collector: Tape:

____Project: Date:

Informant:

Birthdate: Birthplace:

Interview Location:

Other Information:

Subject/Title:

Time	Topic

Stielow OH 1984

FOLKLORE SOUND ARCHIVE WORKSHEET

		Tape No:

Collector--Last Name	First	Other Collector: Last Name, First

Day-Mo-Yr	Project Class	Project or Class title

Informant no. of	Informant: Last Name	First		pseudonym , maiden name

Address--home	City		State	Sex	Race/ethnicity

Birthdate	Birth place: city, state, country	other residences

religion	Occupation (s)	education: ___grade sch. ___high sch. yrs___:N/A ___college

Other Details:

Descriptive Title:

Time	Genre	Contents
0:00		

WKSH: 1984

SEAL

WORKSHEET/FICHE DE TRAVAIL

Collector--Collectionneur:
Date:
Place--Lieu:
Parish--Paroisse:
Informant--Informateur: (_others, append sheets for each)
Name--Nom: Petit Nom:
Sex--Sexe: Maiden Name--Nee:
Race: Dialect:
Birthdate--Date de Naissance: Place--Lieu:
Education: Religion:
Occupation(s):
Residences--Deplacements:

Other--Autres Details:

Folk:___-__-___
Speed--Vitesse:_____
Format (cassette,reel)

Equipment:_____

Length--Duree:_____
_Additional materials

Descriptive Title--Titre Populaire:

Time--Temps	Genre	Classification	Title/contents--Titre/Matiere

APPENDIX 2

FOOT METER CONVERSION

TIME	1-7/8 ips	3-3/4 ips	7-1/2 ips
0:00	0000	0000	0000
0:10	0018	0037	0075
0:20	0037	0075	0150
0:30	0056	0112	0225
0:40	0075	0150	0300
0:50	0093	0187	0375
1:00	0112	0225	0450
2:00	0225	0450	0900
3:00	0377	0675	1350
4:00	0450	0900	1800
5:00	0562	1125	2250
6:00	0675	1350	2700
7:00	0787	1575	3150
8:00	0900	1800	3600
9:00	1012	2025	4050
10:00	1125	2250	4500

TIME	1-7/8 ips	3-3/4 ips	7-1/2 ips
15:00	1682	3375	6750
20:00	2250	4500	9000
25:00	2812	5625	11250
30:00	3375	6750	13500
35:00	3937	7875	15750
40:00	4500	9000	18000
45:00	5062	10125	20250
50:00	5625	11500	22000
55:00	6187	12375	24750
60:00	6750	13500	27000

APPENDIX 3

OCLC FIXED FIELD DESCRIPTORS

Mnemonic	Definition	Default	Notes/Sample Codes
OCLC	OCLC control no.	New	
Rec. Stat:	Record status	m = new record	automatically set
Entrd:	Entry date	current date	six digit yr, mo, dy
Used:	Date last used	current date	automatically set
Type:	Type of record	j = sound record	j = non-musical
Bib lvl:	Bibliographical level	m = monograph	c = collection
Lang:	Language	N/A – no spoken text	eng = English; fre = French; ita = Italian, etc. (see 041)
Source:	Catalog source code	u = unknown	ƀ = Library of Congress d = OCLC, member library
Accom mat:	Accompany material	ƀƀƀƀ = none (up to six codes)	a = discography; b = bibliography e = biography; k = ethnological etc.
Repr:	Reproduction	ƀ = n/a	leave default only
Enc /v/:	Encoding level	no default	I = full cataloging; k = less than full catalog
Ctry:	Country of production	xxƀ = unknown	2 character for most US & Can – 3 digit with 1st S = state; ƀƀU – US, ƀƀC = Can

Dat tp:	Type of date:	u – none supplied	s = single date; m = unk;
			m = multiple; g = doubtedly;
			r = reissue (see Dates)

Mnemonic	Definition	Default	Notes/Sample Codes
MEBE:	Main Entry	Ø – not in body of entry	Ø also of main entry = title; 1 = in body of entry
Mod rec:	Modified record code	ƀ = note modified	indicates date change for machine readability
Comp:	Form of composi-tion	ƀƀ = no info	ma = music (see Ø47); nm = n/a; non-musical
Format:	Format of music manuscript	n = n/a	use default only
Prts:	Existence of parts	n = n/a	use default only
Desc:	Descriptive cat-aloging form	no default	ƀ = non ISBD; a = AACR2 cataloging
Int lvl:	Intellectual level	ƀ = not known or not juvenile	j = juvenile work
LTxt:	Literary text	m = n/a	1 or 2 digits. ƀƀ = musical; h = history; 1 = lectures; o = Folktales; c = conference; a = autobiography; b = biography; t = interviews
Dates:	Date 1, Date 2	ƀƀƀƀ, ƀƀƀƀ = not known	year dates (see Dat tp)

ƀ = blank, Ø = zero

SELECT LIST OF FIELDS IN MARC AMS FORMAT

001	Control Number
035	Local control number ƀƀ
	$a Local system control number
	$b Canceled/invalid control number
040	Cataloging source ƀƀ
	$e Description convention followed
041	Language code ƀƀ
	$a Language code of text
043	Geographic area code ƀƀ
	$a Geographic area code
090	Local Call Number
100	Main Entry – Personal Name
	$a Name
	$b Numeration
	$c Titles & other words assoc. w/name
	$d Dates (birth, death, flourishing)
	$e Relator

0 Forename only ƀ
1 Single surname
2 Multiple surname
3 Name of family

SOURCE: from a handout from Virginia Purdy.

$$ Qualification of a name
(fuller form)
$u Affiliation
110   Main Entry – Corporate Name O
Surname (inverted) Ƅ
    $a Name   1   Place name
    $b Each subordinate unit   2   Name (direct order)
       hierarchy
    $c Place
    $d Dates
    $n Number of part/section
    $p Name of part or section
    $u Affiliation
245   Title Statement   0 No title added entry
    $a Title   1   Title added entry
    $f Inclusive dates   9-0 Nonfiling characters
    $g Bulk dates
    $h Medium
    $n Numer of part/section
    $p Name of part/section
300   Physical Description Ƅb
    $a Extent/quantity
    $b Other physical details
    $c Dimensions/measure
    $f Packaging units
    $3 Materials specified
340   Medium Ƅb
    $a Material base and
       configuration
    $b Dimensions
    $c Material applied to surface
    $d Information recording
       technique
    $e Support
    $f Production/rate/ratio
    $i Technical specifications
       of medium
    $ Materials specified
351   Organization and Arrangement Ƅb
    $a Organization of material
    $b Arrangement

       $c   Archival level
       $3   Materials specified

500   General note ƀƀ
       $a   General note

502   Dissertation note ƀƀ
       $a   Dissertation note

505   Contents note (formatted)   0   Contents (complete) ƀ
       $a   Table of contents   1   Contents
                                            (incomplete) ƀ

506   Restriction of Access ƀƀ
       $a   Terms governing access
       $b   Jurisdiction
       $c   Physical access
            provisions
       $d   Authorized users
       $3   Materials specified

510   Citation note/References   3   Specific location in
       $a   Name of source   source cited not given
       $b   Date of source   4   Specific location in
       $c   Location within source   source cited given
       $3   Materials specified

520   Summary, abstract, annotation,
          scope note ƀƀ
       $a   Note
       $b   Expansion of summary note

524   Preferred citation of described
          materials ƀƀ
       $a   Preferred description

530   Additional Physical Form
          Available Note ƀƀ
       $a   Additional physical form
       $b   Availability source
       $c   Availability conditions
       $d   Order number
       $3   Materials specified

533   Reproductions note ƀƀ
       $a   Type of reproduction
       $b   Place of reproduction
       $c   Agency responsible for
            reproduction
       $d   Date of reproduction

$e Physical description of
 reproduction
$3 Materials specified

535 Location of originals/duplicates   1 Holder of original ƀ
$a Custodian   2 Holder of duplicate
$b Postal address
$c Country of repository
$d Telecommunications address
$3 Materials specified

540 Terms governing use ƀƀ
$a Terms governing use
$b Jurisdiction
$c Authorization
$3 Materials specified

541 Source of acquisition ƀƀ
$a Immediate source of
 acquisition
$b Address
$c Method of acquisition
$d Date of acquisition
$e Action identification
 (accession number)
$f Owner
$g Purchase price
$3 Materials specified

544 Location of Associated
 Materials ƀƀ
$a Custodian
$b Address
$c Country of repository
$d Title (of associated materials)
$e Provenance
$3 Materials specified

545 Biographical or historical note ƀƀ
$a Note
$b Expansion of note

546 Language Note ƀƀ
$a Note
$b Source
$c Information code or alphabet
$3 Materials specified

555   Cumulative index/finding
          aids note ƀƀ
          $a   Finding aids note
          $b   Availability source
          $c   Degree of control
          $d   Bibliographic references
          $3   Materials specified

561   Provenance ƀƀ
          $a   Provenance
          $b   Time of collation
          $3   Materials specified

580   Linking entry complexity note ƀƀ
          $a   Note
          $b   Source

581   Publications Note ƀƀ
          $a   Note
          $3   Materials specified

583   Actions
          $a   Action
          $b   Time identification
          $c   Time of action
          $d   Action interval
          $e   Contingency for action
          $h   Jurisdiction
          $i   Method of action
          $j   Site of action
          $k   Action agent
          $1   Status/condition
          $f   Authorization

584   Accumulation and frequency
          of use ƀƀ
          $a   Accumulation
          $b   Frequency of use
          $3   Materials specified

600–752 Subject Added Entries/Index Terms Type of entry Source

773   Host item entry   0 Display a note
                                       1 Do not Display a note
          $a   Main entry ƀ
          $g   Relationship information
          $k   Record group or collection (?)
          $t   Full title

$w Control number
$3 Materials specified

851   Location ƀƀ
$a Custodian
$b Institutional division
$c Street address
$d Country of repository
$e Location of packaging units
$f Item number
$3 Materials specified

ASSIGNMENT OF HEADINGS FOR FOLKLORE MATERIALS FROM LIBRARY OF CONGRESS MANUAL H16Z7

1. GENERAL RULE. Assign as appropriate a combination of the following headings to the works governed by this memo. The subdivision *Folklore* is free-floating.

[ethnic, national, or occupational group] – [local subdivision] – Folklore

[theme] – Folklore

[headings (s) for specific folklore genres, with local subdivision]

Folklore – [local subdivision]

[locality] – Social life and customs

[other topics, as applicable]

2. SPECIAL PROVISIONS

a. *Collections.* Assign to a collection of folklore texts, if possible, the first three categories of headings listed above. Assign other headings as appropriate for the work.

b. *Works Which Discuss Folklore.* Assign headings for the first three categories of headings above if a specific genre is involved, further subdividing the genre heading by *History and criticism* (or by more specific subdivisions such as *Classification; Themes; Motives;* etc., if appropriate).

If a specific genre is not involved, assign in the place of the third category, the fourth and fifth categories of headings (the subdivision *History and criticism* is not used).

EXPLANATION OF THE CATEGORIES OF HEADINGS

1. ETHNIC, NATIONAL OF OCCUPATIONAL GROUP.

a. If possible, assign headings of the type [*ethnic, national or occupational group*]*—[local subdivision, if appropriate]—Folklore*, e.g.

Afro-Americans—Louisiana—Folklore

Jews—Folklore

Italians—Austria—Folklore

Chimney sweeps—Netherlands—Folklore

b. Do not assign headings of this type for individual nationalities within their own country. The use of *Folklore* or headings for individual genres with local subdivision is sufficient, e.g., *Folklore—Italy*, not *Italians—Folklore*.

c. For Indian groups, assign two headings, one for the individual tribe, if any, and the other for the major group to which the tribe belongs, e.g.

Cree Indians—Folklore

Indians of North America—Canada—Folklore

For Indian legends and tales, see Par. 3d and 3e below.

d. If the work in hand discusses the folklore of an occupational group within a single ethnic group, assign headings for both groups, e.g.

Weavers—Morocco—Folklore

Berbers—Morocco—Folklore

e. Designate the influence exerted on the folklore of an ethnic group by another group by establishing headings of the following type: [*ethnic group*]*—Folklore—[. . .] influences*, e.g., *Finno-Ugrians—Folklore—Slavic influences.*

2. SPECIAL THEMES IN FOLKLORE. If the work in question has a special theme, designate the theme by means of the free-floating subdivision *Folklore*, e.g.

Stars—Folklore

Lizards—Louisiana—Folklore

This use of the subdivision *Folklore* under topics should not be confused with the use of *Legends* under topics. The latter subdivision is used under a few religious topics, e.g. *Grail—Legends*, but only in connection with published medieval romances and legends (H 1730). In all other instances the subdivision *Folklore* should be used for both folkloric texts or for criticism.

The use of the subdivision *Legends and stories* under types of animals has been discontinued (*see* H 1720).

3. HEADINGS FOR SPECIFIC FOLKLORE GENRES. The following is a list of typical folklore genre headings, grouped according to treatment category.

Folklore (General)
  Fairy tales
  Legends
  Tales
Musical
  Ballads
  Folk music
  Folk-songs
Literary
  Fables
  Folk-drama
  Folk literature
  Folk poetry
  Nursery rhymes
  Proverbs
  Riddles

a. To a collection of folkloric texts in one genre, assign the appropriate genre headings, subdividing it by local subdivision, if appropriate. See *Conflict of systems,* p. 1 for an explanation of the various treatments described below.

For works of criticism, subdivide the genre heading, or the heading with local subdivision, by the subdivision *History and criticism,* e.g. *Tales—Arizona—History and criticism.*

Qualify the literary genre headings (the headings of the third column above) by language or nationality in accordance with normal literary form heading practice. Also in the case of translations, subdivide these literary genre headings by the appropriate translation subdivisions, e.g. *— Translations into English; — Translations from German.*

Do not qualify the headings of the first column above by language or nationality. For the special rules for qualifying musical genre headings, and use of the subdivision *Texts, see* H 2067. Do not use translation subdivisions under headings listed under either of the first two columns above.

A few guidelines for assigning particular genre headings are given below.

b. *Fairy tales.* Assign to collections of traditional narratives that typically deal with supernatural beings (such as fairies, ogres, dragons)

or supernatural events, and which are often created for the amusement
of children. It is sometimes difficult, however, to distinguish between
Tales, which represent the blanket term for traditional narratives, and
Fairy tales. If in doubt, prefer *Tales.*

   c. *Folk literature.* Assign this heading to collections containing three or
more folklore genres. For collections of two genres, assign the
appropriate headings for the genres.

   d. *Legends.* Assign to collections of traditional narratives generally
regarded by their tellers as true. They may include narratives which are
religious (such as those associated with the lives of saints or martyrs,
religious objects or beings), supernatural (e.g. vampires, werewolves, or
ghosts), about individuals (e.g. national figures or heroes), or about
specific places (such as those emphasizing place name origins, or folk
histories).

   To works of legends associated with historical persons known to have
existed assign a heading in the form: [*name of person*]−*Legends,* e.g.
Crockett, Davy, 1786-1836−Legends.

   For works concerning legendary figures, use headings in the form:
[*name of person*] (*Legendary character*), e.g. *Pecos Bill* (*Legendary
character*).

   To medieval legends involving religious objects, assign the subdivisions *Legends,* e.g. *Grail−Legends.* (*see* Par. 2, p. 3)

   Qualify the heading *Legends* by the names of religions to designate the
legends of individual religions, e.g. *Legends, Christian; Legends, Buddhist.*

   For legends of American Indian groups, use the subdivision *Legends*
under individual tribes and major groups instead of the subdivision
Folklore. Assign an additional heading for *Legends−[place].*

   Subdivide the heading *Legends* or the form subdivision *Legends* by the
subdivision *History and criticism* for works which discuss the genre.

   e. *Tales.* Assign this heading to collections of traditional narratives that
are for the most part fictitious and are told primarily for entertainment,
e.g. *Tales−Nebraska.*

   For works which are collections of a single tale type, assign a heading
in the form: [Tale name] (*tale*) e.g. *Dragon slayer (Tale).* For works which
discuss the type, subdivide by *History and criticism.*

   For tales of American Indian groups, use the subdivision *Legends*
under individual tribes and major groups instead of the subdivision
Folklore. Assign an additional heading for *Tales−[place].*

   4. *Folklore.* Assign this heading with local subdivision if appropriate,
to works which discuss folklore in general or which discuss folklore as a
discipline.

Do not assign the heading to a work which deals only with one or more folklore genres (i.e. a work which has been assigned a heading for a specific genre with subdivision *History and criticism* or the heading *Folk literature* with the subdivision *History and criticism*).

Do not subdivide the heading by *History and criticism*.

Whenever the heading is subdivided by place, assign also an additional heading of the type *[locality] – Social Life and customs*.

In the case of national groups in other countries than their own (but not American ethnic groups), assign two headings for folklore to bring out the folklore of the place of origin and the present location, e.g.

Germans – Romania – Folklore
Folklore – Romania
Folklore – German

5. Assign other headings as needed, e.g.

Folk medicine
Folk dentistry
Oral Tradition
Literature and folklore
Psychoanalysis and folklore
Folklore and history
Story-telling
   etc.

APPENDIX 6
SAMPLE ASSOCIATIONS

American Association for State and Local History (AASLH)
708 Berry Road
Nashville, TN 37204

American Folklore Society (AFS)
Maryland State Arts Council
15 W. Mulberry St.
Baltimore, MD 21201

Association of Recorded Sound Collections (ARSC)
c/o Les Waffen
P.O. Box 1643
Manassas, VA 22110

Audio Engineering Society (AES)
60 E. 42nd St., Room 449
New York, NY 10017

Folklore Archivist Network
c/o Richard Thill
College of Arts and Sciences
University of Nebraska at Omaha
Omaha, NE 68182

International Association of Sound Archives (IASA)
Open University Library

Walton Hall
Milton Keynes
Bucks, NK76AA
England

International Tape/Disc Association (ITA)
Ten Columbus Circle, Suite 2270
New York, NY 10019

Oral History Association (OHA)
P.O. Box 23734
North Texas State University
Denton, TX 76203

Oral History in the Mid-Atlantic Region (OHMAR)
P.O. Box 266
College Park, MD 20740

Professional Audio Retailers Association (PARA)
9140 Ward Parkway
Kansas City, MO 64114

Society of American Archivists (SAA)
600 S. Federal, Suite 504
Chicago, IL 60605

Society for Ethnomusicology (SEM)
P.O. Box 2984
Ann Arbor, MI 48106

Special Libraries Association (SLA)
235 Park Ave., South
New York, NY 20003

INDEX

About the Author

FREDERICK J. STIELOW is Coordinator of the History and Library Science dual masters program in Archives and Records Management and Assistant Professor of Library Science at the University of Maryland, College Park. A former Fulbright Fellow in Italy and Jameson Fellow at the Library of Congress, he is the author of *A Collection of the Executive Orders of the Governors of Louisiana, Grand Isle of the Gulf,* and articles appearing in *A Guide to the History of Louisiana* (Greenwood Press, 1982), *Louisiana History, Journal of Library History, American Archivist, Provenance,* and in several anthologies. In addition, he is currently coediting *Library Activism in the Sixties* (Greenwood Press, forthcoming with Mary Lee Bundy).